Cyberthreats

Cyberthreats: The Emerging Fault Lines of the Nation State

SUSAN W. BRENNER

UNIVERSITY PRESS

Oxford University Press, Inc., publishes works that further Oxford University's objective of excellence in research, scholarship, and education.

Oxford New York
Auckland Cape Town Dar es Salaam Hong Kong Karachi Kuala Lumpur Madrid Melbourne
Mexico City Nairobi New Delhi Shanghai Taipei Toronto

With offices in
Argentina Austria Brazil Chile Czech Republic France Greece Guatemala Hungary Italy
Japan Poland Portugal Singapore South Korea Switzerland Thailand Turkey Ukraine
Vietnam

Published by Oxford University Press, Inc.
198 Madison Avenue, New York, New York 10016

Library of Congress Cataloging-in-Publication Data
Brenner, Susan W., 1947-
Cyberthreats : the emerging fault lines of the nation
state / Susan W. Brenner.
p. cm.
Includes bibliographical references and index.
ISBN 978-0-19-538501-4 ((hardback) : alk. paper)
1. Computer networks—Law and legislation—Criminal provisions.
2. Political crimes and offenses. 3. War (International law)
4. Transnational crime. I. Title.
K5250.B74 2009
345'.0268--dc22
2008039579

---

3 4 5 6 7 8 9

Printed in the United States of America on acid-free paper

Note to Readers
This publication is designed to provide accurate and authoritative information in regard to the subject matter covered. It is based upon sources believed to be accurate and reliable and is intended to be current as of the time it was written. It is sold with the understanding that the publisher is not engaged in rendering legal, accounting, or other professional services. If legal advice or other expert assistance is required, the services of a competent professional person should be sought. Also, to confirm that the information has not been affected or changed by recent developments, traditional legal research techniques should be used, including checking primary sources where appropriate.

(Based on the Declaration of Principles jointly adopted by a Committee of the American Bar Association and a Committee of Publishers and Associations.)

cha·os n. complete disorder and confusion

Concise Oxford American Dictionary

Contents

ONE

Introduction

[An] invasion force . . . of digital signals marched across the border into Estonia. . . .[1]

ON APRIL 26, 2007, WHAT would become a two-week series of sustained digital attacks began on various components of Estonia's infrastructure.[2] The attacks all took the form of what is known as a Distributed Denial of Service (DDoS) attack.

In a DDoS attack, attackers overwhelm websites and servers by bombarding them with data, or "traffic." The attackers use a network of compromised computers—known as "zombies"—to send massive bursts of data at the targets of the attack. The zombies are computers that have been captured by "bots"—software that subtly and usually invisibly infiltrates an individual's or a business' computer or one used by a governmental, educational, or other agency.[3] The owners of computers recruited into bot networks, or "botnets," usually have no idea their equipment is moonlighting as a minion of some more-or-less sinister force.[4]

Because bot programs give attackers remote control of compromised computers, zombies can be anywhere. As we will see, geography is irrelevant

---

1 Robin Bloor, *Large-scale DOS Attack Menace Continues to Grow*, The Register (June 11, 2007), http://www.theregister.co.uk/2007/06/11/dos_security_cyberwarfare/.

2 *See* Mark Lander & John Markoff, *Digital Fears Emerge after Data Siege in Estonia*, New York Times (May 29, 2007), http://www.nytimes.com/2007/05/29/technology/29estonia.html?ex=1182484800&en=ac3eadbe88fdb21c&ei=5070.

3 *See* Nicholas Ianelli & Aaron Hackworth, Botnets as a Vehicle for Online Crime, CERT Coordination Center (December 1, 2005), http://www.cert.org/archive/pdf/Botnets.pdf. For a good overview of DDoS attacks and botnets, *see* MacAfee North America Criminology Report: Organized Crime and the Internet 2007, http://www.mcafee.com/us/local_content/misc/na_criminology_report_07.pdf.

4 *See, e.g.*, Brian Krebs, *Bringing Botnets out of the Shadows*, Washington Post (March 21, 2006), http://www.washingtonpost.com/wp-dyn/content/article/2006/03/21/AR2006032100279.html.

in cyberspace. So is size: Bot software is currently being used to assemble enormous transnational networks of slave computers. Botnets have grown almost exponentially since they appeared a little more than a decade ago.[5] Early botnets averaged a few hundred computers, but by 2005, the average had risen to 1,000.[6] In 2006, experts reported that the average botnet consisted of 20,000 zombies, with the median size being 45,000 computers; they noted, however, that they had also tracked at least a dozen botnets encompassing more than 100,000 slave computers.[7]

In the Estonian attacks, an estimated 1 million zombie computers were used. If true, it is far from unprecedented; Dutch authorities reportedly encountered a 1.5 million botnet a few years ago.[8] And the size only continues to increase; some foresee "super botnets" comprising millions of slave computers, while others say they are already here.[9] In January 2007, one expert called botnets a "pandemic," estimating that 25% of the world's networked computers—150 million computers—could be zombies.[10]

Changes have been made in how botnets are structured. One innovation is "tiered" command and control, in which these functions are distributed across many different, geographically dispersed computer servers. A tiered botnet operates more like a modern army—with distinct, distributed bot units and command structures—than like the earlier versions, which had a single point of command.[11] And as with an army, the botnet's distributed command structure

---

5 *See Peer-to-Peer Botnets a new and Growing Threat*, CSO (April 17, 2007), http://www2.csoonline.com/blog_view.html?CID=32852; Alexander Gostev, Malware Evolution: January–March 2005, Viruslist.com (April 18, 2005), http://www.viruslist.com/en/analysis?pubid=162454316#botnets.

6 *See* Robert Vamosi, *What Good Are 1,000 Remote-Controlled PCs?*, CNET News (November 24, 2005), http://www.cnet.com.au/software/security/0,239029558,240058520,00.htm.

7 *See The Botnet Trackers*, Washington Post (February 16, 2006), http://www.washingtonpost.com/wp-dyn/content/article/2006/02/16/AR2006021601388.html.

8 *See* Lander & Markoff, *Digital Fears Emerge after Data Siege in Estonia*, *supra*. *See also* Gregg Keizer, *Dutch Botnet Suspects Ran 1.5 Million Machines*, TechWeb (October 21, 2005), http://www.techweb.com/wire/security/172303160.

9 *See* Ryan Vogt & John Aycock, Attack of the 50 Foot Botnet, Technical Report 2006-846-39, Department of Computer Science, University of Calgary (August 2006), http://pages.cpsc.ucalgary.ca/~aycock/papers/50foot.pdf. *See also* Bot Counts, Shadow Server, http://www.shadowserver.org/wiki/pmwiki.php?n=Stats.BotCounts (yearly botnet size for April-June 2007 ranged between 2.5 and 3 million zombies).

10 *See* Tim Weber, *Criminals "May Overwhelm the Web,"* BBC News (January 25, 2007), http://news.bbc.co.uk/2/hi/business/6298641.stm.

11 *See* Scott Berinato, *Attack of the Bots*, Wired (November 2006), http://www.wired.com/wired/archive/14.11/botnet.html?pg=3&topic=botnet&topic_set=.

makes it more resilient to attacks from its opponents. The most recent innovation is using peer-to-peer technology to replace centralized command and control structures with a nonhierarchical, purely distributed system; this means the botnet is not vulnerable to attacks on central control nodes, so the only way to shut it down is to neutralize each of its zombies.[12]

The increasing size and dispersed command structure of botnets make it very difficult for law enforcement officers to find and nullify these armies of slave computers and their masters.[13] Botnets have evolved into massive, amorphous, moving targets that exist transiently in the unbounded regions of cyberspace, as the targets of the Estonian attacks learned to their frustration.

The first tentative intrusions began on April 26, 2007, and increased thereafter; by April 29, a flood of data shut down the Estonian Parliament's email server.[14] In an apparently related incident on the same day, intruders hacked the Reform Party's website and posted a fake political message on it. At that point, the director of Estonia's Computer Emergency Response Team (E-CERT) assembled security experts from the country's Internet service providers, banks, government agencies, and police forces; he also reached out to government agencies in other countries for assistance in tracking and blocking sources of the attack.

By the end of that first week, the Estonian security forces were having some success in frustrating the attack, but they knew the worst was to come. The attacks seemed to emanate from Russia, and May 9 was an important Russian holiday, the anniversary of its defeating Nazi Germany and the day on which the country honors its fallen soldiers. As I explain below, the Estonian defenders knew their attackers intended to shut down the country's computer network and tried to prepare for the worst. The E-CERT director urged the members of his security team to try to keep their sites and services operating; he was under orders to keep an important government site online, but he was told that other government sites, including the Estonian president's website, could be sacrificed, if necessary.

On May 9, data traffic to Estonian servers increased to thousands of times its normal flow; a representative of Estonia's Defense Ministry reported that

---

12 *See* Matt Hines, *Experts: Botnets Add Fault Tolerance*, InfoWorld (June 7, 2007), http://www.infoworld.com/article/07/06/07/Botnets-get-fault-tolerant_1.html.

13 *See* Berinato, Attack of the Bots, *supra*.

14 *See* Lander & Markoff, *Digital Fears Emerge after Data Siege in Estonia*, *supra*. The information in the rest of this paragraph and in much of the next paragraph is taken from this same source.

sites that usually received 1,000 hits a day were being bombarded with 2,000 hits a second.[15] The traffic increased still further on May 10, shutting down Estonia's largest bank. The May attacks also shut down the president's site and other government sites and then shifted to civilian targets: newspapers, television stations, phone systems, schools, and businesses and other financial institutions.[16] To maintain at least some internal Internet service, Estonian authorities had to block most access to Estonian sites by people outside the country; this meant, among other things, that Estonians traveling abroad could not access their email, bank accounts, or other resources.

The DDoS attacks began to wane on May 10, but would continue sporadically for weeks. Estonia's largest bank, which lost at least $1 million in the attacks, was still dealing with intermittent assaults three weeks after the attacks began to subside on May 10. Other victims had similar experiences.[17] The last major wave of after attacks finally ended on May 18.

Security experts in Estonia and abroad agreed that the country's defenders had done an excellent job of dealing with the attacks. Indeed, some said few countries could have defended themselves as skillfully. The attacks were, though, still devastating in many ways for a "wired" country that likes to call itself E-stonia.[18] They revealed how very vulnerable even sophisticated computer systems can be to DDoS attacks; Estonia's defenders were able to react as effectively as they did only because of their unusually sophisticated expertise and because their attackers recklessly put their plans for the attacks online.

At the beginning, as they began to hear rumors about the upcoming attacks, Estonian security experts found detailed plans for the attacks posted in Russian-language forums and chat groups.[19] Those who would become their country's first line of defense in the May cyber assaults watched the

---

15 *See* Steven Lee Myers, *Estonia Computers Blitzed, Possibly by the Russians*, New York Times (May 19, 2007), http://www.nytimes.com/2007/05/19/world/europe/19russia.html?ex=1337227200&en=4817e43658c91382&ei=5088.

16 *See* Lander & Markoff, Digital Fears Emerge after Data Siege in Estonia, *supra*; Myers, Estonia Computers Blitzed, Possibly by the Russians, *supra*.

17 *See* Tony Halpin, *Estonia Accuses Russia of "Waging Cyber War,"* Times Online (May 17, 2007), http://www.timesonline.co.uk/tol/news/world/europe/article1802959.ece.

18 *See* Myers, *Estonia Computers Blitzed, supra*. At the time of the attacks, in Estonia you could "pay for your parking meter via cell phone, access free Wi-Fi at every gas station, and . . . vote in national elections from your PC." Cyrus Farivar, *Cyberwar I: What the Attacks on Estonia Have Taught Us about Online Combat*, Slate (May 22, 2007), http://www.slate.com/id/2166749/.

19 *See* Lander & Markoff, Digital Fears Emerge after Data Siege in Estonia, *supra*.

attacks being orchestrated online in real-time, which gave them an obvious advantage when it came time to respond.

Afterward, Estonian authorities blamed Russia for the attacks, which they referred to as "cyberwarfare."[20] The Estonians based that allegation on several factors, one of which was that the attacks seemed to have been launched in retaliation for the government's removing a statue of a World War II Soviet solider from a park in Tallinn not long before the attacks began and not long before the May 9 holiday. (Officials had expected the removal to trigger street protests from the country's Russian-speaking minority.) Estonian authorities also noted that Russian-language sites were used to plan the attacks and claimed that a member of the Russian security service was "one of the masterminds" of the attacks.[21] Finally, the Estonian authorities claimed that security experts analyzing the DDoS attacks allegedly traced Internet addresses used in the attacks to Russian government agencies, including the office of President Vladimir Putin.[22]

The Estonians also relied on an inferential circumstance in condemning Russia for the attacks. The inferential circumstance was the premise that the attacks were of such magnitude that mere civilians could not have carried them out; according to this theory, only a state could have been responsible for the attacks.[23] The Russian government vehemently denied any involvement in them, a denial many outside Estonia found credible. Some foreign experts said it might be impossible ever to ascertain precisely who was responsible for the attacks. As one put it, "'The Internet is perfect for plausible deniability.'"[24]

As time passed, even the Estonian authorities abandoned the idea that the attacks were Russian cyberwarfare.[25] By June, Estonia's prime minister was describing them as "'criminal activity'" and asking the Russian government

---

20 *See* Halpin, *Estonia Accuses Russia of "Waging Cyber War," supra* (Estonia's Foreign Minister "accused the Kremlin" of being directly involved in the attacks).

21 *See* Ian Traynor, *Russia Accused of Unleashing Cyberwar to Disable Estonia*, Guardian Unlimited (May 17, 2007), http://www.guardian.co.uk/russia/article/0,,2081438,00.html. Other experts "were divided" as to whether it was possible to identify those involved in the attacks. *See id.*

22 *See* Lander & Markoff, Digital Fears Emerge after Data Siege in Estonia, *supra.*

23 *See* Traynor, Russia Accused of Unleashing Cyberwar to Disable Estonia, *supra.*

24 Lander & Markoff, *Digital Fears Emerge after Data Siege in Estonia, supra* (quoting Gadi Evron, Israeli computer security expert who investigated the attacks).

25 *See* John Schwartz, *When Computers Attack*, New York Times (June 24, 2007), http://www.nytimes.com/2007/06/24/weekinreview/24schwartz.html?ex=1183694400&en=bb2b9a4c1a7bfc84&ei=5070.

for assistance in finding those responsible.[26] And the international community had come to the sobering realization that the Estonian attacks were not, after all, something beyond the capacity of mere civilians. The technology and techniques used in the attacks were not new, and they did not involve tactics that only a nation-state would be able to implement.[27]

The Estonian attacks are instructive for our purposes not merely because they illustrate the kind of evils that are emerging from the dark side of cyberspace. More important for this discussion, they demonstrate how evolving threats emerging from cyberspace challenge the conceptual categories we have so far used to avoid chaos and maintain order in our societies and in our lives. What was never ambiguous was that the country of Estonia was attacked, repeatedly and maliciously. What was, and remains, ambiguous is what kind of attack it was, and who was responsible. The two issues—the nature of the attack and the identity of the attackers—are actually interrelated.

To understand why, let us consider the possibilities. The first set of possibilities goes to the nature of the attack and encompasses three alternatives: cyberwarfare, cyberterrorism, or cybercrime. We will parse these three alternatives in detail in the next chapter, so here I will simply note the basic distinctions among them.

Cyberwarfare, like its real-world[28] counterpart, consists of a military conflict between two nation-states. At a minimum, therefore, for the attacks to constitute cyberwarfare, they would have to have been launched by another country, such as Russia, which ultimately proved to be blameless. No evidence ever pointed to any other country's being responsible for the attacks.

---

26 *Estonia Asks For Russia Help to Find Web Criminals*, Reuters (June 6, 2007), http://www.javno.com/en/world/clanak.php?id=50606. A few weeks after the attacks ended, some non-Estonian experts concluded they had been launched by globally coordinated political activists. *See* Michelle Price, *Political Activists Blamed for Russian Cyber Assaults*, Information Age (June 4, 2007), http://www.information-age.com/infolog/ia_today/2007/06/political_activists_blamed_for.html.

27 *See* Schwartz, When Computers Attack, *supra*.

28 In this and succeeding chapters, I use the term "real-world" to denote the default reality we inhabit, i.e., the reality that is not cyberspace. In calling it the "real-world," I am not implying that what goes on in cyberspace is not "real," either in terms of our immediate experience of it or in terms of the consequences activity in cyberspace may have for our external environment. I use the term "real-world" to denote the physical, tangible environment in which we spend the better part of our lives. Here, "real" has the essentially the same connotation it has when coupled with "property"; in the law, "real property" is understood as being something tangible and physical, as opposed to intangible commodities such as rights. *See*, e.g., "Abatement and Revival" § 150, Corpus Juris Secundum (Thomson West 2007).

Cybercrime, like its real-world counterpart, consists of private citizens' intentionally engaging in activity that threatens a society's ability to maintain internal order. Therefore, for the attacks to constitute cybercrime, they would have to have been launched by individuals who were pursuing typical criminal goals, such as enriching themselves through theft or extortion. There was never, however, any evidence that the attacks were undertaken for the purpose of theft, extortion, or any of the motives typical of mere criminals. The attackers seemed bent on destruction for destruction's sake, and that brings us to the third and final possibility.

Cyberterrorism, like its real-world counterpart, consists of private citizens' engaging in terrorist activity. Empirically, terrorism is often indistinguishable from criminal activity because, like criminals, terrorists cause death and injury to people and damage to property; the distinction between the two consequently lies not in the conduct involved or the result achieved but in the motivations for the conduct and for its result. Terrorists act to promote ideological principles, not to enrich or otherwise benefit themselves as individuals. Therefore, for the Estonian attacks to constitute cyberterrorism, individuals who sought to promote ideological principles would have launched them, in this instance by shutting down various websites and generally crippling the country's Internet access. The fact that the attacks apparently sought destruction and disruption for their own sake inferentially supports the conclusion they were cyberterrorism, but terrorists usually take responsibility for their actions. That, after all, is the point: Terrorism—including, presumably, cyberterrorism—is havoc for political reasons. Unless the political motives for a terrorist attack are acknowledged and publicized, the attack has no purpose. And no one ever attributed political motivations to the Estonian attacks.

Where does this leave us? It leaves us with the new reality of a wired world. Estonia will never know who was responsible for the attacks or why they were launched, something that is not possible in the unwired, physical world. In the physical world, when a country is at war, it knows it is at war and, most likely, with whom. So when the German army invaded Poland in 1939, Poland knew it was at war with Germany; and when the Japanese air force bombed Pearl Harbor in 1941, the United States knew it was at war with Japan.

Activity in the physical world is visible and therefore transparent. The immediate visibility of armies invading and planes bombing translates into war, the responsibility for which is usually also apparent. Murder, theft, rape, and all the other crimes we have traditionally encountered generally reveal themselves as what they are upon commission, though their authorship may remain obscure, at least for a time. The same is true for terrorism: When two planes flew into the World Trade Center, it was immediately apparent this was terrorism, not war, not crime, and not accident. Nation-states do not use commercial airliners to wage war, and criminals do not engage in destruction for the sake of destruction. Once the nature of the attack was clear, the focus shifted to identifying those responsible through a combination of inference and crediting the eventual claims of responsibility.

None of this is true for conduct vectored through cyberspace. In the next few chapters, we will explore in detail why and how cyberspace erodes, and eliminates, our ability to identify the nature of an attack and/or the identity of those responsible. The Estonian episode was far from an isolated event; other countries, including the United States, have been the objects of similar, though rather more targeted, attacks. In each instance, as we shall see, the nature of the attack remained ambiguous and the identity of those responsible was never ascertainable.

This undeniable reality is a matter of great import for all of us—for private citizens and governments alike—because it undermines the conceptual, legal, and practical strategies we rely on to defeat chaos and maintain order within and among our societies. We have never been able to eliminate chaos in the real, physical world, but we have learned how to keep it under enough control that it does not threaten the fabric of our lives.

Cyberspace changes that. The problem we confront is that the tactics we use to control chaos in the real, physical world are generally ineffective when it comes to the cyberworld. If the chaos evolving in the cyberworld stayed in that virtual environment, we would have little or no reason to be concerned; we could simply quarantine cyberspace and isolate the problem. Unfortunately, what happens in the cyberworld does not stay in the cyberworld; it migrates out into our world because cyberspace is not a true externality. It is simply a vector for human activity, both good and bad. Cyberspace lets the worst of everyplace leak out into anyplace, and that is part of our problem.

As we shall see, the concepts and strategies we use to maintain order in the physical world are all based on the concept of "place," of geographical territory. Our notions of security are enclave notions; we control chaos by limiting its ability to manifest itself in a particular area. Our world is made up of a patchwork of enclaves in which chaos is being controlled more or less successfully. We have complicated rules and strategies for controlling chaos within and among these enclaves. And, as we shall see, all of those rules and strategies, even the ones designed to control chaos among enclaves, are based on this notion of sovereign spaces.

Cyberspace presents us with what is, in essence, a fourth (or maybe fifth) dimension—a behavioral dimension rather than a spatial dimension. Cyberspace is not "real" in any tangible sense, but as we saw with the Estonian attacks, it can have very real effects in the spatial world we inhabit. But because cyberspace is neither a "real" place nor is situated in a "real," tangible space, it is not subject to the terrestrial rules and strategies we use to control chaos within and among our physical enclaves.

The individuals whose conduct manifests itself through cyberspace are, of course, located in a terrestrial enclave, which *can* mean they are subject to these rules and strategies. Unfortunately, as we saw with the Estonian attacks, this is not inevitable: If we cannot identify those responsible for chaos emanating from cyberspace, we cannot subject them to the terrestrial rules and strategies we use to discourage this type of activity. And as we shall see, even if we can identify the perpetrators they may still be beyond the reach of these measures; they may, for example, be operating from Country A, which discourages internal chaos but has no problem with allowing its residents to prey on those residing in Countries C-Z.

In the next several chapters, I explain in detail precisely why the terrestrial rules and strategies are not effective for conduct vectored through cyberspace. The problem, essentially, is that they assume visible, identifiable activity. So, if we identify activity as war, we use our war rules and strategies to deal with it; if we identify activity as crime, we use our crime rules and strategies to deal with it; and the same is also true if we identify activity as terrorism because we currently treat terrorism as a variety of crime.[29] As I explain in the next chapters, the way we control the incidence of chaos is to react to outbreaks of war, crime, and terrorism in a way that is designed to discourage such events.

As we have seen, this system breaks down when neither the nature of the activity nor the identity of those responsible is apparent. Not only do we not know whom to target with our reactive efforts, we do not know what kind of reaction is appropriate. The Estonian authorities believed they were engaged in cyberwarfare with Russia, but their belief did not rise to the level of certainty that would have warranted an offensive counterattack with real-world weapons. Their belief was, apparently, erroneous, but what if it had not been? What if Russia had really been engaging in cyberwarfare against Estonia? What if (hypothetically) the attacks were the first of a series of

29 This assumes we are able to identify those who are responsible for the war, crime, or terrorism, which is usually true for activity in the physical world.

cyberwarfare assaults by Russia? Uncertain whether it is, in fact, at war, Estonia passively tries to fend off the never-ending, increasingly sophisticated attacks until its economy and society are so weakened they collapse, at which point Russia kindly offers to send troops to stabilize the situation.

That scenario may seem absurd, but it probably is not. As we will see in later chapters, countries are preparing for cyberwarfare, and it appears cyberwarfare will look nothing like its real-world counterpart. Real-world warfare is overt and destructive; cyberwarfare will be subtle and erosive. China, for example, has already articulated plans for cyberwarfare that involve using civilians and civilian entities in attacking foreign corporate and financial institutions. In the real, physical world, warfare is like professional football: only the designated players participate. In the cyberworld, warfare will be much more catholic; civilians are likely to be prime players and prime targets.

That creates at least the potential for conflating war, crime, and terrorism. As we all know, crime and terrorism are civil—civilian on civilian affairs—while war is the exclusive province of the military. Armies, which are often composed of erstwhile civilians, fight wars; "pure" civilians do not. Civilians are, of course, caught up in warfare, but we have developed an elaborate set of rules for how "noncombatants" are to be treated; and we refer to unavoidable harms to civilians as "collateral damage" because it is a byproduct of the purely military effort.

In cyberwarfare, it seems, there may be no room for noncombatants. And this brings me back to the point I made earlier: If civilians are legitimate targets in cyberwarfare, then how can a country tell whether it is dealing with war, crime, or terrorism? The distinctions are of profound importance in the world in which we currently live because they determine . . . everything. They determine who will respond to an attack and how they will respond. We do not, for example, use nuclear devices or other military weapons against bank robbers or terrorists. In the United States, anyway, our law bars the military from participating in civilian law enforcement; we have an absolute, unbreachable partition between civil and military threat response strategies. And we, like every other functioning country, have a carefully calibrated hierarchy of threats and an equally carefully calibrated hierarchy of threat responses for the real world.

Because it is becoming increasingly apparent that these threat and response hierarchies are not effective against cyberthreats, we must develop a new approach for cyberthreats. We must devise principles and strategies that are effective in this new threat environment. Logically, the cyberthreat strategies can either supplement our real-world threat and response hierarchies, or replace them. If we decide cyberthreats are merely a new and distinct category of threats—an analogue of crime, war, and terror—then the approach we devise will be additive; that is, it will supplement the principles and strategies we employ for these traditional, real-world threats. If, on the other hand, we decide that cyberthreats are not a distinct category of threats but are, instead, evolving variations of the three traditional threat categories, then we will either need to upgrade our current discrete threat and response hierarchies with new expanded versions or implement, instead, an entirely new, holistic approach to controlling chaos offline and online.

That is what this book is about. I am not presumptuous enough to attempt to resolve all of these issues here. My goal is rather to explain why they are issues we must confront and to offer some modest suggestions as to how we might go about resolving them. Though that may sound unambitious, it actually is not. The law-abiding, stable societies most of us enjoy are the product of centuries of struggle against chaos in its various forms. Some of the methods we employ to deal with chaos are ancient; others, such as professional policing, are relatively new. All, however, are well rooted in history, tradition, and culture; we are so accustomed to having the military deal with war and a professional police force deal with crime and terrorism that it is difficult for us to imagine anything different. What we have seems "right"—inevitable.

And so it may be. But because what we have is clearly not enough for the world we are beginning to confront, we need to think about what we can do differently to make that world as safe as possible for us and for those who come after us. And that is what we do here.

The next several chapters parse the real-world threat categories and the principles and strategies we have devised to deal with them. They also demonstrate why and how these principles and strategies are not effective against cyberthreats. We then consider how societies can improve their ability to control cyberthreats without encroaching on individual liberties or the constructive anarchy of cyberspace.

Before we begin, I need to clarify one point: Later, particularly Chapter 7, I will analyze what I call "cyb3rchaos"[30]—the potential disruption attributable to new, elusive threats emerging from cyberspace. When I discuss this potential disruption, I by no means intend to suggest that we are on the brink of a complete social and cultural meltdown. As far as I can tell, the Cyber-Vandals are not at the gate and we are not the Roman Empire in the early fifth century AD.

---

30 I define this term more precisely in Chapter 7.

But meltdown is not the only hazard that evolved civilizations face; contumacious, erosive threats can ultimately prove to be, if not equally devastating, devastating enough to present cause for concern. The British Empire, after all, never fell; it declined, to a shadow of what it had been.

In 2004, three years before I write this, the Federal Bureau of Investigation estimated that cybercrime cost U.S. citizens about $400 billion,[31] and in July of 2007, FBI Director Robert Mueller said he believes only about one-third of the cybercrime in the United States is actually reported to the FBI.[32] I have heard cybercrime cost estimates that are much, much higher than the figure cited for 2004; and as everyone involved with cybercrime knows, it has dramatically increased in the last three years and will continue to increase unless and until governments begin to create realistic disincentives for cybercriminals. I also believe, based on reliable anecdotal evidence, that the reporting rate for cybercrime in the United States, anyway, is much less than Director Mueller estimates. As we will see, victimized businesses do not report cybercrime to law enforcement for many reasons, and I suspect similar forces often influence individual victims, as well.

So, while twenty-first-century western civilization is obviously not on the brink of a Cyber-Decline and Fall, cybercrime and other threats emanating from cyberspace are, in my opinion, a very legitimate cause for concern. The reason I am writing this is because I believe the problem is solvable, but not if we continue trying to use old solutions for new evils.

---

31 *See* Alice Lipowicz, *Rentable Crime Networks Latest Security Threat*, Washington Technology (July 5, 2005), http://www.washingtontechnology.com/online/1_1/26546-1.html.

32 *FBI Director Encourages Businesses to Report Cybercrime*, Security Solutions (July 10, 2007), http://securitysolutions.com/news/security_fbi_director_encourages/.

TWO

Order & Disorder

Crime, War, and Terrorism

With the establishment of the . . . armed forces, the police . . . and prisons, the . . . structure of the modern state was . . . complete.[33]

HUMAN SOCIETIES, LIKE ALL SELF-ORGANIZING systems,[34] must maintain order if they are to survive.[35] "Order" is the subjugation of chaos; it means a sufficient measure of control has been established over the environment within which a system operates and the entities that compose it so the latter can successfully discharge the tasks necessary for the perpetuation of the system (and themselves).[36]

If a self-organizing system is to survive, it must at a minimum ensure the continuity of a populace of the entities of which it is composed. For a biological system, such as a human tribe or society, this means ensuring that (a) its constituent entities have the necessities (e.g., food, water, shelter) they need to survive and to reproduce; (b) their offspring achieve adulthood and are successfully incorporated into the system; and (c) these discrete, constituent entities and the system itself are protected from the depredations of competitors and predators.[37]

---

33 Martin van Creveld, The Rise and Decline of the State 169 (Cambridge University Press 1999).

34 The concept of self-organizing systems emerged in biology to describe structure, such as ant colonies and schools of fish, in which organization develops from interactions within the system. *See*, e.g., Steven Johnson, Emergence: The Connected Lives of Ants, Brains, Cities, and Software 17–20, 29–40, 53–57 (Scribner 2002). The principle of self-organization has since evolved to encompass a variety of systems, including human societies and organizations. *See*, e.g., Johnson, Emergence, *supra*.

35 *See* Susan W. Brenner, *Toward a Criminal Law for Cyberspace: Distributed Security*, 10 Boston University Journal of Science & Technology Law 1, 8–11 (2004). In this and later discussions, I use the term "society" as a global, generic term designating humans who live in a stable grouping of some kind. I will, in other words, use "society" and "system" essentially interchangeably to denote organized human groupings.

36 *See* Brenner, *Toward a Criminal Law for Cyberspace, supra*.

37 *See*, e.g., Edward O. Wilson, Sociobiology: The New Synthesis 37–62 (Harvard University Press 2000).

Order in human societies therefore has both an internal and an external component:[38] Internal order governs the constituent individuals' activities and relationships with each other and their relationships with the system as a whole. As I explain below, if systems are to establish and maintain internal order, they must implement rules which define these activities and relationships so that, for instance, an individual knows whether he is a worker or a soldier, a commoner or a member of the governing royal family. The rules must also specify the rights, duties, and obligations associated with the status and relationships assigned to each member of the system.

External order governs a system's relationship with its environment. For human systems, "environment" includes both (a) the physical context within which a system functions and (b) biological agents, including other humans, who can threaten the system's survival directly by attacking its constituent members and/or indirectly by competing with them for food or other essential resources. Every human system will consequently also implement rules that structure its interactions with the environment, as I explain in more detail below.

The rules that structure internal and external order do not operate independently. They instead interact and evolve, allowing the system to adapt to changes in its environment.

Rules are an absolute necessity for the emergence of self-organizing systems, including human societies. Without rules to order activities and relationships, there is no "system"; at best, there is a shifting congeries of entities, an assemblage lacking coherence and function. But though rules are a constant, the ways they manifest themselves are not. This is particularly true for societies composed of intelligent entities, such as humans. Human systems must establish and enforce rules that maintain the necessary threshold of internal and external order, but the rules they adopt can be more or less idiosyncratic, depending primarily upon the nature of the challenges the members of a particular system must confront.

As we will see below, the rules modern human social systems devise to maintain system order are, almost exclusively, territorially based. That is, each system, or society, is situated in a territorially circumscribed space to which it lays claim by the tenets of some applicable law, by the force of weapons, and/or by whatever other standard applies. This is a trait human social systems share with the systems established by other biological species, but,

---

38 *See* Brenner, *Toward a Criminal Law for Cyberspace, supra.* The following discussion of order and challenges to the process of maintaining order is drawn from this source.

as I explain below, the modern nation-state has taken the systemic focus on territory to a new level.[39]

External Order

As noted above, external order encompasses a system's relationship with its physical and biological environment. Like other systems, human societies must organize and implement the efforts of their members to deal with physical threats (earthquakes, droughts, fires) and with threats competing societies pose.[40]

For thousands of years, humans lived in tribes that were organized around kinship and had, at most, a very rudimentary division of labor.[41] The task of maintaining external order—including the task of fending off threats from competing tribes—generally fell exclusively to the male members of the tribe and was generally adventitious.[42] That is, the males belonging to a given tribe fought with neighboring or invasive tribes if and when the need arose.[43] No formal rules established a specialized occupational class for maintaining external order.

As our remote ancestors developed agriculture, they settled into what became far more complex, geographically fixed societies; the state, with its more intricate division of labor, had begun to emerge.[44] As part of this division of labor, these larger and more complex societies eventually created a separate institution—the military—to deal with threats from "outsiders," that is, from competing societies.[45] Unlike their predecessors, the members of this institution were professional soldiers; their only function was to wage war,

---

39 *See*, e.g., Walter C. Opello Jr. & Stephen J. Rosow, The Nation-State and Global Order: A Historical Introduction to Contemporary Politics 1–8 (Lynne Rienner 1999).

40 *See* Brenner, *Toward a Criminal Law for Cyberspace, supra.*

41 *See*, e.g., Martin van Creveld, The Rise and Decline of the State 2–12 (Cambridge University Press 1999).

42 *See*, e.g., *id.*

43 *See*, e.g., *id.*

44 *See*, e.g., *id.* at 1–126 tracing the evolution of human societies from tribes through city-states and empires to nation-states). *See also* Brenner, *Toward a Criminal Law for Cyberspace, supra.*

45 *See*, e.g., van Creveld, The Rise and Decline of the State, *supra. See also* Brenner, *Toward a Criminal Law for Cyberspace, supra.*

defensively or offensively.[46] The Roman Empire was perhaps the most extreme example of this trend in the ancient world.[47]

Our reliance on a professional military to deal with external human threats increased during the post-medieval eras, when societies evolved into territorially based nation-states.[48] The nation-state differed from the preceding forms of human social organization in a number of ways, but its most distinctive characteristic is the centralization and monopolization of certain "sovereign" functions and prerogatives, which it exercises within a specified territory over which its declared authority is "both exclusive and all-embracing."[49]

With the triumph of the nation-state, therefore, "territory" became the absolute point of demarcation between external threats arising from competing states and challenges to internal order, which we will examine below.[50] The nation-state assumed sole responsibility for dealing with both as part of its sovereign functions, which meant it enjoyed a monopoly on the legitimate use of force.[51] As I explain below, structural constraints dictated that the use of force to maintain internal order would be governed by system-specific rules; for centuries, the conduct of warfare was a more or less ad hoc affair,[52] but the triumph of the nation-state would change this.

Though idiosyncratic rules governing aspects of warfare evolved earlier, it was really not until the nineteenth century that nations began to develop a standardized law of warfare.[53] The goal was to establish "internationally recognized legal conventions" that governed the conduct of warfare; the purpose was to "'humanize war'" by specifying which activities were, and which

---

46 *See*, e.g., van Creveld, The Rise and Decline of the State, *supra*. *See also* Brenner, *Toward a Criminal Law for Cyberspace, supra*.

47 *See*, e.g., van Creveld, The Rise and Decline of the State, *supra*.

48 *See*, e.g., *id.* at 133 (noting that territory is perhaps "the most important characteristic" of the modern nation-state). *See also* Brenner, *Toward a Criminal Law for Cyberspace, supra*.

49 van Creveld, The Rise and Decline of the State, *supra* at 1.

50 *See*, e.g., *id.* at 133 (noting that territory is perhaps "the most important characteristic" of the modern nation-state). *See also* Brenner, *Toward a Criminal Law for Cyberspace, supra*.

51 *See*, e.g., van Creveld, The Rise and Decline of the State, *supra* at 155–170.

52 *See*, e.g., Doyne Dawson, The Origins of Western Warfare: Militarism and Morality in the Ancient World 13–77 (Westview Press 1996).

53 *See*, e.g., Karma Nabulsi, Traditions of War: Occupation, Resistance, and the Law 4–18 (Oxford University Press 1999). *See also* Theodor Meron, Henry's Wars and Shakespeare's Laws: Perspectives on the Law of War in the Later Middle Ages 7–46 (Clarendon Press 1993).

were not, acceptable.[54] This effort eventually led to the adoption of the Geneva and Hague Conventions and related instruments, which we examine in the next chapter.[55] Another product of this effort was the development of laws prohibiting those who were not "official" members of the military—those who did not wear the uniform of their country—from participating in conflicts between nation-states.[56] And as we will also see in the next chapter, nation-states began to develop law that attempted to protect noncombatants—"civilians"—from the injury, death and destruction endemic to these conflicts.[57]

Internal Order

As noted above, human systems achieve internal order by using rules to structure the relationships and activities of those who comprise a society in predictable, productive ways.[58] These "civil" rules structure important relationships (e.g., ruler-ruled, husband-wife, employer-employee) and allocate essential tasks (e.g., farmer, teacher, mayor). They set legitimate social expectations (e.g., emancipation, adulthood, voting, property ownership) and establish a baseline of order by defining the behaviors that are "appropriate" in that society.

The members of a society tend to abide by its civil rules because they are socialized to believe in those rules; they also come to perceive that conforming their behavior to the civil rules as the "right" thing to do. As they mature, the members of a society learn they will gain the approval of others and avoid their disapproval by conforming to the dictates of the civil rules. Civil rules, alone, suffice to maintain order in biological systems populated by individually nonintelligent entities, such as ants or bees, but human beings are more problematic. We are more problematic because we, as individuals, are highly intelligent. Unlike members of nonintelligent collective species, we have the capacity to deviate, that is, to deliberately violate the civil rules that are meant to maintain internal order.

---

54 *See*, e.g., Nabulsi, Traditions of War, *supra*.

55 *See*, e.g., *id*.

56 *See*, e.g., Brenner, *Toward A Criminal Law for Cyberspace, supra*. *See also* van Creveld, The Rise and Decline of the State, *supra* at 169.

57 *See*, e.g., *id*. at 164–165.

58 *See* Brenner, *Toward A Criminal Law for Cyberspace, supra*. The discussion in the next several paragraphs is all drawn from this source.

That capacity for deviance creates the need for an additional, external mechanism to enforce the civil rules. Societies have, so far, done this by implementing a secondary set of rules—"criminal" rules—that reinforce the need to obey civil rules. Every human system will have a set of criminal rules, because every society must deal with some level of deviance on the part of its members.

Every society has, for example, civil rules that define property rights and criminal rules that prohibit violating these property rights and prescribe sanctions that will be imposed upon those who commit such violations. Every society also has a more or less extensive repertoire of criminal rules and associated sanctions that buttress the civil rules' prohibition on inflicting various types of "harm" on persons and on other protected interests. These rules proscribe the infliction of physical "harm" on persons (e.g., murder, rape, and assault); they may prohibit the infliction of reputational "harm" (e.g., libel) and intangible "harms" (e.g., adultery, gambling and other "immoral" conduct). As we will see in Chapter 3, this repertoire of criminal rules encompasses the infliction of "harms" that are specifically directed at undermining the stability of the society; these "harms" include terrorism, treason, riot, and obstructing justice. The sanctions include, but are not limited to, societal disapproval, or denunciation; in most modern societies, they also emphasize punishment in the form of incarceration, death, fines, supervised release, and, in some instances, banishment.

Nation-states let individuals sort out disagreements over the application of civil rules (civil litigation), but they maintain exclusive control over the enforcement of their criminal rules because the violation of such a rule is a direct threat to the state's ability to maintain internal order. No modern nation-state can survive if its members are free to prey upon each other by violating the civil rules, and concomitant criminal rules, that protect their safety, their lives, and their property. As I explain below, modern nation-states have therefore evolved a distinct, and essentially uniform, strategy for enforcing their criminal rules.

Crime Control

Criminal rules are meaningless unless they are ***enforced;*** states consequently must ensure that these rules are being obeyed if they are to maintain internal order. Violating a criminal rule is generically known as committing a "crime," and the commission of crimes is a complex, enduring aspect of human social life.

Because the capacity for committing crimes is inherent in human nature (or, perhaps more accurately, in human intelligence), nation-states understand that they cannot eliminate crime and so strive to control it.

Historically, their efforts to control the incidence of crime have been predicated on the assumptions that publicly sanctioning those who violate criminal rules does all of the following: (a) expresses societal condemnation of the violations (and the violators); (b) exacts punishment for the affront to society; and most importantly for our purposes; (c) controls crime by deterring future violations. This last assumption incorporates two subsidiary assumptions: (a) criminal sanctions deter violations by presenting us with a simple choice—obey the rules or suffer the intentionally unpleasant consequences; and (b) rule violators will be identified, apprehended, and sanctioned.

The first subsidiary assumption is based on the premise that inflicting punishment should increase the "cost" of violating a criminal rule; this premise, in turn, is based on the assumption that prospective offenders will weigh the benefits to be gained from committing crime against the "cost" that will be imposed if they are apprehended, convicted, and sanctioned.[59] When the "cost" becomes high enough, so the logic goes, individuals will refrain from violating criminal rules.[60] Studies, however, show that the deterrent effect of punishment is a joint function of (a) the severity of the punishment and (b) the likelihood of being punished. It appears, then, that even the prospect of the death penalty may not deter someone from committing murder if they believe there is a slight chance, if any, that they will be apprehended and punished.

Crime control therefore requires that there be some system in place which ensures that criminal rule violators are identified, apprehended, and sanctioned with a fair degree of efficacy. There must, in other words, be a credible threat of retaliation for violating criminal rules; absent such a threat, the rules and their attendant sanctions cannot effectively deter crime and maintain internal order.[61] The absence of a credible threat of enforcement was a major factor in the United States' failed experiment with alcohol Prohibition in the 1920s; it

---

59 *See id.*

60 *See id.* Incarceration and capital punishment have the additional effect of incapacitating convicted offenders from re-offending—permanently when capital punishment or life imprisonment are imposed or for a defined period of time when a limited term of incarceration is imposed.

61 *See id.* There would, in other words, be no disincentive—no "cost"—associated with violating a criminal rule.

was against the law to possess or consume alcohol, but the extraordinary lack of enforcement produced what one historian calls "the toothless law."[62]

Enforcement, therefore, is critical. Until relatively recently, societies tended to rely on citizen enforcement to sustain this threat; individuals were required to apprehend criminals or face fines and other punishments. In colonial America, for example, law enforcement was the ". . . duty of every citizen. Citizens were expected to be armed and equipped to chase suspects on foot, on horse, or with wagon. . . . "[63]

Though this system may have been adequate for primarily rural societies, its effectiveness eroded as urbanization increased during the Industrial Revolution; the attendant rise in urban crime led to various efforts to develop an alternative system, all of which failed.[64] The model of crime control—or "law enforcement"—used almost exclusively in the modern world emerged in 1829, when Sir Robert Peel created the London Metropolitan Police. The Metropolitan Police was something new: an independent, quasi-military agency staffed by full-time, uniformed professionals whose sole task was to react to crimes and apprehend the perpetrators, who would then be appropriately punished. Peel's model quickly migrated to America and then spread around the world; it has been the dominant approach to crime control for at least a century. As a result, citizens in the twenty-first century assume no responsibility for crime; they see that as the sole province of professionalized police forces who maintain internal order by reacting to completed crimes.

Crime Control Model

The crime control model described above evolved to deal with real-world crime: with the commission of crime that occurs wholly in a physical environment. Four characteristics of real-world crime shaped the way this model approaches crime: (a) proximity, (b) scale, (c) physical constraints, and (d) patterns.

As to the first characteristic, in real-world crime, the perpetrator and victim are physically proximate when a crime is committed (or attempted). It is, for instance, not possible to rape, or realistically attempt to rape someone,

---

62 Andrew Sinclair, Prohibition: The Era of Excess 178–229 (Little, Brown 1962).

63 Roger Roots, *Are Cops Constitutional?*, 11 Seton Hall Const. L.J. 685, 692 (2001). *See also* Saul Cornell, A Well-Regulated Militia 12–13 (Oxford University Press 2006).

64 *See* Brenner, *Toward a Criminal Law for Cyberspace, supra.* The discussion that follows is all drawn from this source.

if the rapist and the victim are fifty miles apart; and in a nontechnological world, it is physically impossible to pick someone's pocket, rob them, or defraud them out of their property if the thief and victim are in different cities, different states, or different countries.

With regard to scale, real-world crime tends to be one-to-one crime; that is, it usually involves one perpetrator and one victim. A crime begins when the victimization of the target starts and ends when it is concluded; during this event, the perpetrator focuses her attention on consummating that distinct crime. When it is complete, she can move onto another crime and another victim. Like proximity, the one-to-one character of real-world crime derives from the constraints that physical reality imposes upon human activity: a thief cannot pick more than one pocket at a time; scam artists defraud one person at a time; and prior to firearms, it was very difficult to cause the simultaneous deaths of more than one person. Real-world crime therefore tends to be serial crime.

Physical constraints have other consequences for real-world crime. Like other areas of human endeavor, real-world crimes, even very simple crimes, require some level of preparation, planning, and implementation if they are to succeed. One who intends to rob a bank must visit it to learn about its layout, security, and routine; this exposes her to scrutiny from witnesses whose observations may later contribute to her being apprehended. As she robs the bank, she leaves trace evidence behind and is again subject to observations (height, weight, accent, skin color, sex) that can result in her being apprehended; the same is true as she flees the scene. She may have obtained a weapon or a disguise before the robbery and may need help disposing of the money afterward. Each step takes time and effort and thereby augments the exertion required to commit the crime and increases the risks involved in its commission.

Finally, patterns emerge in the real-world crimes that are committed in a given society. Victimization tends to fall into patterns for two reasons. One is that because of the combined efficacy of the civil rules discussed above and the deterrent influence of crime control efforts, only a small segment of a society will persistently commit crimes; those who do are statistically likely to be from economically deprived backgrounds (they have more to gain and less to lose) and reside in areas that share demographic characteristics (the risk of apprehension is likely to be less). These offenders will be inclined to concentrate their depredations on people who live in these areas because they are convenient victims; consequently, much routine crime in a society will be concentrated in identifiable areas. The other reason patterns emerge is that societies have a repertoire of crimes that range from more to less serious in terms of the "harm" each inflicts. Rape produces nonconsensual sexual intercourse, theft results in a loss of property, murder causes a loss of life, and so on. In societies that are maintaining the necessary baseline of internal order, serious crimes will occur much less often than minor crimes.

These characteristics became embedded assumptions about the nature of real-world crime that shaped our current approach to law enforcement. The assumption of proximity added a basic dynamic: victim-perpetrator proximity and victimization; perpetrator efforts to evade apprehension; investigation; identification, and apprehension of the perpetrator.

This dynamic reflects a time when crime was local, when victims and perpetrators lived in the same neighborhood or village. A victim might know the perpetrator by name or by reputation; if she did not know him, there was still a good chance he could be identified by witnesses or by his ties in the community. If the perpetrator was a stranger, this enhanced the likelihood of his being apprehended; he would "stand out" as someone who did not belong. Law enforcement deals effectively with this type of crime because its spatial limitations mean investigations are limited in scope; investigations still focus on the physical scene of the crime.

The model incorporates one-to-one victimization as its default assumption and that, in conjunction with another assumption, yields the proposition that crime is committed on a limited scale. The other assumption is that law-abiding conduct is the norm and crime is unusual. The assumption that crimes is unusual derives from the factors noted above: Individuals are socialized to accept civil rules as prescribing the "correct" standards of behavior; criminal rules reinforce this by emphasizing that the behaviors they condemn are outside the norm and are likely to be sanctioned. The result is that crime becomes a subset—usually a small subset—of the total behaviors in a society; the limited incidence of criminal behavior, coupled with one-to-one victimization as the default crime mode, means law enforcement personnel can focus their efforts on a limited segment of the conduct within a given society. Essentially, it means crime is "manageable" ***within*** a given nation-state.

Finally, the model incorporates the premise that crime falls into patterns. It assumes crime will be limited in incidence and in the types of "harms" it inflicts; it also assumes that an identifiable percentage of crime will occur in geographically and demographically demarcated areas. The combined effects of localized crime, and the differential frequency with which various crimes are committed, gives law enforcement the ability to concentrate its resources in areas where crime is most likely to occur, further enhancing its ability to react to completed crimes.

Summary

By the mid-twentieth century, nation-states discretely controlled most of the world's territory. Nation-states also monopolized the legitimate use of force to maintain order, both internally and externally. The systems nation-states used for this purpose were effectively mirror images of each other: They relied on professional, uniformed, hierarchically organized warriors to resolve external conflicts arising with other nation-states; and they relied on professional, uniformed, hierarchically organized law enforcement officers to maintain internal order by reacting to the commission of crimes within the territory the nation-state controlled.

The activities of a nation-state's military and its law enforcement personnel were each exclusive; those who were not authorized to participate—"civilians"—played no legitimate role in either activity. The rules governing the conduct of warfare treated civilians essentially as part of the context in which warfare occurred; they were not allowed to participate, combatants were to make every legitimate effort to avoid injuring them but, if this occurred, it could be dismissed as collateral damage. The rules governing the conduct of law enforcement activities were to some extent analogous. Here, civilians played a more significant role because they were the ostensible purpose of the whole effort; law enforcement officers' role was "to protect" civilians from the depredations of those who flouted the society's civil and criminal rules. But civilians could become part of the context of law enforcement activities, such as when civilians were unwillingly trapped at the scene of a shoot-out between criminals and police officers.

This system seems eminently reasonable to us for at least two reasons: One is that we have grown up with it, been socialized into it, seen it mythologized in movies and on TV and therefore regard it as not only reasonable but inevitable; never having experienced a different system, we cannot imagine why we would do things differently. The other reason we find this system eminently reasonable is that it seems to work quite well, or at least well enough, for real-world warfare and real-world crimes. It is, as we would presumably concede, far from perfect, but as long as it keeps our lives orderly, we cannot imagine any need for change.

And that brings us to the cyberworld and the different kinds of threats it produces. We take up those threats and their effect on this two-pronged system for maintaining order in the next chapter.

THREE

Cyberthreats

New Issues, New Challenges

Law-breaking . . . goes in the direction of the largest profits. Crime . . . tends to flourish in the legal categories and the geographical areas where enforcement is weak.[65]

HISTORY IS OFTEN INSTRUCTIVE IN analyzing the impact new technologies can have on a society's ability to maintain internal order. Roughly a century ago, law enforcement in the United States (and elsewhere) was being forced to confront automobile technology, which altered the landscape upon which crimes were committed in various ways, some of which are analogous to the issues cyberspace creates for modern law enforcement.

Unlike the challenges cyberspace creates, those resulting from criminals' use of automobiles did not result from the inadequacy of existing "legal categories." The crimes automotive technology facilitated in the early twentieth-century were well-established, traditional offenses, such as theft, robbery, and kidnapping. The challenges arising from criminals' use of automotive technology resulted from their using the technology to manipulate "geographical areas." Automobiles gave criminals a unique advantage in federal systems such as the United States. They could, as one author noted, "plan a crime in one state, execute it in another, and then return to the first state or hurtle into some other remote locality for the hiding-out . . . period."[66] Bank robbers, car thieves, kidnappers, pimps, and other criminals quickly learned that if they used motor vehicles to flee a state after committing a crime there, they could frustrate law enforcement efforts to find and apprehend them.[67] As late as 1934, Clyde Barrow, of the Bonnie and Clyde gang, wrote a letter to

---

65 Arthur C. Millspaugh, Crime Control by the National Government 278 (The Brookings Institution 1937).

66 *Id.* at 46. Criminals in nonfederal systems could accomplish something similar by committing a crime in one country and then crossing the border into another country.

67 *Id.*

Henry Ford, thanking him for his "steel-bodied V-8 automobiles" because they made it so much easier for the gang to elude police after they committed a robbery.[68]

There were several reasons why automobiles frustrated law enforcement and encouraged the commission of crimes. The most obvious, perhaps, is the one noted above: Once a criminal who had, say, committed a crime in Indiana crossed the border into Illinois, Indiana law enforcement officers no longer had jurisdiction to arrest him or to investigate the crime,[69] and Illinois law enforcement had no jurisdiction because no crime had been committed in their state. There were legal procedures by which an offender, if located, could be extradited back to Indiana for prosecution,[70] but they were complex and took time;[71] and if Indiana began such a proceeding, in the interim, the perpetrator could move into another state or disguise himself so completely that Illinois police would not be able to find him. And Illinois police might not be particularly interested in spending their time investigating a crime that had been committed outside their jurisdiction.

Another problem, at least in the earlier years of automobile technology, was that law enforcement officers might not have motor vehicles. In 1917, Hinton D. Clabaugh, Supervisor of the U.S. Department of Justice's office in Chicago, had no automobiles for his federal agents.[72] The agents were, among other things, forced to use streetcars to conduct "high-speed" chases of felons

---

68 E. R. Milner, The Lives and Times of Bonnie and Clyde 135 (Southern Illinois Press 1996). Barrow liked the V-8 Fords because they were "very fast," could travel great distances without needing to be refueled, and were so common he could "change license plates and appear to be another law-abiding citizen." *Id.*

69 *See*, e.g., Hon. Frederic S. Berman & Jay M. Lippman, *The Fugitive in New York: Can Law Enforcement Cross State Lines and Act under Color of Its Office?*, 39 New York Law School Law Review 637, 648–649 (1994). While state law generally agrees that an officer who is in hot pursuit of a suspect can arrest the suspect in another jurisdiction, out-of-state officers who are not in hot pursuit have neither the authority to make an arrest nor to investigate in another state. *See id. See also* John J. Murphy, *Revising Domestic Extradition Law*, 131 University of Pennsylvania Law Review 1063. 1064 (1983) ("Crimes . . . against the laws of any State can only be defined, prosecuted and pardoned by the sovereign authority of that State").

70 The Constitution includes a provision governing state-to-state extradition of those in flight after committing "Treason, Felony or other Crime." *See* U.S. Const. art. iv § 2 cl. 2. In 1793, Congress enacted the implementing legislation needed for this provision to become effective. *See* Act of Feb. 12, 1793, ch. 7, 1 Stat. 302 (current version codified as 18 U.S. Code § 3182). For a review of domestic extradition law and procedure, *see* Murphy, *Revising Domestic Extradition Law, supra.*

71 *See id.* at 1090–1092 (noting that state-to-state extradition procedure is "woefully cumbersome").

72 *See* Joan M. Jensen, The Price of Vigilance 18 (Rand McNally 1968).

fleeing in automobiles.[73] Clabaugh's problem was, as it so often is for law enforcement, a budgetary issue; Congress simply refused to appropriate funds for the equipment the agents needed.[74] Local police met with similar problems, at least when the technology was still new; in a 1909 speech, Louisville Police Chief J. H. Haager said citizens seemed to regard it as an "extravagance" when his department bought its first motor vehicle.[75]

Over the years, citizens came to accept that police needed automobiles and the funds to buy them; motor vehicles eventually became a basic tool of law enforcement. In at least some instances, citizen-approved funding was supplemented by other means. In an effort analogous to efforts that would occur many decades later, Chicago businessmen donated seventy-five automobiles to the Department of Justice in 1917, suggesting they be distributed for federal agents' use in various cities, which they were.[76] And as we will see in a later chapter, private sector entities have donated computer equipment to contemporary law enforcement agencies to improve their ability in dealing with cybercrime.[77]

The challenges automotive technology created for law enforcement in the United States (and elsewhere) were actually resolved rather quickly, for two reasons. One was that citizens realized that unless law enforcement officers had motor vehicles, criminals could commit crimes with relative impunity. They therefore supported the appropriation of funds for the purchase of the vehicles, and by the late 1920s, most of the "medium-to-large" police departments in the United States, as well as U.S. federal agents, "utilized the automobile for patrol service."[78] Smaller departments followed suit, and motorized police long ago became a firmly embedded assumption in our culture.

The other reason was that Congress acted quickly and decisively to address the cross-jurisdictional legal problem. Because there really was nothing states could do about the problem, Congress chose to federalize much of what had heretofore been purely state crime. Between 1910 and 1934, Congress used its

---

73 *See id.*

74 *See id.* at 17–18.

75 Cynthia Morris & Brian Vila, The Role of Police in American Society: A Documentary History 84 (Greenwood Press 1999).

76 *See* Jensen, The Price of Vigilance, *supra* at 20.

77 *See, e.g., Police Heartened by Donations from Public,* L.A. Times (May 27, 1994), 1994 WLNR 4169684 (bank donated 13 computers to Los Angeles police department); *Metro Report,* Dallas Morning News (November 6, 1992), 1992 WLNR 4797635 (company donated computer equipment to the Dallas police department).

78 Morris & Vila, The Role of Police in American Society, *supra* at 82.

power under the Commerce Clause[79] to adopt legislation that made it a federal crime to transport the following across state lines: prostitutes (Mann Act of 1910), stolen motor vehicles (Dyer Act of 1919), kidnap victims (Lindbergh Act of 1932), and other stolen property (National Stolen Property Act of 1934).[80] Congress also made it a federal crime for someone to cross state lines while fleeing prosecution for any of a list of felonies (the Fugitive Felon Act of 1934).[81] A host of similar statutes followed over the years; the twentieth century saw a vast expansion (some would say overexpansion) of federal criminal law, most of it based on the premise that only federal authorities could deal effectively with interstate criminal activity.[82] And though some believe this expansion of federal criminal jurisdiction to be doctrinally problematic,[83] it has been effective; the availability of federal investigation and prosecution means that robbers, kidnappers, and others who commit what are essentially garden-variety state crimes can no longer evade justice simply by moving from one state to another.

Unfortunately, the rather superficial analogy between the challenges automotive technology and computer technology have posed for law enforcement breaks down at this point. As I explain in more detail below, computers constitute both a much more complex and much less stable technology than automotive technology. Once someone learns to drive a motor vehicle, which is not a particularly complex task, she can essentially operate any noncommercial vehicle manufactured in, say, the last half-century. Once someone learns to use a computer, she has acquired a base level of expertise for that hardware and for the software it uses, but this is not the end of the matter. Unlike automobile technology, computer hardware and software change rapidly, which means there is an essentially continuous learning curve for computer technology; this

---

79 The Commerce Clause gives Congress the power to "regulate commerce with foreign nations, and among the several states." U.S. Constitution art. 1, § 8, subdiv. 2. In 1903, the Supreme Court upheld Congress' use of the Clause to make it a federal crime to transport lottery tickets from state to state. *See* Champion v. Ames, 188 U.S. 321, 322–324 (1903). That decision established Congress' authority to use the Commerce Clause to criminalize what had up until then been "mere" state crimes. *See* Craig M. Bradley, *Racketeering and the Federalization of Crime*, 22 American Criminal Law Review 213, 218–222 (1984). Congress used this authority to react to the increasing "culture of mobility" in the United States. *See*, e.g., Kathleen F. Brickey, *Criminal Mischief: The Federalization of American Criminal Law* 46 Hastings Law Journal 1135, 1141 (1995).

80 *See*, e.g., *id.* at 1142–1144.

81 *See*, e.g., *id.* at 1144.

82 *See*, e.g., *id.* at 1142–1144.

83 *See*, e.g., *id.* at 1172–1174. *See also* Bradley, *Racketeering and the Federalization of Crime*, *supra* at 262–266.

is, as we will see, especially true for the hardware and software law enforcement must use in investigating cybercrime and other cyberthreats, as well. It also means that those who use computer technology must have the resources to continue to upgrade the hardware and software they use.

So, as we shall see, law enforcement immediately encounters two intractable empirical phenomena: The computer hardware and software they need to investigate cybercrime and other cyberthreats must be upgraded with great frequency, which is expensive. The officers who use this hardware and software must be trained and retrained in its use as it evolves in novelty and complexity; this tends to be not only expensive but also time consuming.[84] The time officers spend learning how to use upgraded equipment takes them away from performing their official duties.[85]

The effects of these difficulties are only enhanced by the certain characteristics of the threats that come from cyberspace. As we will see, cyberspace allows criminals, terrorists and warriors to conceal their identities while launching attacks from almost any point on the globe to almost any other point.[86] Cyberspace therefore almost exponentially increases the complexity of the cross-jurisdictional investigative challenges early twentieth-century law enforcement encountered when criminals began using motor vehicles.

In the remainder of this chapter we will develop a taxonomy of cyberthreats; that is, we will define cybercrime, cyberterrorism, and cyberwarfare. In Chapters 4 and 5, we return to the issues I raised above, examining in more detail precisely how and why these cyberthreats create unprecedented challenges for law enforcement and the military in countries around the world. In Chapters 6 and 7, we consider how governments around the world can respond to these challenges.

Cybercrime

An online dictionary defines "cybercrime" as "a crime committed on a computer network."[87] That not-uncommon definition immediately prompts me

---

84 *See*, e.g., U.S. Government General Accountability Office, Cybercrime: Public and Private Entities Face Challenges in Addressing Cyber Threats 39–40 (Government Printing Office, 2007) [hereinafter, "GAO Cybercrime Report"].

85 *See id.*

86 *See id.* at 41.

87 Cybercrime—definitions from Dictionary.com, http://dictionary.reference.com/browse/cybercrime.

to ask several questions: Is cybercrime, then, different from regular "crime"? If so, how? If not—if cybercrime is merely a boutique version of regular "crime"—then why do we need a new term for it?[88]

The first step in answering these questions is parsing out what cybercrime is and what it is not. When we do this, we see that the definition quoted above needs to be modified for two reasons. The first reason is that it implicitly assumes every cybercrime constitutes nothing more than the commission of a traditional crime by nontraditional means, that is, committing theft by using a computer network instead of, say, a gun. As I have argued elsewhere,[89] that is, in fact, true for much of the cybercrime we have seen so far.

For example, online fraud such as the notorious "419 scam"[90] is nothing new as far as law is concerned; it is "old wine in new bottles."[91] People have been defrauding each other for centuries, probably for millennia. Centuries ago, English common law developed the concept of fraud to address conduct that deprived the victim of her property but could not be prosecuted as theft.[92]

Basically, "theft" consists of the perpetrator's—the thief's—"taking" property from the victim against her will or without her knowledge; in other words, "theft" is someone's taking my property without my permission.[93] Fraud victims voluntarily, even eagerly, hand their property over to the

---

88 As these questions may demonstrate, the problem I *see* with this definition is that lawyers need to be able to fit the concept of "cybercrime" into the specific legal framework used in their country and into the more general legal framework that ties together legal systems around the world in their joint battle against cybercrime. We simply cannot accomplish this if we have not precisely defined the phenomenon with which we are concerned.

As to international efforts, it might be more accurate to cite the *evolving* framework that is intended to unite legal systems in the battle against cybercrime. *See* Convention on Cybercrime, Council of Europe, November 23, 2001, C.E.T.S. No. 185, http://conventions.coe.int/Treaty/en/Treaties/Html/185.htm [hereinafter Convention on Cybercrime Treaty]; Convention on Cybercrime, Council of Europe, Signatures and Ratifications, November 23, 2001, C.E.T.S. No. 185, http://conventions.coe.int/Treaty/Commun/ChercheSig.asp?NT=185&CM=8&DF=12/11/2006&CL=ENG.

89 *See* Susan W. Brenner, *Is There Such a Thing as Virtual Crime?*, 4 California Criminal Law Review 1 ¶¶ 120–29 (2001), http://www.boalt.org/CCLR/v4/v4brenner.htm [hereinafter Brenner, *Virtual Crime*].

90 *See* Advance fee fraud—Wikipedia, http://en.wikipedia.org/wiki/Advance_fee_fraud; Nigeria—The 419 Coalition Website, http://home.rica.net/alphae/419coal/.

91 *See* Advance fee fraud, *supra.*

92 *See*, e.g., Jeffrey S. Parker & Michael K. Block, *The Limits of Federal Criminal Sentencing Policy; or, Confessions of Two Reformed Reformers*, 9 George Mason Law Review 1001, 1045 n. 176 (1001).

93 *See id.*

fraudster because he has deceived them into believing they are giving him the property as part of a legitimate transaction—a transaction from which they believe they will profit handsomely.[94] So, because fraud victims voluntarily hand their property over to the fraudster, this type of activity is not "theft" in the traditional sense.[95] English common law consequently developed a new crime—"fraud" or "larceny by trick"—that encompassed a criminal's using deception (or fraud) to induce the victim voluntarily to hand over their property.[96] The premise of this crime is that although the victim consents to transferring her property to the criminal, she does so under false pretenses, which means the transfer is, in effect, theft—the fraudster wrongfully obtains the victim's property.[97]

Until the twentieth century, criminals had only two ways to defraud others: They could do so face to face by orally offering to sell someone the Brooklyn Bridge for a *very* good price; or they could do the same thing in writing, by using snail mail.[98] The proliferation of telephones in the twentieth century added a third option: scam artists could now use the telephone to sell the Bridge, again at a *very* good price.[99] At the end of the twentieth century, fraud began migrating online. Online versions of fraud schemes are often known as "419 scams" because so many of them have come from Nigeria and section 419 of the Nigerian criminal code outlaws fraud.[100]

The same thing that happened to fraud is happening with other traditional crimes, such as theft, extortion, harassment, and trespassing.[101] These crimes, and others, are also migrating online. And there is no reason to believe that many, if not most, of our traditional crimes will not migrate online in some fashion. A few traditional crimes—such as rape and bigamy—probably will not make the transition, at least not in their current form, because the commission of these crimes requires physical activity that

---

94 *See*, e.g., David W. Maurer, The Big Con 31–102 (Anchor Books, 1999).

95 *See id.*

96 *See id.*

97 *See*, e.g., John Wesley Bartram, *Pleading for Theft Consolidation in Virginia: Larceny, Embezzlement, False Pretenses and § 19.2-284*, 56 Washington & Lee Law Review 249, 261 (1999).

98 *See*, e.g., Maurer, The Big Con 31–102, *supra*.

99 *See*, e.g., Federal Trade Commission, Putting Telephone Scams . . . On Hold (2004), http://www.ftc.gov/bcp/conline/pubs/tmarkg/target.htm.

100 *See* Advance fee fraud—Wikipedia, *supra*.

101 *See* Brenner, *Virtual Crime*, *supra* at ¶¶ 39–50, 61–68.

cannot occur online (unless we were to revise our definition of rape and bigamy to encompass virtual conduct).[102]

Unfortunately, the same is not true of homicide: While we so far have no documented cases in which computer technology was used to take human life, this is conceivable and will no doubt occur one of these days (if, indeed, it has not).[103] Those who speculate on such things postulate cases in which a killer hacks the database of a hospital and kills patients by altering the dosage of their medication or, in one ingenious scenario, altering the blood type of a patient about to undergo surgery.[104] A killer would probably find this a particularly clever way to commit murder, for several reasons. One is that the crime might never be discovered; the deaths might well be ascribed to negligence on the part of hospital staff.[105] Even if the deaths were identified as homicide, it could be very difficult to determine who the "true" intended victim was, especially if the killer had altered information on multiple patients.

But though most of the cybercrime we have encountered to this point in our cyberhistory consists of merely committing traditional crimes by new means, this has not been true of ***all*** the cybercrime we have seen. We have already encountered at least one entirely new cybercrime: the Distributed Denial of Service attacks I described in Chapter 1. As we saw there, a DDoS attack overloads servers and "make[s] a computer resource unavailable to its intended users."[106] I described the Estonian DDoS attacks in Chapter 1; another notorious

---

102 *Id.* at ¶¶ 104–26. Elements of rape and bigamy can certainly migrate online as when a rapist uses the Internet to track a victim or to trick her into meeting him at the place he plans to attack.

103 There are reports of attempts to use computer technology to cause injury or death. As one author explains, "hackers have infiltrated hospital computers and altered prescriptions . . . . [A] hacker prescribed potentially lethal drugs to a nine-year old boy who was suffering from meningitis. The boy was saved only because a nurse caught the deviation prior to the drug being administered." Howard L. Steele, Jr., *The Prevention of Non-Consensual Access to "Confidential" Health-Care Information in Cyberspace*, 1 Computer Law Review & Technology Journal 101 (1997), http://www.smu.edu/csr/Steele.pdf. This same interloper had also prescribed unnecessary antibiotics to a seventy-year-old woman. *Id.*

104 *Stealing the Network: How to Own a Continent* outlines a creative cyber-homicide scenario: *Uber*-hacker Bob Knuth tricks Saul, a student, into hacking into a hospital's wireless network to alter the archived record of a man's blood type. FX et al., Stealing the Network: How to Own a Continent 39–75 (Syngress Publishing 2004). Saul thinks he is altering a dummy record used for administrative purposes only. *Id.* When the man whose record was altered comes in for surgery some time later, he is given the wrong type of blood and dies; in the book, anyway, no one realizes why he died, nor does anyone realize his death was murder.

105 *See id.*

106 Denial of service attack—Wikipedia, http://en.wikipedia.org/wiki/Denial-of-service_attack. *See also* Chapter 1.

episode occurred in February of 2000, when a fifteen-year old Canadian known as "Mafiaboy" launched a series of DDoS attacks that effectively shut down websites operated by CNN, Yahoo!, Amazon.com, and eBay, among others.[107]

Although this was not true of the Estonian attacks, DDoS attacks are increasingly used for extortion.[108] Someone launches a DDoS attack on a website or a company's network, then stops the attack and sends an email to the victim explaining that the attack will continue unless and until the website owner or the company pays the author of the email to be "protected" from such attacks.[109] This is the twenty-first version of an old crime; it is, for example, substantively indistinguishable from a tactic the American Mafia was using more than half a century ago, though they relied on arson rather than on interrupted data streams.[110]

A DDoS attack can be used for extortion (and, as we shall see, for terrorism and for warfare), but that is not inevitable. A "pure" DDoS attack, such as Mafiaboy's 2000 attacks on CNN, Yahoo!, Amazon.com, and eBay, denies access to a website or a network but does so for no consequential reason; that is, the attacker wreaks havoc merely for the sake of doing so. She does not use the DDoS attack as a tool for committing theft, fraud, extortion, burglary, or any crime that was within a pre-twentieth-century prosecutor's repertoire.[111]

---

107 *See*, e.g., Pierre Thomas & D. Ian Hopper, *Canadian Juvenile Charged in Connection with February "Denial of Service" Attacks*, CNN.com, April 18, 2000, http://archives.cnn.com/2000/TECH/computing/04/18/hacker.arrest.01/.

108 *See*, e.g., MacAfee North America Criminology Report: Organized Crime and the Internet 2007 12 (2007), http://www.mcafee.com/us/local_content/misc/na_criminology_report_07.pdf.

> In the Internet version of a protection racket, criminal gangs will threaten companies with disruption of their networks, DoS attacks, or the theft of valuable information unless they pay a ransom. Reputation threats, in which a hacker threatens to deface a company's web site, are often part of an extortion scheme.

See also McAfee NA Virtual Criminology Report 6–19 (2005), http://www.softmart.com/mcafee/docs/McAfee%20NA%20Virtual%20Criminology%20Report.pdf; Paul McNamara, *Addressing "DDoS Extortion,"* Network World, May 23, 2005, http://www.networkworld.com/columnists/2005/052305buzz.html; Jose Nazario, *Cyber Extortion, A Very Real Threat*, IT-Observer, June 7, 2006, http://www.it-observer.com/articles/1153/cyber_extortion_very_real_threat/.

109 *See*, e.g., Erik Larkin, *Web of Crime: Enter the Professionals*, PC World, August 22, 2005, http://pcworld.about.com/news/Aug222005id122240.htm. *See also* MacAfee North America Criminology Report: Organized Crime and the Internet 2007, http://www.mcafee.com/us/local_content/misc/na_criminology_report_07.pdf.

110 *See*, e.g., Robert W. Winslow, Crime in a Free Society: Selections from the President's Commission on Law Enforcement and Administration of Justice 192–209 (Dickinson Publishing 1968).

111 *See* Brenner, *Virtual Crime*, *supra* at ¶¶ 73–76.

A "pure" DDoS attack is therefore a new type of crime: a "pure" cybercrime.[112] By "pure" cybercrime, I mean that it represents the infliction of "harm" of a type and in a fashion that could not be replicated in the real, physical world.[113] As such, it cannot be addressed with existing criminal law; we must, instead, adopt new law that makes it a crime to launch such an attack.[114]

Therefore, the first reason I find the definition quoted at the beginning of this section unsatisfactory is that it is too narrow. It does not encompass the reality that cybercrime can consist of committing "new" crimes—engaging in conduct which inflicts "harms" we have not seen before and have therefore not outlawed—as well as "old" crimes.

---

112 *See id.*

113 For the proposition that a "pure" DDoS attack does inflict "harm," *see* Susan W. Brenner, *Law in an Era of Pervasive Technology*, 15 Widener Law Journal 667, 773 (2006):

> When a DDoS attack takes a commercial site offline, the company is "harmed" because it loses actual revenue, revenue opportunities, and opportunities for advertising and public relations. The same holds for attacks that target . . . sites maintained by government agencies, educational institutions, charities, religious institutions or other groups; agency and educational sites cannot provide services and charities . . . lose opportunities to solicit financial . . . support. There is a "harm" even if an attack shuts down a personal website an elderly gentleman uses to share information about his grandchildren and hobby; he has paid whatever fees are necessary to allow him to create the website and has the right to have it remain online and undisturbed.

As to the difficulty, if not impossibility, of replicating a DDoS "harm" in the real-world, *see*, e.g., Marc D. Goodman & Susan W. Brenner, *The Emerging Consensus on Criminal Conduct in Cyberspace*, 10 International Journal of Law and Information Technology 139, 152 (2002):

> One can analogize a denial of service attack to using the telephone to shut down a pizza delivery business by calling the business' telephone number repeatedly, persistently and without remorse, thereby preventing any other callers from getting through to place their orders. Now, while it may be possible for someone to execute such a scheme, it would be very onerous to do so because it would require a great deal of physical effort and concentration on the perpetrator's part; he would have to be constantly dialing, maintaining the connection until it was broken and then redialing quickly to prevent any other calls from coming in. It would also involve a significant risk of apprehension because the victim could contact the authorities, who would presumably have no difficulty tracing the calls to the perpetrator, since he would presumably be using his personal or business telephone. (Aside from the increased risk of apprehension, the mechanics involved would make using a public telephone to conduct such a maneuver a daunting, if not impossible endeavor.)

114 The United Kingdom's 1990 Computer Misuse Act outlawed hacking and other online variants of traditional crime, but it did not address DDoS attacks. DDoS attacks were not a crime in the United Kingdom until new legislation to that effect was adopted in November of 2006. Tom Espiner, *U.K. Outlaws Denial-of-Service Attacks*, CNET News.com (November 10), 2006, http://news.com.com/2100-7348_3-6134472.html.

The other reason I find the definition unsatisfactory is that it equates the commission of cybercrime with the use of a computer network.[115] The use of networks is certainly characteristic of cybercrime; indeed, it is the default mode of cybercrime (which is why some experts say that leaving an "air gap"—not connecting a computer to a network—is the best security measure).[116] It is possible, though, to use a non-networked computer for illegal purposes. A non-networked computer can, for example, be used to counterfeit currency or to forge documents.[117] In either instance, a computer—but not a computer network—is being used to commit a crime. Here, the computer is being used to commit an "old" crime, but it is also possible that a non-networked computer could be used to commit a "new" crime of some type.

A more accurate definition of cybercrime is that it involves using computer technology to engage in activity that has been proscribed by a society's criminal law. In other words, as we saw in Chapter 2, cybercrime consists of using computer technology to engage in activity that threatens a society's ability to maintain internal order.[118] This definition eliminates the limiting assumptions

---

115 *See* Cybercrime—definitions from Dictionary.com, http://dictionary.reference.com/browse/cybercrime.

116 *See*, e.g., Michael Bobbitt, *(Un)Bridging the Gap*, Information Security (July 2000), http://infosecuritymag.techtarget.com/articles/july00/cover.shtml.

117 *See*, e.g., United States v. Nettles, 476 F.3d 508, 511–512 (7th Cir. 2007) (defendant used computer to counterfeit currency); United States v. Reddick, 185 Fed. Appx. 817, 818–819 (11th Cir. 2006) (defendant used computer to forge checks); State v. Brewster, 2004 WL 1284008 (Ohio App. 2004) (same). *See also* United States Secret Service: Know Your Money—Counterfeit Awareness, http://www.secretservice.gov/money_technologies.shtm.

118 In this chapter and in the chapters that follow, we will assume that cyberthreats of whatever type undermine a society's—a discrete nation-state's—ability to maintain internal and external order. *See* Chapter 2.

It is at least conceptually possible that at some point nation-states will cease to become the defining institution responsible for maintaining order of either type. It is, for example, possible that nation-states will merge into larger, perhaps regional entities so they can combine resources and more effectively control cyberthreats and other evolving threats to social order. We *see* what may be the beginnings of such a strategy in the Organization of American States and the Asia-Pacific Economic Cooperation initiative. Indeed, while this seems unlikely at the present, we might someday see the emergence of a global government, which would assume these tasks.

I offer no opinion on the likelihood that the default institution upon which we rely for the maintenance of social order will evolve beyond the nation-state, at any point in the future. I do not think it matters for the analysis I present in this book. If all nation-states merge into a single global government, that government will still need to maintain internal order . . . unless and until technologies fundamentally alter human nature and eradicate the possibility of deviating from the rules that maintain order. And that government might still grapple with the need to maintain external order in the face of, say, threats from rogue stateless groups.

that (a) equate cybercrime with the commission of traditional "crimes" and (b) link it with networked computers. It consequently encompasses (a) both traditional and emerging cybercrimes and (b) *any* use of computer technology that poses a threat to the maintenance of internal order.[119]

This, though, is a conceptual, rather than an operational, definition of cybercrime; that is, it answers the questions I posed at the beginning of this discussion. It defines cybercrime as a category of criminal activity that is distinct from traditional "crime" and, in so doing, provides an essential conceptual foundation for policymakers who are charged with allocating resources for law enforcement, setting operational priorities, and enacting cybercrime-appropriate legislation.

The definition given above is not an operational definition because it does not provide the legal predicate law enforcement needs to respond to cybercrimes. An operational definition of cybercrime encompasses the aggregate body of legal definitions a society uses to proscribe the commission of cybercrimes, old and new.[120] An operational definition furthers the goal of maintaining order within a nation-state by informing the populace as to what uses of computer technology are, and are not, acceptable. And by doing this, it defines the parameters within which the nation-state's law enforcement officers operate; in nation-states that have established an adequate operational definition of cybercrime, law enforcement has the doctrinal tools it needs to react to, and thereby attempt to deter, this type of criminal activity.

Because cybercrime is clearly still evolving, and because cybercriminals exploit gaps in law that result from this evolution, operational definitions of cybercrime must also evolve. Nation-states must monitor online activity to identify emerging but as yet unproscribed activities that threaten their ability to maintain internal order. Once they identify a threat, countries should

---

119 It may seem peculiar that this definition does not include the intent to inflict "harm," because we use intent to differentiate between the criminal infliction of "harm" and lesser, albeit still "harmful" conduct that we address via the administrative or civil justice processes. It is certainly possible that the inadvertent or negligent use of computer technology—especially networked computer technology—could cause "harm" of a type and a scale commensurate with that which would trigger the imposition of criminal liability if it were done intentionally. I did not include intent in the definition given above because my goal here is to articulate an adequate conceptual definition of cybercrime. I *see* intent as an issue that arises when we move from the conceptual to the operational stage, i.e., when we begin to define particular types of cybercrime.

120 *See* Chapter 2.

criminalize it, just as the United Kingdom rather belatedly criminalized DDoS attacks.[121]

Cyberterrorism

Conceptually, cyberterrorism consists of using computer technology to engage in terrorist activity.[122] This definition mirrors the conceptual definition of cybercrime I proposed in the previous section, which is appropriate given that societies generally treat terrorism as a type of crime.[123]

In modern history,[124] the move to criminalize terrorism began in the 1930s as a reaction to the 1934 assassination of King Alexander I of Yugoslavia

---

121 *See*, e.g., Espiner, U.K. Outlaws Denial-of-Service Attacks, *supra*.

122 *See* Clay Wilson, Congressional Research Service, Computer Attack and Cyberterrorism: Vulnerabilities and Policy Issues for Congress (Library of Congress, Updated April 1, 2005). *See also* John Rollins & Clay Wilson, Terrorist Capabilities for Cyberattack: Overview and Policy Issues 3–5, Congressional Research Service (Library of Congress, 2007).

123 *See*, e.g., 18 U.C. Code §§ 2332a & 2332b (federal terrorism offenses). As one author notes, the law enforcement approach to terrorism "is grounded in international criminal law and extant core anti-terrorism conventions." Mark D. Kielsgard, *A Human Rights Approach to Counter-terrorism*, 36 California Western International Law Journal 249, 253 (2006) (citing a number of United Nations and regional antiterrorism conventions). As this author also notes, the model is "carried out by civilian authorities and treats terrorism as a criminal matter." *Id.* at 254.

124 Terrorism is at least two thousand years old: In the first century BC, the Zealots Sicarii, a violent branch of an ancient Jewish political group, used violence in an attempt to expel their Roman rulers from Judea. *See*, e.g., "Zealotry," Wikipedia, http://en.wikipedia.org/wiki/Zealotry. *See also* M. Cherif Bassiouni, *Assessing "Terrorism" into the New Millennium*, 12 DePaul Business Law Journal 1, 1 (2000). One group of Zealots—the Sicarii—assassinated Roman citizens and Jews who did not support their effort to end Rome's colonial rule. *See id.* The ancient and medieval eras saw the rise of many other terrorist groups, perhaps the most notable being the eleventh century Order of the Assassins, a Muslim sect that was active in Persia, Syria, and Palestine. *See*, e.g., Walter Laqueur, The New Terrorism: Fanaticism and the Arms of Mass Destruction 11 (Oxford University Press 1999). As one author notes, their major historical contribution was pioneering the use of "suicide terrorists," in their case, suicide assassins. *See id.* For other ancient and medieval terrorist groups, *see id.* at 10–13.

Authorities in the ancient and medieval worlds seemed to have responded to terrorist acts in the same way they responded to the commission of other, more traditional crimes. *See*, e.g., Bassiouni, Assessing "Terrorism" into the New Millennium, *supra* at *1 (Roman governor of Palestine responded to Zealot Sicarii attacks by "ordering the death of all zealots," among other things). In the eighteenth and nineteenth centuries British officials responded in the same way to the thugs, organized bands of what seem to have been a type of terrorist operating in India. *See*, e.g., Percival Griffths, The British Impact on India 204–206 (MacDonald 1952). *See also* David J. Whittaker, Terrorists and Terrorism in the Contemporary World 19–20 (Routledge 2004).

by terrorists.[125] The effort culminated in the issuance of the 1937 League of Nations' Convention for the Prevention and Punishment of Terrorism that required countries ratifying the Convention to adopt legislation criminalizing terrorism.[126] Article 1 of the Convention defined terrorism as "criminal acts directed against a State and intended or calculated to create a state of terror in the minds of particular persons, or a group of persons or the general public."[127] Article 2 defined the "criminal acts" encompassed by the Convention as acts (a) causing death, serious injury or loss of liberty to heads of state, their spouses and other public officials; (b) willfully causing the destruction of public property and (c) endangering the lives and safety of the general public.[128]

Members of the League of Nations, especially countries that had dealt with terrorism, supported the drafting of the Convention.[129] Czechoslovakia, for example, asserted that criminalizing terrorism was "necessary to protect 'security of life and limb, health, liberty and public property intended for the

---

> Terrorism acquired its contemporary connotation from its use during the French Revolution. The French legislature led by Maximilien Robespierre, . . . ordered the public execution of 17,000 people ('regime de la terreur') to educate the citizenry of the necessity of virtue. Robespierre's supporters . . . accused him of using terrorism in an attempt to identify the illegitimate use of terror. Terrorism, initially associated with state-perpetrated violence, shifted to describing non-state actors following its application to the French and Russian anarchists of the 1880s and 1890s.

Reuven Young, *Defining Terrorism: The Evolution of Terrorism as a Legal Concept in International Law and Its Influence on Definitions in Domestic Legislation*, 29 Boston College Comparative & International Law Review 23, 27–28 (2006).

125 *See* Ben Saul, *The Legal Response of the League of Nations to Terrorism*, 4 J. Int'l Crim. Just. 78, 79 (2006):

> King Alexander I of Yugoslavia was assassinated by Croatian separatists while on a state visit to France; the French Foreign Minister . . . and two by-standers were also killed. The suspects fled to Italy, and France requested their extradition under a treaty . . . which excluded political crimes from extradition. The Court of Appeal of Turin refused to surrender the accused because regicide and related offences were politically motivated and thus non-extraditable.

The League of Nations came under political pressure to address this problem. *See id.* Its response was shaped by earlier proposals that sought to criminalize terrorism. *See id.* at 80.

126 *See* Young, *Defining Terrorism, supra* at 35–36.

127 Convention for the Prevention and Punishment of Terrorism, November 16, 1937, art. I(2), 19 League of Nations O.J. 23 (1938).

128 *See* Lee Bruce Kress, Marius H. Livingston, & Marie G. Wanek, International Terrorism in the Contemporary World 364 (Greenwood Press 1978).

129 *See* Saul, *The Legal Response of the League of Nations to Terrorism, supra* at 82–83.

common use.'"[130] Other nations agreed, asserting that terrorism should be criminalized because it inflicted "harm" of type surpassing that caused by regular "crime;" several claimed that whereas "crime" is an offense against individuals, terrorism is an offense against the sanctity and security of the state.[131] The League of Nations' Convention never went into effect, but its approach ultimately proved influential; its successor, the United Nations, has consistently defined terrorism as criminal activity.[132]

Societies conflate crime and terrorism because both threaten their ability to maintain internal order.[133] The implicit assumption is that all threats to internal order should be dealt with in the same way, that is, as distinct varieties of "crime." This assumption is the product of pragmatic considerations: As we saw in Chapter 2, societies have historically relied on two sets of internal rules ("civil" and "criminal") and a unique institution (the military) to maintain internal and external order, respectively. Since terrorist threats have historically come from actors within a society, and since they threatened internal order, the logical solution was to treat these threats as but another type of crime, rather than developing a distinctive new approach for terrorism.[134]

Although societies conflate crime and terrorism, we must distinguish the two because they differ in ways that are relevant to how societies respond to these threats. We, of course, are concerned with how societies can respond effectively to threats once they migrate online.[135] Basically, crime is personal,

---

130 *Id.* (quoting J. Starke, The Convention for the Prevention and Punishment of Terrorism, 19 BYBIL (1938) at 60). As one author noted, "ordinary criminal offences aim to achieve the same object." *Id.*

131 *See id.* at 80–83.

132 *See* Young, *Defining Terrorism, supra*, 29 Boston College Comparative & International Law Review at 36–40. *See, e.g.*, G.A. Res. 49/60, U.N. Doc. A/RES/49/60 (December 9, 1994), http://www.un.org/documents/ga/res/49/a49r060.htm.

133 *See* Chapters 1 & 2.

134 Some claim the terrorism-as-crime model is inadequate for dealing with today's transnational terrorists and should therefore be supplemented with a terrorism-as-war model, in appropriate instances. *See, e.g.*, Fionnuala Ní Aoláin, *Hamdan and Common Article 3: Did the Supreme Court Get It Right?*, 91 Minnesota Law Review 1523, 1554–1555 (2007); Kim Lane Scheppele, *We Are All Post-911 Now*, 75 Fordham Law Review 697, 615–616 (2006); Noah Feldman, *Choices of Law, Choices of War*, 25 Harvard Journal of Law & Public Policy 257, 258 (2002). These arguments often derive, at least in the United States, from the greater restrictions U.S. law places on evidence gathering in criminal cases than in military and intelligence endeavors. *See, e.g.*, Scheppele, *We Are All Post-911 Now, supra*.

135 *See* Chapter 2.

while terrorism is political.[136] Crimes are committed for individual and personal reasons, the most important of which are personal gain and the desire or need to harm others psychologically and/or physically.[137]

Terrorism often results in the infliction of harms indistinguishable from those caused by certain types of crime (such as death, personal injury, or property destruction), but the harms are inflicted for very different reasons.[138] A federal statute, for example, defines "terrorism" as (a) committing acts that *constitute crimes* under the laws of the United States or another country (b) in order to do any of the following: intimidate a civilian population; influence government policy by intimidation or coercion; or affect the conduct of a government by "mass destruction, assassination, or kidnapping."[139] This formulation, like many others, defines the unique crime of terrorism by incorporating the definition of other, more traditional offenses. That is, one commits terrorism by committing any of a set of specified traditional crimes for the explicit purpose of advancing a political agenda.

As the statutory definition summarized above suggests, terrorism is usually intended to demoralize a civilian population, directly or indirectly.[140] We are unfortunately familiar with terrorist acts that were intended to directly demoralize a civilian population, such as the 9/11 attacks in the United States and the 3/11 Madrid bombings.[141] In both instances, violence was used for symbolic purposes, the goal being to terrify and demoralize the civilian populace of societies with which Al-Qaeda deems itself to be at war—an ideological war aimed at allowing the restoration of the "ancient

---

136 *See*, e.g., Paul R. Pillar, Terrorism and U.S. Foreign Policy 13–14 (Brookings Institution Press, 2004). "Terrorists' concerns are macroconcerns about changing a larger order; . . . criminals are focused on the microlevel of pecuniary gain and personal relationships." *Id.* at 14.

137 *See id.* at 13–15.

138 *See*, e.g., Pippa Norris, Montague Kern & Marion Just, *Introduction: Framing Terrorism*, *in* Framing Terrorism: The News Media, the Government, and the Public 3, 8 (Pippa Norris, Montague Kern & Marion Just eds., Routledge, 2003) [hereinafter Framing Terrorism] (distinguishing terrorism from "crimes motivated purely by private gain, such as blackmail, murder, or physical assault directed against individuals, groups, or companies, without any political objectives").

139 18 U.S.C. § 2331(1) (2000). For more definitions, *see*, e.g., Mohammad Iqbal, *Defining Cyberterrorism*, 22 John Marshall Journal of Computer & Information Law 397 (2004).

140 *See* John Rollins & Clay Wilson, Terrorist Capabilities for Cyberattack: Overview and Policy Issues 3, Congressional Research Service (Library of Congress, 2007).

141 *See*, e.g., "September 11, 2001 attacks," Wikipedia, http://en.wikipedia.org/wiki/September_11,_2001_attacks; "2004 Madrid train bombings," Wikipedia, http://en.wikipedia.org/wiki/Madrid_bombing.

Islamic caliphate."[142] The goal in these and similar attacks is to demoralize civilians by demonstrating their vulnerability, which, of course, is the result of their government's inability to protect them from seemingly random violence.[143]

In the real, physical world, therefore, terrorism usually achieves its primary goal[144] of demoralizing civilians by destroying property and injuring or killing civilians.[145] This focus on "harming" civilians and civilian-owned property distinguishes terrorism from warfare, which, as we will see at the end of this chapter, is not supposed to target civilians.[146]

The world's experience so far has been primarily, perhaps exclusively, with the real-world terrorism that focuses on achieving civilian demoralization by ***directly*** demonstrating civilians' vulnerability to attack. The tactics modern, Al-Qaeda-style terrorists use approach a theater of violence in

---

142 *See* Lawrence Wright, The Looming Tower: Al-Qaeda and the Road to 9/11 175, 234–35 (Knopf 2006). *See also* Norris, Kern, & Just, *Framing Terrorism, supra* at 7–8.

143 One source explains how this demoralization ties into the terrorists' goals:

> Terrorists may create . . . fear . . . to influence their negotiations with . . . governments, but fear has secondary consequences that further undermine government authority. . . . [F]ear fragments and isolates society into anxious groups of individuals concerned only with their personal survival . . . "Terrorism destroys the solidarity, cooperation, and interdependence on which social functioning is based, and substitutes insecurity and distrust."

Huddy et al., *Fear and Terrorism, in* Norris, Kern, & Just, *Framing Terrorism, supra* (quoting Martha C. Hutchinson, *The Concept of Revolutionary Terrorism*, 6 Journal of Conflict Resolution 288 (1973)). *See also* Information Operations: Warfare and the Hard Reality of Soft Power 92 (Leigh Armistead ed., Brassey's Inc. 2004) [hereinafter Information Operations] ("terrorism is an attack on the legitimacy of the established order").

144 It is important to realize—especially when we begin to analyze cyberterrorism—that terrorists also have secondary goals. Their secondary goals involve the successful conducting of activities that sustain and promote their ability to work toward achieving their primary goal. These goals include disseminating propaganda; recruiting new members of a terrorist group and retaining existing members; fund-raising to support terrorist activities and the terrorists themselves; training terrorists in attack strategies; coordinating attacks; and researching attack targets. *See, e.g.*, Eben Kaplan, Council on Foreign Relations, Terrorists and the Internet (2006), http://www.cfr.org/publication/10005/. *See generally* U.S. Department of State, Country Reports on Terrorism 2006 17 (2007), http://www.state.gov/s/ct/rls/crt/2006/ [hereinafter Country Reports on Terrorism 2005].

145 "[A]cts done to advance an ideological . . . cause and to induce terror in any population . . . are terrorism if they cause . . . death or serious injury; serious risk to public health or safety; destruction or serious damage to property." Young, *Defining Terrorism, supra* at 86 (summarizing Terrorism Suppression Act, 2002, § 5 (N.Z.)).

146 *See* U.N. Office of the High Comm'r for Human Rights, *Geneva Convention Relative to the Protection of Civilian Persons in Time of War*, August 12, 1949, http://www.unhchr.ch/html/menu3/b/92.htm. *See also* Terrorism—Wikipedia, http://en.wikipedia.org/wiki/Definition_of_terrorism.

which death, injury and destruction become at once an object lesson and a threat.[147] They are, as I noted above, an object lesson in the government's inability to keep its populace safe, and they are a concomitant threat of more death and destruction to come. As one author noted, the violence "is intended to produce effects beyond any immediate physical damage."[148] The immediate victims of violence therefore serve merely as "attention getting devices."[149]

Below, I argue that cyberterrorism seeks to achieve the same ends by different means. It approaches the task more subtly: Instead of relying on theatrical violence, cyberterrorism seeks to erode our confidence in the constituent infrastructures that are essential for our security and survival in modern, urban societies. This indirect approach does not rely on inculcating fear that we or our loved ones will become the victims of random violence. Instead, it seeks to achieve something that may be even more devastating in the long run. While the fear that we or those we love will become the victims of sudden, random violence can certainly be demoralizing, there is always the possibility that people will come to believe random violence is a controllable threat. "If I don't ride the commuter train, I'll be safe." "If I don't shop at the central marketplace, I'll be safe." We have seen these kinds of adaptive behaviors in Israel and in sectarian-violence plagued Iraq, among other places.[150]

As I explain below, the indirect approach I ascribe to cyberterrorism focuses on using technology to erode our confidence in the information and the systems we necessarily rely on to function in our modern, urban environments. The fear it seeks to inculcate is not the fear of death, maiming or explosive property destruction; it is the fear of losing control. If systems do not work—if there is no electricity, no water, no phone service, or no ATM service—I cannot function in the only way I, as a modern urbanite, know to conduct my life. My world begins to come apart, in frightening and erratic ways, and that can, in its own way, be as demoralizing as the 9/11 attacks. We will, as I said, return to this issue below.

---

147 *See*, e.g., Mohammed M. Hafez, Suicide Bombers in Iraq 11–14 (United States Institute of Peace 2007).

148 Steven Livingston, The Terrorism Spectacle 3 (Westview Press, 1994).

149 *Id.*

150 *See*, e.g., Liza M. Wiemer & Benay Katz, Waiting for Peace: How Israelis Live with Terrorism 159–61 (Gefen Publishing, 2005); *Life in Iraq*, BBC News, October 31, 2006, http://news.bbc.co.uk/2/hi/in_depth/middle_east/2006/life_in_iraq/.

To date, there have apparently been no known instances of cyberterrorism.[151] There have been cases the media has incorrectly described as cyberterrorism: In 2000, for example, an Australian man hacked into a municipal waste-management system and dumped "millions of litres of raw sewage" into parks, rivers, and businesses.[152] And in 1997, a Massachusetts hacker shut down all communications to a Federal Aviation Administration control tower at an airport for six hours.[153] But these and other cases cited as cyberterrorism actually constitute cybercrime. In each instance, the perpetrator acted out of individual motivations—a desire for revenge or power—rather than out of a desire to advance a particular ideology by demoralizing segments of a civilian population.[154] The motivations, in other words, were personal, not political.

To understand what cyberterrorism can—and likely will—be, we need to parse out how terrorists can use computer technology to achieve their goals, that is, to demoralize civilians and undermine a society's ability to sustain

---

151 *But see* Arquilla & Ronfeldt, *Waging War Through the Internet, supra* at E1. *See also* GAO Cybercrime Report at 21–22:

> In March 2005, security consultants . . . reported that hackers were targeting the U.S. electric power grid and had gained access to U.S. utilities' electronic control systems. Computer security specialists reported that, in a few cases, these intrusions had "caused an impact." While officials stated that hackers had not caused serious damage to the systems . . . the constant threat of intrusion has heightened concerns that electric companies may not have adequately fortified their defenses against a potentially catastrophic strike.

152 Tony Smith, *Hacker Jailed for Revenge Sewage Attacks*, Register, October. 31, 2001, http://www.theregister.co.uk/2001/10/31/hacker_jailed_for_revenge_sewage/.

153 Bill Wallace, *Next Major Attack Could Be Over Net*, San Francisco Chronicle A1 (November 12, 2001), http://www.sfgate.com/cgi-bin/article.cgi?file=/chronicle/archive/2001/11/12/MN29929.DTL.

154 Probably the closest thing we have to a reported cyberterrorist attack came in 1998 when:

> Tamil guerrillas swamped Sri Lankan embassies with 800 e-mails a day over a two-week period. The messages read, "We are the Internet Black Tigers and we're doing this to disrupt your communications." Intelligence authorities characterized it as the first known . . . attack by terrorists against a country's computer systems.

Rohas Nagpal, *Cyber Terrorism in the Context of Globalization*, 2 World Congress on Informatics & Law 22 (2002), http://www.ied.org/congreso/ponencias/Nagpal,%20Rohas.pdf. *See also* Council on Foreign Relations, Liberation Tigers of Tamil Eelam (2006), http://www.cfr.org/publication/9242/ (Tamil Tigers as terrorists).

Some might argue that this attack does not qualify as cyberterrorism because it targeted computer systems at embassies located in countries other than Sri Lanka, and therefore did not impact Sri Lanka's civilian populace. But the attack did shut down embassy computers and "had the desired effect of generating fear in the embassies." Dorothy E. Denning, *Activism, Hacktivism, and Cyberterrorism: The Internet as a Tool for Influencing Foreign Policy*, *in* Networks and Netwars: The Future of Terror, Crime, and Militancy 236, 239 (John Arquilla & David F. Ronfeldt eds., RAND 2002).

internal order.[155] Conceptually, computer technology's use for this purpose falls into three categories: (a) weapon of mass destruction; (b) weapon of mass distraction; and (c) weapon of mass disruption.[156] We will examine each, in order.

Weapon of Mass Destruction

This is a conceptual option, but not, I submit, a real possibility.[157] The notion that computer technology can be a weapon of mass destruction is based on a flawed premise: the notion that computers, alone, can be used to inflict the kind of demoralizing carnage the world saw in New York and Washington, D.C., on 9/11 or in Madrid on 3/11.[158] Computers, as such, cannot inflict physical damage on persons or property; that is the province of real-world implements of death and destruction.[159]

---

155 While some dismiss the possibility of cyberterrorism, others correctly understand that it is not merely a possibility, but probably an inevitability. *See* Arquilla, *Waging War Through the Internet*, *supra* at E1:

> Despite . . . al Qaeda's long-standing interest in cyber terror, we have been . . . dismissive of this burgeoning threat. In part, that's because we doubt terrorists will focus on using computers to attack computer systems, believing instead that "real terrorists" want to kill people and blow things up. . . .
>
> From a purely psychological point of view, this idea makes sense, as traditional terrorists have been leg-breakers. . . . But . . . we have made it very hard for al Qaeda to mount new attacks within the United States. So, if Osama bin Laden wants to pursue his goal of attacking our economy, disruptive cyber-terror strikes via the Internet are likely to be an increasingly important element in his offensive.

156 The discussion that follows focuses on terrorists' use of computer technology to further their *primary* goal of advancing an ideological agenda. It does not address the use of computer technology to further the secondary goals noted earlier.

157 In testifying before Congress, USSTRATCOM Commander General Cartwright said that "regret factors associated with a cyber attack could . . . be in the magnitude of a weapon of mass destruction." U.S.-China Economic & Security Review Commission, 110th Cong., Report to Congress 13 (2007) [hereinafter, "U.S-China Economic & Security Review Commission Report 2007"], http://www.uscc.gov/annual_report/2007/07_annual_report.php. Because General Cartwright's "regret factors" referred to the "psychological effects that would be generated by the sense of disruption and chaos caused by a cyber attack", I think he actually meant that a cyberattack can constitute a weapon of mass disruption, which I discuss later in this chapter.

158 2004 Madrid Train Bombings—Wikipedia, http://en.wikipedia.org/wiki/11_March_2004_Madrid_train_bombings.

159 The erroneous assumption that computer technology is merely another mode of mass destruction accounts for the skepticism many express about the prospects of a "digital Pearl Harbor" or a "digital 9/11." *See*, e.g., Drew Clark, *Computer Security Officials Discount Chances of "Digital Pearl Harbor*," GovExec.com (June 3, 2003), http://www.govexec.com/dailyfed/0603/060303td2.htm.

But computers ***can*** be used to set in motion forces that produce physical damage. Instead of hacking into a municipal waste-management system for revenge, cyberterrorists could disable the systems that control a nuclear power plant and cause an explosion such as the one at Chernobyl in 1986.[160] By claiming responsibility for the catastrophe, the cyberterrorists could exploit the resulting illness, death, and radioactive contamination to undermine citizens' faith in their government's ability to protect them and maintain order.

This is a viable terrorism scenario, but it is not a ***cyber***terrorism scenario. Though computer technology is used to trigger the explosion, the victims would recall this as a ***nuclear*** catastrophe, not as a ***computer*** catastrophe. Here, as in other computer as weapon of mass destruction scenarios, computer technology plays an incidental role in the commission of a terrorist act. It acts as a detonator, not as a weapon. To characterize scenarios such as this as cyberterrorism is as inappropriate as describing the 1998 U.S. embassy bombings carried out by Al-Qaeda as automotive-terrorism because vehicles were used to deliver the bombs to the target sites.[161]

Weapon of Mass Distraction

This is both a conceptual and a real possibility. Here, computer technology plays a pivotal role in the commission of a terrorist act, an act that differs in essential ways from the real-world terrorism to which we are accustomed. Computer technology is used to manipulate a civilian population psychologically. This manipulation saps civilian morale by undermining citizens' faith in the efficacy of their government.[162] Depending on the type of manipulation involved, it can also result in the infliction of personal injury, death, and property destruction.

To understand how computer technology could be used purely for psychological manipulation, consider this scenario: on September 11, 2001, as

---

160 *See*, e.g., Chernobyl disaster—Wikipedia, http://en.wikipedia.org/wiki/Chernobyl_accident. *See also* Barton Gellman, *Cyber-Attacks by Al-Qaeda Feared*, Washington Post A1 (June 27, 2002), http://www.washingtonpost.com/ac2/wp-dyn/A50765-2002Jun26?start=24&per=24 (reporting that in 1998 a twelve-year-old hacker "broke into the computer system that runs Arizona's Roosevelt Dam" and could have released 489 trillion gallons of water, which would have flooded the cities of Mesa and Tempe).

161 *See*, e.g., Wright, The Looming Tower, *supra* at 270–72.

162 Susan W. Brenner & Marc D. Goodman, *In Defense of Cyberterrorism: An Argument for Anticipating Cyber-Attacks*, 2002 University of Illinois Journal of Law, Technology & Policy 1, 31–40.

planes crashed into the World Trade Center and the Pentagon, millions of Americans watched the events unfold on television; many also used the Internet to try to find out more about what was happening.[163] The CNN site experienced particularly heavy traffic that day.[164] What if, instead of finding CNN-generated content, these visitors had encountered a Web page that announced—in appropriately terrifying graphics—"World War—Nuclear Holocaust in Europe and Australia, Japan Devastated by Chemical Attack"?[165]

Because this was 2001, an over-the-top Orson Welles "War of the Worlds" reaction would be unlikely;[166] for at least the last decade people have typically obtained their news from several types of media and from various sources within each type. But the posting of such a falsified page could have acted as a terror multiplier, enhancing the unnerving effects of the day's real-world terrorist events.[167] It could also have left lingering doubts in the public's mind as to whether "the government" had actually "covered up" the extraterritorial disasters once reported on CNN. These doubts could have provided the predicate for a long-term campaign of eroding public confidence in public officials and news outlets.

Now consider a scenario coupling psychological manipulation with injury, even death. At 1:00 pm on a Wednesday in San Francisco, the local Office of Emergency Services and Homeland Security receives messages via a secure government computer system informing them that a "suitcase nuclear device" is on the Bay Area Rapid Transit (BART) system, the public transportation system that serves San Francisco and surrounding cities.[168] The officials are told the device is in the hands of terrorists who will detonate it in two hours, at 3:00 pm. If such a device were detonated, the death and destruction would be unimaginable—far greater than that inflicted on 9/11.

---

163 *See*, e.g., September 11, 2001 timeline for the day of the attacks—Wikipedia, http://en.wikipedia.org/wiki/September_11,_2001_timeline_for_the_day_of_the_attacks ("8:49:34 a.m.—CNN and MSNBC's websites receive such heavy traffic that . . . many servers collapse.").

164 *See id.*

165 For analogous, but much less dramatic attacks, *see* Brenner & Goodman, *In Defense of Cyberterrorism*, *supra* at 32–34. For more a fictional account of analogous but far more dramatic attacks, *see* Jack Henderson, Circumference of Darkness 398–399 (Bantam Books 2007).

166 The War of the Worlds (radio)—Wikipedia, http://en.wikipedia.org/wiki/The_War_of_the_Worlds_%28radio%29.

167 *See* Brenner & Goodman, *In Defense of Cyberterrorism*, *supra* at 26.

168 Bay Area Rapid Transit—Wikipedia, http://en.wikipedia.org/wiki/Bay_Area_Rapid_Transit.

The officials issue an immediate evacuation order for the San Francisco area. This produces chaos as panicked citizens desperately try to flee an impending nuclear disaster. People trying to flee by automobile clog the streets and accidents ensue; aside from anything else, this impedes people's ability to leave the area and further increases their panic. Those who do not have vehicles cannot use the BART so they turn to other means of mass transit, such as buses; this produces a stampede for access to buses, in which the vulnerable are trampled. Some of those who do not have access to vehicles try to seize automobiles from their owners and drivers; this results in deaths and further injuries. Some of those who cannot find any means of leaving the city quickly take their own lives; others strike at out whoever or whatever is nearby.

Death, injury, and property damage ensue . . . but there is no impending disaster, no suitcase nuke. Terrorists hacked the government computer system and sent credible, fake messages, which the local officials reasonably believed. The net result is that the terrorists achieve injury, death, and destruction as well as a dramatic erosion in the public's confidence in the government's ability to ensure their security without having to deploy an actual weapon.[169]

In these and other computer as weapon of mass distraction scenarios, computer technology is used primarily for psychological manipulation. The first scenario is a "true" computer as weapon of mass distraction scenario; the second scenario tends to blend weapon of mass distraction with hypothesized weapon of mass disruption effects (which we examine below). The point, though, is that neither scenario involves the actual use of real-world weapons; in both, the computer is the only implement the terrorists employ.

Weapon of Mass Disruption

When terrorists use computer technology as a weapon of mass disruption, their goal is to undermine a civilian populace's faith in the stability and reliability of essential infrastructure components such as mass transit, power supplies, communications, financial institutions, and health care services.[170] Though the weapon of mass disruption and weapon of mass distraction

---

169 For a similar, equally-fictive account of how false information can be used to create confusion and a resulting risk of injury, *see* Chris Suellentrop, *Sim City: Terrortown*, Wired 103, 103–104 (October 2006), http://www.wired.com/wired/archive/14.10/posts.html?pg=2.

170 *See* Brenner & Goodman, In Defense of Cyberterrorism, *supra* at 26.

alternatives both target civilians' faith in essential aspects of their society, the two differ in how computer technology is used to corrode civilian confidence in societal infrastructures and services.

As we saw in the previous section, terrorists launch a psychological attack when they use computer technology as a weapon of mass distraction; the goal is to undermine civilians' confidence in one or more of the systems they rely on for essential goods or services. The cyberterrorists accomplish this by making citizens ***believe*** a system has been compromised and is no longer functioning effectively. The terrorists do not actually impair the functioning of the system. Their goal is to inflict psychological, not systemic, damage.

When computer technology is used as a weapon of mass disruption, terrorists' goal *is* the infliction of systemic damage on one or more target system. This version of cyberterrorism is closer to the scenarios that sometimes appear in the popular media in which cyberterrorists shut down an electrical grid or the systems supplying natural gas or petroleum to a particular populace.[171]

Like the weapon of mass distraction alternative, this scenario is both a conceptual and a real possibility. Here, terrorists use computer technology in a fashion that is analogous to, but less devastating than, their use of real-world weapons of mass destruction. Their goal is not to inflict the catastrophic, theatrical carnage and destruction we saw on 9/11. Rather, it is more insidious: to demoralize a civilian populace by making civilians question the government's ability to keep things working.

In pursuing this alternative, terrorists seek to exploit the unacknowledged but inherent fragility of modern urban life. As some have noted, the increasingly urbanized, increasingly technologized lifestyle many of us enjoy in the twenty-first century makes us more vulnerable to this type of terrorism than are those in more traditional, rural societies. One author pointed out, for example, that hostile forces could unlock the "disruptive potential" of a city by attacking key points in its infrastructure; such attacks would halt economic and other activity and erode the city's ability to provide the services that are an essential component of its political legitimacy.[172] He also noted

---

171 *See*, e.g., Dan Verton, Black Ice: The Invisible Threat of Cyber-Terrorism 1–16 (McGraw-Hill, 2003). *See also* Jeremy Kirk, *Russian Expert: Terrorists May Try Cyberattacks*, InfoWorld.com (December 13, 2006), http://www.infoworld.com/article/06/12/13/HNcyberterroralert_1.html?sour.

172 John Robb, *The Role of Cities*, Global Guerrillas, October 21, 2006, http://globalguerrillas.typepad.com/globalguerrillas/2006/10/the_role_of_cit.html.

that certain attributes of modern urban areas have the unintended but ***de facto*** effect of engineering them "to radiate instability."[173] This makes them particularly vulnerable to attack.[174] The attributes he cites are (a) "[e]xtreme mobility and interconnectedness," such as widespread use of automobiles, cell phones, and other computer technology; (b) "[c]omplete reliance on high volume infrastructure networks"; and (c) "[c]omplex and heterogeneous social networks that are held together under pressure."[175] The combined effect of these and other attributes of modern urban enclaves is to make them vulnerable in ways older cities were not.

Though this author was analyzing urban vulnerability to physical attacks, cyberterrorists can exploit the same vulnerabilities. A 2006 exercise the U.S. Secret Service and Department of Homeland Security conducted proves that. In February 2006, more than three hundred public and private sector participants from the United States, Australia, Canada, New Zealand, and the United Kingdom conducted a simulated cyberterrorism exercise—"Cyber Storm"—on various government agencies and businesses.[176] The participants were in sixty different geographical locations, and the four-day exercise had three goals: (a) disrupting "critical infrastructure," which would lead to "cascading effects" within the countries' "economic, societal, and

---

173 *Id.*

174 *Id.*

175 *Id.*

176 U.S. Department of Homeland Security, National Cyber Exercise: Cyber Storm 1 (2006), www.automationalley.com/MiRSA/Studies/prep_cyberstormreport_sep06.pdf [hereinafter Cyber Storm Report]:

> Cyber Storm was a coordinated effort between international, Federal and State governments, and private sector organizations to exercise their response, coordination, and recovery mechanisms in reaction to . . . cyber events. . . . Over 100 public and private agencies, associations, and corporations participated in the exercise from over 60 locations and 5 countries. . . . The . . . scenario simulated a large-scale cyber campaign affecting or disrupting . . . critical infrastructure elements primarily within the Energy, Information Technology (IT), Telecommunications and Transportation sectors. The exercise was conducted primarily on a separate exercise network without impacting real world information systems.

The U.S. government participants included 8 federal departments and 3 agencies plus 4 states (Michigan, Montana, New York, and Washington). *See* U.S. Department of Homeland Security, Presentation, National Cyber Exercise: Cyber Storm, New York City Metro ISSA Meeting 11 (June 21, 2006), http://www.cryptome.org/cyberstorm.ppt [hereinafter Cyber Storm powerpoint]. The private sector participants included "9 major IT firms," 6 electric utility firms, 2 major air carriers. *See id.* The IT (information technology) firms included Microsoft, Verisign, Cisco, Symantec, and MacAfee. *See id.* The airlines were Delta and Federal Express. *See id.*

governmental structures";[177] (b) hindering "governments' ability to respond to the cyber attacks"; and (c) undermining "public confidence in the governments' ability to provide and protect services."[178]

Before we analyze the exercise itself, we need to consider its impact. Although the Department of Homeland Security (DHS) described Cyber Storm as a success, it actually revealed a number of problems. One was that the defenders, who were trying to fend off the simulated attacks, had difficulty coordinating attack response efforts between (a) the public and private sectors and (b) different agencies within the public sectors.[179] As a result, the

---

177 *Id.* at 1.

> The exercise simulated a sophisticated cyber attack campaign through . . . scenarios directed against critical infrastructures. The intent . . . was to highlight the interconnectedness of cyber systems with the physical infrastructure and to exercise coordination . . . between the public and private sectors. Each of the scenarios was . . . executed in a closed and secure environment.

Id. at 11.

178 *See id. See also* U.S. Department of Homeland Security, Fact Sheet: Cyber Storm Exercise (September 13, 2006), http://www.dhs.gov/xnews/releases/pr_1158340980371.shtm.

In March 2008, the Department of Homeland Security ran the second iteration of Cyber Storm: Cyber Storm II. Cyber Storm II was conducted on a much larger scale: It involved 18 federal agencies, 5 countries (Australia, Canada, New Zealand, the United Kingdom, and the United States), 9 states, and over 40 U.S. companies. *See*, e.g., Anne Broache, *Homeland Security "Cyber Storm" Simulates Crisis*, CNet News (March 14, 2008), http://news.cnet.com/8301-10784_3-9894236-7.html. *See also* Fact Sheet: Cyber Storm II, Department of Homeland Security, http://www.dhs.gov/xprepresp/training/gc_1204738760400.shtm. It involved simulated attacks on four infrastructure sectors: "chemical, information technology (IT), communications and transportation (rail/pipe)." Cyber Storm: Securing Cyberspace, Department of Homeland Security, http://www.dhs.gov/xprepresp/training/gc_1204738275985.shtm. All we know of the scenario is that it was "executed by persistent, fictitious adversaries with a distinct political and economic agenda" who used "sophisticated attack vectors to create a large-scale incident requiring players to focus on response." Fact Sheet: Cyber Storm II, *supra*.

The Department of Homeland Security announced it would issue a report on Cyber Storm II later in 2008, but because much of the information concerning the exercise is classified, many doubt the report will be particularly revealing. *See*, e.g., Anne Broache, *Homeland Security "Cyber Storm" Simulates Crisis, supra. See also* William Jackson, *High-Stakes War Games*, Government Computer News (April 14, 2008), http://www.gcn.com/print/27_8/46115-1.html. Reports from Australia indicate that the 55 Australian organizations that participated in Cyber Storm II were not particularly successful in dealing with the simulated attacks. *See*, e.g., Liam Tung, *Australia Crumbles under Cyber Storm Attack*, ZDNet Australia (May 21, 2008), http://www.zdnet.com.au/news/security/soa/Australia-crumbles-under-Cyber-Storm-attack/0,130061744,339289145,00.htm. Speaking at a conference, the head of Australia's Cyber Storm effort reported that chains of command crumbled as organizations became overwhelmed with attacks (and the damage they inflicted). *See id.*

179 *See* Cyber Storm Report at 6–9. This problem seems to have been particularly severe when the coordination efforts involved participants from different countries.

defenders tended to deal with each attack as an "individual and discrete" event; they had great difficulty in coordinating their efforts into a cohesive, cross-sector response, especially when the attacks increased in number and frequency.[180] Another problem for the defenders was the "relatively modest resources" available to them.[181] The DHS report notes, for example, that the National Cyber Response Coordination Group (NCRCG) became so overwhelmed by the attacks that developing "an accurate situational picture was challenging."[182] This had several effects, one of which involved the "weapon of mass distraction" scenario described above.

As the report explains, the NCRCG is, among other things, responsible for maintaining "a clear public communications channel" during attacks or other disasters; the goal is to "deliver key technical messages at a layman's level . . . in a timely . . . manner and establish a confident . . . media presence."[183] The DHS report notes the importance of maintaining public confidence in the government's ability to deal with such a crisis: "More damage could have occurred as a result of erroneous and panicked public responses to incorrect media coverage than by actual attacks by the adversary."[184] It also notes that the "current NCRCG composition and staffing was significantly challenged by the adversary's robust media campaign that accompanied the cyber attacks."[185] In other words, the Cyber Storm exercise showed that a few talented, determined attackers can inflict serious damage on a modern society's critical infrastructures.[186]

Cyber Storm involved a series of attacks launched by the fictitious "Worldwide Anti-Globalization Alliance" (WAGA), a "loose coalition of well financed organizations led by a single world-wide group, who sought the use of the Internet as its primary means of causing damage."[187] WAGA was at once a free-standing organization and a coalition encompassing five equally

---

180 *See* Cyber Storm Report at 6–9. Essentially the same problem arose in Cyber Storm II, the successor exercise noted above. *See supra* note 114. *See, e.g.*, Tung, *Australia Crumbles under Cyber Storm Attack, supra.*

181 *See* Cyber Storm Report at 6–9.

182 *See id.* at 6–7.

183 *See id.* at 7.

184 *See id.* at 7.

185 *See id.* at 7.

186 *See DHS Releases Report on Cyber Storm Exercise,* Security Focus (September 14, 2006), http://www.securityfocus.com/brief/303. See *also* Cyber Storm Report at 6–9.

187 *Id.* at 14 ("The simulated adversaries did not represent a specific . . . terrorist group. . . . The[y] . . . were a loose coalition of well financed 'hacktivists.'"). *See also* Cyber Storm powerpoint.

fictive organizations, each with "a different political agenda."[188] As its name implies, WAGA opposed globalization; the other four organizations opposed nuclear technology and weapons and other aspects of modern society.[189] The WAGA coalition's efforts were facilitated by the opportunistic contributions of certain individuals: "Auggie Jones, Cyber Saboteur," self-interested hackers of unknown origin, an anonymous but disgruntled airport employee, and "The Tricky Trio," three activist hackers from Berlin.[190]

The effort generally focused on attacking "cyber systems used to control both physical infrastructures and digital commerce and services."[191] The goal was to maximize economic harm and generate "public distrust by disrupting services and by misleading news media."[192]

Over the four days the exercise was live, the disparate attackers accomplished a variety of things, such as crashing the Federal Aviation Association's computer control system, causing power and Internet outages, attacking media sites, shutting off the heat, planting logic bombs in computer systems, compromising medical data, posting a false Amber alert, altering one "No Fly" list and posting the contents of another online, launching DDoS attacks on various systems, shutting down commuter trains, and altering account balances in financial institutions.[193] The opportunistic hackers of unknown origin apparently pursued their own interests by distributing malware, purchased personally identifying information for use in identity theft and engaged in online extortion, all presumably out of self-interest.[194]

As we have already seen, the responses from the various defenders were . . . not particularly effective. The DHS report attributed the less-than-stellar performance to various things, such as the lack of coordination and resources

---

188 *See* Cyber Storm powerpoint. The organizations were WAGA itself, Freedom not Bombs, the Black Hood Society (an offshoot of Freedom not Bombs), the People's Pact, and the Internet Technopolitic Front. *See id.* After the DHS report became public, some criticized the designers of the exercise for casting "far left" groups as the cyberterrorists. *See*, e.g., Kevin Poulsen, *When Hippies Turn to Cyber Terror*, Wired (August 15, 2006), http://blog.wired.com/27bstroke6/2006/08/when_hippies_tu.html?entry_id=1539952 (noting that the fictive "Freedom not Bombs" group is "suspiciously similar to the real Food Not Bombs, which provides vegan meals to the homeless."

189 *See* Cyber Storm powerpoint.

190 *See id.*

191 *See id.*

192 *See id.*

193 *See id.*

194 *See id.*

I noted earlier. It also cited a need for improved training, planning, risk assessment, and tools.[195]

I am sure all of these contributed to the success WAGA and its collaborators enjoyed in Cyber Storm, but I also think other, more general factors played a role in that success. We will examine those factors in detail in Chapter 4. At this point, I merely want to note one factor I think exacerbated the difficulties the Cyber Storm defenders faced.

And that factor is the structure of an attack: As we saw in Chapter 2, societies divide threats into internal (crime) and external (war). Both crime and war are singular transactions; in each there is one threat source—one opponent, in effect—and one defender that is attempting to nullify that threat source. The opponent can, of course, have various facets; in World War II, for example, there were two "sides"—the Axis and the Allies—each of which encompassed multiple countries. The same can be true of crime; a crime may be perpetrated by a single criminal, by two or three criminals working together, or even by an organized gang of criminals. But in all of these scenarios there are still only two "sides:" the opponent (the attacker) and the defender (the entity attempting to repel the attacker).

This creates a straightforward dynamic: A nation-state nullifies the threat (crime or war) by successfully responding to that opponent. In warfare, the encounter with the opponent can occur over a long period of time and can span a wide geographical area, as in World War II; but in wars, such as World War II, each combatant knows precisely the opponent it must overcome to prevail and ensure the continuation of external order. And something similar can occur in crime; serial killers and organized gangs, for example, engage in the repetitive commission of crimes over a period of time and, often, over a geographical area. But here, too, the state responsible for maintaining internal order confronts what is still only a single opponent; here, as in warfare (but on a lesser scale), the state concentrates its resources on nullifying the criminal opponent to ensure the continuation of internal order.

In Cyber Storm, there was no singular transaction and no single, identifiable opponent. Instead, there was a cacophony of unknown opponents in various geographical locations and who were driven by equally unknown, idiosyncratic motives. In Cyber Storm, those charged with responding to the attacks confronted an asymmetrical, ambiguously heterogeneous cohort of anonymous attackers. This characteristic, which the Cyber Storm attacks share with other

195 *See id.* at 1–2.

cyberattacks, has a number of implications for defenders, one of which is that they cannot focus on "a" threat. I think that factor accounts for many of the problems Cyber Storm identified in our current response structures. As the DHS report noted, the defenders tended to respond to each incident of the Cyber Storm assault as if it were an isolated, free-standing attack. This is neither surprising nor exceptionable because this is the way attacks are structured in the real-world and the Cyber Storm defenders, like other defenders, were trained to respond to real-world threats. That must change, at least in part. Defenders can no longer routinely assume that threats will be identifiable, singular, and sequential; they must also be able to respond to aggregated threats that can be labyrinthine in structure and discontinuous in occurrence.

As I said, we will return to this issue in Chapter 4. Before we take up the conceptual and other difficulties defenders face, we need to complete our taxonomy by examining cyberwarfare.

Cyberwarfare

Before we can analyze "cyberwarfare," we must define "warfare." We all generally understand what "war" is. As the Supreme Court said almost a century and a half ago, "war" is "the exercise of force by bodies politic . . . against each other, for the purpose of coercion."[196] One author describes "warfare" as having four essential constituent elements: (a) There must be "a contention between at least two" nation-states; (b) the nation-states must use their armed forces in this "contention"; (c) each nation-state's goal must be to overpower the opposing state(s) (the enemy) and impose "peace on the victor's terms"; and (d) the contending states will have "symmetrical, although diametrically opposed, goals."[197]

These conceptual definitions of warfare capture its essence: War has historically been (a) a struggle involving the use of "armed force" (b) between opposing sovereignties, nation-states in the last few centuries, sovereign cities, tribes, and groups before that.[198] The conduct of war—the external threat to social order—has been the exclusive province of the sovereign, the nation-state in modern history; crime—the internal threat—has been left to individuals.[199]

---

196 The Brig Amy Warwick (The Prize Cases), 67 U.S. 635, 652 (1863).

197 Yoram Dinstein, War, Aggression and Self-Defense 4–5 (Cambridge University Press, 2005).

198 One has only to contrast the American public's attitude toward World War II and toward the Vietnam war, particularly in its later stages, to appreciate the importance of legitimacy.

199 *See* Chapter 2, *supra.*

We will return to the inevitability of nation-states' involvement in warfare in a moment, but first we must consider an operational definition of war. The conceptual definitions give us two indispensable characteristics of warfare—(a) a struggle between opposing sovereignties (b) that entails the use of armed force—but there is another, often-overlooked characteristic we must include in our analysis of cyberwarfare.

Warfare

Operationally, a historian of warfare defines it as having "three essential dimensions: violence; legitimacy; and legality."[200] We already know war involves violence, and I suspect we intuitively understand the need for its legitimacy.[201] Legitimacy is the ideological device nation-states use to motivate warrior-citizens to fight for their country—to convince them that killing their fellow man in battle is, in fact, the "right" thing to do. War therefore differs from crime, which can also involve the use of violence, because war "derives legitimacy from a political, societal, or religious source. Men are . . . given license to ignore . . . accepted societal conventions against killing and destroying."[202]

"Legality," the historian's third dimension, is the additional characteristic we need for our analysis. I suspect most of us give little thought to the premise that wars must be "legal," as well as legitimate. This is an ancient requirement that has greatly evolved in sophistication over, say, the last millennium.[203] As the historian cited above explains, wars are fought according to

> understood sets of rules. In Medieval Europe, special courts enforced these rules, specifying, for example, that the knight who seized his enemy's right gauntlet was his rightful captor and thus deserving of whatever

---

200 *See generally*, Michael S. Neiberg, Warfare in World History (Routledge, 2001).

201 See note 198, *supra*.

202 Michael Neiberg, Warfare in World History, *supra* at 2.

203 *See id.* at 9–20, 46–58. *See*, e.g., Chris Jochnick & Roger Normand, *The Legitimation of Violence: A Critical History of the Laws of War*, 35 Harvard International Law Journal 49, 60 (1994) ("In the second millennium B.C., the wars between Egypt and Sumeria were governed by a complex set of rules obligating belligerents to distinguish combatants from civilians and providing procedures for declaring war, conducting arbitration, and concluding peace treaties"). *See also* Lieutenant Commander Gregory P. Noone, *The History and Evolution of the Law of War Prior to World War II*, 47 Naval Law Review 176, 182–187 (2000).

> ransom he could negotiate. . . . Often, these codes and laws remained unwritten, though . . . they could be extremely elaborate. War is not, then, a free-for-all. . . . It is carefully governed and regulated by both armed forces and the societies they serve.[204]

The modern rules of war have their origins in Hugo Grotius's 1625 treatise on "the law of war and peace"—*De Jure Belli ac Pacis*."[205] Grotius argued that war should be "governed by a strict set of laws."[206] He was particularly concerned with the way war was being conducted.[207] According to Grotius, "when arms have . . . been taken up there is no longer any respect for law . . . ; it is as if, in accordance with a general decree, frenzy had openly been set loose for the committing of all crimes."[208]

Grotius, and others who would later express similar sentiments, was part of a general reaction to the way wars had been waged. Until "well into" the eighteenth century, the armies nation-states fielded "were composed largely of mercenaries, whose pay was intermittent and who, for lack of a regular supply service, had to 'live off the country.'"[209] These untrained and undisciplined "soldiers" brutalized civilians and essentially razed farms and towns in the areas they passed through; during the Thirty Years War in the early seventeenth century, "over half the German-speaking population was wiped out" and most of Europe was left a "shambles."[210]

Grotius's writings and the devastation left by the Thirty Years War resulted in a number of reforms, perhaps the most important of which was that soldiering was professionalized: Troops were trained, organized in a "chain of command" consisting of "battalions, regiments and other standard units," and were fed, clothed, and paid regularly.[211] Armies added staff to handle

---

204 Neiberg, Warfare in World History, *supra* at 2.

205 *See* Hugo Grotius, The Law of War and Peace (De Jure Belli ac Pacis) (1625), http://www.lonang.com/exlibris/grotius/index.html. *See also* Noone, *The History and Evolution of the Law of War Prior to World War II, supra* at 187.

206 Jochnick & Normand, *The Legitimation of Violence, supra* at 60.

207 *See* Hugo Grotius, The Law of War and Peace, *supra* Prolegomena ¶ 28. *See also* Noone, *The History and Evolution of the Law of War Prior to World War II, supra* at 187–188.

208 *See* Grotius, The Law of War and Peace, *supra* Prolegomena ¶ 28 *See also* Noone, *The History and Evolution of the Law of War Prior to World War II, supra* at 187–188.

209 Telford Taylor, The Anatomy of the Nuremburg Trials: A Personal Memoir 6 (Little, Brown, 1992).

210 *Id.*

211 *Id.*

supply and transport and established procedures to maintain discipline among their troops.[212] From all this, customs and rules governing soldiers' relationships with civilians and their conduct while occupying foreign territory developed.[213]

These practices influenced "the philosophers and publicists of the eighteenth century Enlightenment," notably Jeremy Bentham, Francois Voltaire, and Jean-Jacques Rousseau, all of whom supported Grotius's argument that international law should govern the conduct of war.[214] Rousseau, for example, argued in ***The Social Contract*** that members of a nation-state's armed forces have the right to kill soldiers fighting for the opposing sovereign, but not if the soldiers lay down their arms and surrender.[215] Other developments along this line occurred in the eighteenth century, but until the nineteenth century, the laws of war remained unwritten.[216]

In the nineteenth century, humanitarian concerns prompted by newspapers' graphic accounts of battlefield violence would play a role in the eventual codification of a law of war, as would the Union Army's commissioning Francis Lieber, a professor, to draft a code setting out the rules governing the conduct of warfare and one governing responses to guerilla activity.[217] Like the treaties it would influence, the code governing the conduct of warfare cited the principle of "military necessity" as the basis for determining what actions were appropriate during military combat. The principle was set out in Article 15:

> Military necessity admits of all direct destruction of life or limb of armed enemies, and of other persons whose destruction is incidentally unavoidable in the armed contests of the war; it allows of the capturing of every armed enemy . . .; it allows of all destruction of property, and obstruction of the ways and channels of traffic, travel, or communication, and of all

---

212 *Id.*

213 *Id.*

214 *See id. See also* Noone, *The History and Evolution of the Law of War Prior to World War II, supra* at 188.

215 *See* Jean-Jacques Rousseau, Discourse on Political Economy and the Social Contract 52 (Christopher Betts, trans., Oxford University Press 1994).

216 *See* Noone, *The History and Evolution of the Law of War Prior to World War II, supra* at 188–189.

217 *See* Francis Lieber, Instruction for the Government of Armies of the United States in the Field, General Order No. 100 (April 24, 1863), http://www.yale.edu/lawweb/avalon/lieber.htm#sec2. *See also* Noone, *The History and Evolution of the Law of War Prior to World War II, supra* at 189–193.

> withholding of sustenance or means of life from the enemy; of the appropriation of whatever an enemy's country affords necessary for the subsistence and safety of the army....[218]

Article 16 qualified the principle, explaining that military necessity "does not admit of cruelty—that is, the infliction of suffering for the sake of suffering or for revenge.... It does not admit of ... wanton devastation.... and ... does not include any act of hostility which makes the return to peace unnecessarily difficult."[219] Other articles in this code specified that solders were not to harm civilians or private property "in hostile countries occupied by them"[220] and that a prisoner of war "is subject to no punishment for being a public enemy, nor is any revenge wreaked upon him."[221]

In 1874, the Union Army's rules governing the conduct of warfare became the basis of a "Declaration Concerning the Laws and Customs of War," which was drafted at a conference in Brussels.[222] The Declaration was never formally adopted and therefore never became effective, but it stimulated a series of efforts that ultimately culminated in the Hague Conference of 1899.[223]

The Conference produced the Hague Convention of 1899, which, though it did relatively little to develop a fully realized law of war, did formally articulate the principle that during war "populations and belligerents remain under the ... the principles of international law."[224] As we have seen, this statement incorporated a concept that had evolved over at least two centuries, namely, that civilians and surrendering combatants should be treated

---

218 Lieber, Instruction for the Government of Armies of the United States in the Field, *supra* at Article 15.

219 *Id.* at Article 16.

220 *See id.* at Article 37. Other Articles prescribe similar treatment for museums, works of art, libraries, "scientific collections," hospitals, churches, charities, and educational institutions. *See id.* at Articles 34–36.

221 *Id.* at Article 56. Other Articles prescribe similar treatment for hostages, chaplains, medical staff and servants. *See id.* at Articles 53–55.

222 *See* Noone, *The History and Evolution of the Law of War Prior to World War II, supra* at 194.

223 *Id.* at 194–196.

224 Preamble of the (Hague II) Convention with Respect to the Laws and Customs of War on Land, The Hague (July 29, 1899), http://www.yale.edu/lawweb/avalon/lawofwar/hague02.htm. *See* Noone, *The History and Evolution of the Law of War Prior to World War II, supra* at 196–197.

as noncombatants.[225] Aside from giving some consideration to noncombatants, the 1899 Hague Convention focused primarily on the methods that could be used to conduct war; it proscribed the use of poison, for example; set certain restrictions on the use of deception; and outlined procedures that should be followed to minimize the death and destruction resulting from "bombardment."[226] The second Hague Conference, held in 1907, produced another Convention that differed very little from its predecessor.[227]

In the aftermath of World War I, various countries adopted pacts outlawing the use of chemical weapons, an effort that seems to have led to the promulgation of the 1929 Geneva Conventions: the Geneva Convention for the Amelioration of the Condition of the Wounded and Sick in Armies in the Field and the Geneva Convention Relative to the Treatment of Prisoners of War.[228] Both Conventions refined principles concerning the treatment of erstwhile combatants that had been articulated in earlier agreements.[229]

In 1949, the two 1929 Geneva Conventions were superseded by four Conventions that have been ratified by many countries: Convention (I) for the Amelioration of the Condition of the Wounded and Sick in Armed Forces in the Field; Convention (II) for the Amelioration of the Condition of Wounded, Sick and Shipwrecked Members of Armed Forces at Sea; Convention (III) Relative to the Treatment of Prisoners of War; and Convention (IV) Relative to the Protection of Civilian Persons in Time

225 Apparently, until the Middle Ages, warring states tended to treat all inhabitants of opposing states as enemies, "including women and children." Jill M. Sheldon, *Nuclear Weapons And The Laws Of War: Does Customary International Law Prohibit The Use Of Nuclear Weapons In All Circumstances?*, 20 Fordham International Law Journal 181, 243 n. 426 (1996) (citing Lester Nurick, *The Distinction Between Combatant and Noncombatant in the Law of War*, 39 American Journal of International Law 680, 681 (1945)). But by 1806, Napoleon's minister, Talleyrand, would write that "the law of nations does not permit that the rights of war, and of conquest . . . should be applied to peaceable, unarmed citizens." Noone, *The History and Evolution of the Law of War Prior to World War II, supra* at 1189 (citing Telford Taylor, The Anatomy of the Nuremburg Trials: A Personal Memoir 7 (1992)).

226 *See* (Hague II) Convention with Respect to the Laws and Customs of War on Land, The Hague (July 29, 1899), http://www.yale.edu/lawweb/avalon/lawofwar/hague02.htm.

227 Preamble of the (Hague II) Convention with Respect to the Laws and Customs of War on Land, The Hague (July 29, 1899), http://www.yale.edu/lawweb/avalon/lawofwar/hague02.htm. *See* Noone, *The History and Evolution of the Law of War Prior to World War II, supra* at 198–199.

228 *See id.* at 199–203.

229 *See id.* at 202–203.

of War.[230] The 1949 Conventions were supplemented in 1977 by two Additional Protocols, which have been also widely adopted.[231]

As must be evident, the primary concern of most of these documents is with establishing rules that govern "how an armed conflict may proceed once it has begun."[232] And that brings us to a basic dichotomy in the law of war: "International jurisprudence makes a distinction between laws governing the resort to force (***jus ad bellum***) and laws regulating wartime conduct (***jus in bello***)."[233]

The documents we just reviewed are for the most part concerned with the ***jus in bello***, which, unlike the ***jus ad bellum***, "operates at the level of the individual."[234] The ***jus in bello*** is concerned with the behavior of individuals (combatants) toward other individuals (combatants and noncombatants). Its goal is to import humanitarian considerations into the conduct of war, once commenced.[235] As we have seen, the ***jus in bello*** is predicated on the doctrine of "military necessity," which essentially "requires a balance between the need to achieve a military victory and the needs of humanity."[236]

The ***jus ad bellum*** operates at the nation-state level and is predicated on the broader principle of necessity.[237] The "central issue of the ***jus ad bellum***" is

---

230 *See* The Laws of War, The Avalon Project at Yale Law School, http://www.yale.edu/lawweb/avalon/lawofwar/lawwar.htm. *See also* Rosa Ehrenreich Brooks, *War Everywhere: Rights, National Security Law, and the Law of Armed Conflict in the Age of Terror*, 153 University of Pennsylvania Law Review 675, 691 (2004) (1949 Geneva Conventions "further rationalized and codified customary and treaty-based norms relating to armed conflict, outlining the rules applicable to civilians, prisoners of war, and wounded and sick members of armed forces").

231 *See* The Laws of War, The Avalon Project at Yale Law School,*supra*. *See also* Brooks, *War Everywhere: Rights, National Security Law, and the Law of Armed Conflict in the Age of Terror, supra* at 691–692.

232 James D. Fry, *The UN Security Council and the Law of Armed Conflict: Amity or Enmity?*, 38 George Washington International Law Review, 327, 330 (2006).

233 Jochnick & Normand, *The Legitimation of Violence: A Critical History of the Laws of War, supra* at 52. "*Jus in bello* is further divided into the Geneva laws (the `humanitarian laws'), which protect specific classes of war victims . . . and the Hague laws (the `laws of war'), which regulate the overall means and methods of combat." *Id.*

234 Craig J.S. Forrest, *The Doctrine of Military Necessity and the Protection of Cultural Property during Armed Conflicts*, 37 Case Western International Law Journal 177, 179 (2007).

235 *See id.* at 177–180. *See also* Alexander C. Linn, *The Just War Doctrine and State Liability for Paramilitary War Crimes*, 34 Georgia Journal of International & Comparative Law 619, 625–626 (2006) (noting that Christian and non-Christian societies pursued "*jus in bello*, rules to limit the horrors of war, for ethical, religious, or utilitarian reasons").

236 Forrest, *The Doctrine Of Military Necessity and the Protection of Cultural Property during Armed Conflicts, supra* at 181.

237 *Id.* at 179–181.

the legality of a nation-state's commencing warfare against another nation-state.[238] The Western version of the ***jus ad bellum*** evolved from Roman and Greek doctrines;[239] it was formally articulated St. Augustine, whose work Thomas Aquinas later refined and expanded.[240]

Aquinas concluded that a just war was a "necessary" war, and that a "necessary" war meets three criteria: It must be waged by a sovereign entity because sovereigns are charged with protecting the common good.[241] Second, the justification for war must be based on "a fault of the external enemy";[242] Aquinas cited self-defense and the desires to avenge a wrong or to regain unjustly seized persons or property as examples of what would justify war.[243] Finally, Aquinas said war must be waged with a "just intention," such as protecting the common good and/or the desire to secure peace.[244] The ***jus ad bellum*** could, therefore, justify defensive war (a war based only on self-defense) or nondefensive war (a war justified on some other basis).[245]

The two principles governed different stages of an armed conflict. As we have seen, the ***jus ad bellum*** determined whether the aggressor nation had

---

238 Dinstein, War, Aggression and Self-Defense, *supra* at 5.

239 *See* Paul D. Marquardt, *Law without Borders: The Constitutionality of an International Criminal Court*, 33 Columbia Journal of Transnational Law 73, 76 (1995). *See*, e.g., The Nicomachean Ethics 339 (F. H. Peters & David Ross trans., Kegan Paul 1904) ("we make war in order that we may enjoy peace"). A similar concept has existed in many other cultures. *See*, e.g., Stephanie Bellier, *Unilateral and Multilateral Preventive Self-Defense*, 58 Maine Law Review 508, 522 (2006) (noting that the just war principle "is not foreign to Judaism, Islam, or Buddhism); *see* Noone, *The History and Evolution of the Law of War Prior to World War II, supra* at 183–185.

240 *See*, e.g., Benjamin v. Madison III, Comment, *Trial by Jury or by Military Tribunal for Accused Terrorist Detainees Facing the Death Penalty? An Examination of Principles That Transcend the U.S. Constitution*, 17 University of Florida Journal of Law & Public Policy 347, 401–402 (2006). *See* 19 Saint Augustine, The City of God Book 19 § 12 (Marcus Dods trans., Random House 1993) ("It is . . . with the desire for peace that wars are waged. . . . And hence it is obvious that peace is the end sought for by war").

241 *See* Madison III, Comment, *supra* at 402 (citing Thomas Aquinas, Summa Theologica, pt. II-II, question 40, art. 1 (The Fathers of the English Dominican Province trans., Benziger Brothers 1947).

242 *Id.*

243 *See id. See also* Linn, *The Just War Doctrine and State Liability for Paramilitary Crimes, supra* at 627 n. 30.

244 *See* Benjamin v. Madison III, Comment, *Trial By Jury Or By Military Tribunal For Accused Terrorist Detainees Facing The Death Penalty?, supra* at 402. *See also* Linn, *The Just War Doctrine and State Liability for Paramilitary Crimes, supra* at 627.

245 *See* Madison III, Comment, *supra* at 402–403. *See also* Linn, *The Just War Doctrine and State Liability for Paramilitary Crimes, supra* at 627 (the sovereign commencing war "must have a rightful intention to advance good and to avoid evil").

justifiably resorted to the use of armed force; this was its only purpose. The ***jus ad bellum*** consequently ceased to apply once hostilities had begun; the ***jus in bello*** took over "at the advent of armed conflict" and applied "to all parties, irrespective of which" state was the aggressor or the justification the aggressor state gave for commencing the conflict.[246]

Implicit in the concept of the ***jus ad bellum*** was the premise that war was an acceptable option for nation-states.[247] *A* war could be legal or illegal, but "war" *per se* was not outlawed.[248] Indeed, according to one scholar the prevailing conviction in the nineteenth and early twentieth centuries was that "every State had a right . . . to embark upon war whenever it pleased."[249]

States were, however, expected to give notice of their intent to do so: For centuries, it was understood that wars were commenced with a declaration of war.[250] "A declaration of war is a . . . formal announcement, issued by the constitutionally competent authority of a State, setting the exact time at which war begins with a designated enemy (or enemies)."[251] In the Western tradition, the expectation that states would declare war before commencing hostilities dates back at least to Roman times; Cicero, for example, made this a threshold requirement for the commencement of a just war.[252] And it was incorporated into the ***jus ad bellum***. Article 1 of the Hague Convention (III) of 1907 states that parties to the Convention "recognize that hostilities . . . must not commence without previous and explicit warning, in the form either of a reasoned declaration of war or of an ultimatum with conditional declaration of war."[253]

---

246 Forrest, *The Doctrine of Military Necessity and the Protection of Cultural Property during Armed Conflicts, supra* at 181.

247 Dinstein, War, Aggression and Self-Defense, *supra* at 69–72.

248 *Id.*

249 *Id.* at 71.

250 *Id.* at 29.

251 *Id.* at 71.

252 *Id.* at 59.

253 Hague Convention (III), Convention Relative to the Opening of Hostilities, art. 1, October 18, 1907, http://www.yale.edu/lawweb/avalon/lawofwar/hague03.htm. Aside from anything else, a declaration of war "`served the legal function of triggering international law governing neutral and belligerent states." William C. Peters, *On Law, Wars and Mercenaries: The Case for Courts-Martial Jurisdiction over Civilian Contractor Misconduct in Iraq*, 2006 Brigham Young University Law Review 367, 404 (quoting Curtis A. Bradley & Jack L. Goldsmith, Foreign Relations Law 177, 178 (2003)).

This changed after World War II, with the creation of the United Nations. The United Nations Charter "abolished the legal institution of war."[254] Article 2(4) of the Charter states that members of the United Nations "shall refrain in their international relations from the threat or use of force against the territorial integrity or political independence of any state."[255] Under current international law, therefore, declarations of war no longer serve a purpose because they cannot even partially justify a state's commencing war with another state.[256]

Article 2(4) of the United Nations Charter may have effected another change in the historical understanding of "warfare." In outlawing war as a legitimate tool of sovereignty, the Charter uses the term "force," rather than the term of art commonly used over the last several centuries—"armed force."[257] Over the years, this has produced a great deal of discussion as to whether, for example, the Charter outlaws the use of "economic pressure" as well as the more traditional use of armed, or military, force.[258] Most experts have concluded that Article 2(4) only encompasses the use of military force, and so does not extend to economic or psychological pressure, unless they are "coupled with the use" or a threatened use of military force.[259]

While it outlaws non-defensive war, the U.N. Charter preserves the legality of defensive war, at least under certain circumstances. Article 51 of the Charter preserves "the inherent right of individual or collective self-defence if an armed attack occurs."[260] Under the Charter, "self-defense is the only

---

254 Stephen Neff, *Towards a Law of Unarmed Conflict: A Proposal for A New International Law of Hostility*, 28 Cornell International Law Journal 1, 15 (1995).

255 United Nations Charter Article 2(4), June 26, 1945, http://www.yale.edu/lawweb/avalon/un/unchart.htm.

256 *See* Neff, *Towards a Law of Unarmed Conflict: A Proposal for A New International Law of Hostility, supra* at 15. *See also* Paul W. Kahn, *War Powers and the Millennium*, 34 Loyola of Los Angeles Law Review 11, 16–17 (2000).

257 *See* United Nations Charter, *supra*. *See also* Dinstein, War, Aggression and Self-Defense, *supra* at 81.

258 *See*, e.g., *id.*

259 *Id. But see* Clinton E. Cameron, Note, *Developing a Standard for Politically Related State Economic Action*, 13 Michigan Journal of International Law 218, 219 (1991) As noted earlier, Article 2(4) outlaws both the use and the threatened use of force.

260 United Nations Charter Article 51, June 26, 1945, http://www.yale.edu/lawweb/avalon/un/unchart.htm. *See also* Definition of Aggression, United Nations General Assembly Resolution 3314, United Nations Document A/9631; Concerning Friendly Relations and Cooperation Among States in Accordance with the Charter of the United Nations, United Nations General Assembly Resolution 2625, United Nations Document A/8028 (1970).

legally recognized justification for the use of force abroad."[261] Article 51 serves as an affirmative defense to charges that a state violated Article 2(4).[262] That is, a nation-state's use of armed force against another nation-state in violation of Article 2(4) will be excused if (a) it was in response to an armed attack from the other nation-state and (b) the responding state reported its actions to the U.N. Security Council, which may intervene.[263] Experts disagree as to whether the affirmative defense established under Article 51 applies when there has been a threatened use of armed force, rather than an actual use of armed force.[264]

The ***jus ad bellum*** currently exists in a very attenuated form; indeed, the U.N. Charter is sometimes known as the "***jus contra bellum***" because it is the law that outlaws war, at least, the use of nondefensive was as a legitimate prerogative of sovereignty.[265] As we saw above, the ***jus in bello*** not only exists, it has been significantly expanded beyond the tentative steps taken in the nineteenth century.

In the twenty-first century, we have two categories of law that govern different, but not mutually exclusive, scenarios: the "war" and "not-war" scenarios. The domestic civil-criminal laws we analyzed in Chapter 2 apply in the "not-war" scenario, when a nation-state is at peace and its only concern is with maintaining internal order. The laws of warfare apply to the "war" scenario: Articles 2(4) and 51 of the U.N. Charter, the vestiges of the ***jus ad bellum***, govern a nation-state's initial use of armed force against another nation-state; they determine if that use of force was justified. The modern ***jus in bello***—primarily the Geneva and Hague Conventions—govern the tactics nation-states employ while engaging in armed struggles with each other.

All of these rules, all of these laws, were formulated with activity in the real, physical world in mind. War is ancient, but cyberwarfare is not and, as we will see in the remainder of this chapter and in the next two chapters,

---

261 United Nations Charter Article 51, *supra*.

262 *See* Michael Y. Kieval, Note, *Be Reasonable! Thoughts On The Effectiveness Of State Criticism In Enforcing International Law*, 26 Michigan Journal of International Law 869, 883 (2005).

263 *See*, e.g., Amy E. Eckert & Manooher Mofidi, *Doctrine Or Doctrinaire—The First Strike Doctrine And Preemptive Self-Defense Under International Law*, 12 Tulane Journal of International and Comparative Law 117, 132 (2004).

264 *See*, e.g., Ryan Schildkraut, Note, *Where There Are Good Arms, There Must Be Good Laws: An Empirical Assessment Of Customary International Law Regarding Preemptive Force*, 16 Minnesota Journal of International Law 193, 202–211 (2007).

265 *See*, e.g., Amy E. Eckert & Manooher Mofidi, *Doctrine Or Doctrinaire*, *supra*.

that disconnect creates difficult issues for lawyers, for diplomats, for warriors, and for civilians.

Cyberwarfare

In 2006, the U.S. Air Force adopted a new mission statement in which it pledges to "fight in Air, Space, ***and Cyberspace***."[266] The new statement recognizes what has been apparent for some time: warfare can and will migrate into cyberspace.[267]

Definitionally, cyberwarfare is the conduct of military operations by virtual means.[268] It consists of nation-states' using cyberspace to achieve essentially the same ends they pursue through the use of conventional military force: achieving advantages over a competing nation-state or preventing a competing nation-state from achieving advantages over them.

Cyberwarfare may already be a reality. Reports have announced that the People's Republic of China has launched cyberattacks in an attempt to cripple Taiwan's infrastructure and paralyze the island nation's government and economy.[269] The attacks allegedly targeted Taiwan's public utility, communications, transportation, and operational security networks.[270]

---

266 Air Force Link—Welcome, http://www.af.mil/main/welcome.asp (emphasis added). The new mission statement added the reference to cyberspace. *See* Michael W. Wynne, Sec'y of the Air Force, Cyberspace as a Domain in Which the Air Force Flies and Fights (November 2, 2006), http://www.af.mil/library/speeches/speech.asp?id=283.

267 *See*, e.g., Clay Wilson, Congressional Research Service, Information Operations and Cyberwarfare: Capabilities and Related Policy Issues CRS-1 to CRS-8 (Library of Congress 2006), http://www.fas.org/irp/crs/RL31787.pdf. *See also* U.S.-China Economic & Security Review Commission, 109th Cong., Report to Congress (2006), http://www.uscc.gov/researchpapers/annual_reports.htm. [hereinafter, "U.S. Economic & Security Review Commission 2006 Report"]; Office of the Secretary of Defense, 110th Cong., Annual Report to Congress: Military Power of the People's Republic of China 21–22 (2007), http://www.defenselink.mil/pubs/china.html; Office of the Secretary of Defense, 109th Cong., Annual Report to Congress: Military Power of the People's Republic of China 35–36 (2006), http://www.defenselink.mil/pubs/china.html.

268 *See*, e.g., Steven A. Hildreth, Congressional Research Service, Cyberwarfare (Washington, DC, Library of Congress 2001), http:// www.fas.org/irp/crs/RL30735.pdf. For a cyberwarfare scenario, *see* John Arquilla, *The Great Cyberwar of 2002*, Wired, February 1998, http://www.wired.com/wired/archive/6.02/cyberwar_pr.html [hereinafter John Arquilla, *The Great Cyberwar*].

269 Bill Gertz, *Chinese Information Warfare Threatens Taiwan*, Washington Times A3 (October 13, 2004).

270 *Id.* As we saw in Chapter 1, Estonia for a time believed it was the target of cyberwarfare attacks launched by Russia.

Regardless of whether these attacks actually occurred, China is clearly committed to cyberwarfare, or "information warfare." A 2007 Pentagon study is one of many to report that the People's Liberation Army ["PLA"] is "building capabilities for information warfare, computer network operations, and electronic warfare, all of which could be used in preemptive attacks."[271] This document noted that the PLA "has established information warfare units to develop viruses to attack enemy computer systems and networks, and tactics . . . to protect friendly computer systems and networks."[272] It also says the PLA is incorporating offensive cyberwarfare into its military exercise, in the form of "first strikes" against enemy networks.[273]

China's efforts in this regard are nothing new. China has for some time been pursing nontraditional methods of waging warfare—including cyber- and space-war—because its military leaders believe "it cannot defeat the United States in a traditional force-on-force match up."[274] A 2006 report from the U.S.-China Economic & Security Review Commission elaborated on China's commitment to cyberwarfare:

> China is actively improving its non-traditional military capabilities. . . . China's approach to exploiting the technological vulnerabilities of adversaries extends beyond destroying or crippling military targets. Chinese military writings refer to attacking key civilian targets such as financial systems.
>
> The Commission . . . has seen clear examples of computer network penetrations coming from China. . . . [T]he PLA, leveraging private sector expertise, steadily increases its focus on cyber-warfare capabilities and is making serious strides in this field. . . . [T]he PLA's cyber-warfare strategy has evolved from defending its own computer networks to attacking the networks of its adversaries. . . .[275]

China is not alone in developing the capacity for cyberwarfare. According to one expert, "at least 20 nations" have "their own cyberattack programs." Dawn S. Onley & Patience Wait, *Red Storm Rising*, Government Computer News (August 21, 2006), http://www.gcn.com/print/25_25/41716-1.html (quoting John Thompson, chairman and chief executive officer of Symantec Corp).

271 Office of the Secretary of Defense, 110th Cong., Annual Report to Congress: Military Power of the People's Republic of China 12 (2007), *supra*.

272 *Id.* at 22.

273 *Id.*

274 *See* U.S.-China Economic & Security Review Commission, 110th Cong., Report to Congress 8 (2007), http://www.uscc.gov/researchpapers/annual_reports.htm.

275 U.S. Economic & Security Review Commission 2006 Report, *supra* at 137.

And testifying before the House Armed Services Committee in March of 2007, General James Cartwright, Commander of the U.S. Strategic Command, said that

> cyberspace has emerged as a war-fighting domain not unlike land, sea, and air, and we are engaged in a less visible, but none-the-less critical battle against sophisticated cyberspace attacks. We are engaging these cyberspace attacks offshore, as they seek to probe military, civil, and commercial systems, and consistent with principles of self defense, defend the [Department of Defense] portion of the Global Information Grid (GIG) at home.[276]

China is, of course, not the only country developing offensive cyberwarfare capabilities: According to one expert, "at least 20 nations" have "their own cyberattack programs."[277] The United States' response to the burgeoning cyberwar threat has not been limited to expanding the Air Force's mission; it is also creating a special "Air Force Cyber Command," which, as I write this, is scheduled to go operational in 2009.[278] In announcing the new Cyber Command, Secretary of the Air Force Michael Wynne said that

> the traditional principles of war . . . apply fully in cyberspace. . . . [H]e stressed that cyberspace operations can include far more than computer network attack and defense. He cited the use of improvised explosive devices in Iraq, terrorist use of Global Positioning Satellites and satellite communications, Internet financial transactions by adversaries, radar and navigational jamming, and attacking American servers as just a few examples of operations that involve the cyberspace domain.[279]

---

276 Statement of General James E. Cartwright, Commander, United States Strategic Command before the House Armed Services Committee 11 (March 21, 2007), http://armedservices.house.gov/hearing_information.shtml. General Cartwright also said, "a purely defensive posture" is likely to fail, in cyberspace as it has in the real-world, and argued for the United States' developing the ability to "take the fight to our adversaries." *Id.* at 12.

277 Dawn S. Onley & Patience Wait, *Red Storm Rising, supra.*

278 *See* Marty Graham, *Welcome to Cyberwar Country, USA*, Wired (February 11, 2008), http://www.wired.com/politics/security/news/2008/02/cyber_command?currentPage=all. The date has been pushed back; the command was originally to have been fully operational by 2007. *See Air Force Cyber Command Online for Future Operations*, Air Force Link (July 17, 2007), http://www.af.mil/news/story.asp?id=123060959.

279 8th Air Force Named as Cyberspace Command, The Information Warfare Site (November 2, 2006), http://www.iwar.org.uk/iwar/resources/cybercommand/index.htm.

Most of the news releases accompanying the announcement of the Cyber Command stressed the importance of the Air Force's "achieving dominance" in the domain of cyberspace, as it has in the domains of "air, space, land and sea."[280] Some also noted that a "[f]ailure to control and dominate the cyber domain could be catastrophic."[281]

Countries are clearly gearing up for cyberwarfare,[282] even though no one knows precisely what that will mean. In the next two chapters we will explore the issues, and the ambiguities, cyberwarfare (and cybercrime and cyberterrorism) will create for those charged with maintaining order, both internal and external, in the world's nation-states. But before we get into specifics, I want to note a few general thoughts on how the emerging constellation of cyberthreats is likely to challenge our understanding of war, crime, and terrorism as distinct threat categories.

As we saw earlier, the distinguishing characteristic of war is that it is a struggle between nation-states.[283] Like all human activity, war is ultimately carried out by individuals, but here the individuals act not out of their own self-interest but on behalf of a nation-state. Like terrorism, warfare tends to result in the destruction of property (often on a massive scale) and in the injury and deaths of individuals (also often on a massive scale).[284] Unlike terrorism, war is limited, at least in theory, to clashes between the aggregations of individuals (armies) that respectively act for warring nation-states. Injuring and killing civilians (those who were not recruited to serve in a combatant nation-state's military) occurs, but like most property damage and destruction, it is supposed to be a collateral event.[285] And as we saw earlier, the primary objective in warfare is to "triumph" over the adversarial nation-state(s), whatever that means in a given context.

---

280 *See*, e.g., Air Force Leaders to Discuss New "Cyber Command," Air Force Print News (November 5, 2006), http://www.8af.acc.af.mil/news/story_print.asp?id=123031988.

281 *See id.*

282 *See* Charles Billo & Weston Chang, Cyber Warfare: An Analysis of the Means and Motivations of Selected Nation-States (Institute for Security Technology Studies 2004).

283 *See* Dinstein, War, Aggression and Self-Defense, *supra* at 5 ("One element seems common to all definitions of war . . .. [W]ar is a contest between states").

284 *See*, e.g., Niall Ferguson, The Pity of War 248–317 (Basic Books, 1999) (describing economic losses and loss of life in World War I).

285 *See* Geneva Convention IV, Relative to the Protection of Civilian Persons in Time of War arts. 3, 28, 53, August 12, 1949, http://www.genevaconventions.org/.

In the real-world, there can be ambiguity as to whether an event is a crime or an act of terrorism,[286] but war is always unambiguous. When the Japanese bombed Pearl Harbor in 1941,[287] it was clearly an act of war; the same was true when Hitler invaded Poland in 1939, and it has also been true throughout recorded history.[288]

War is unambiguous in the real-world because it is unique; only nation-states can summon the resources needed to launch a physical land, sea, or air attack on another nation-state. The clarity of war is enhanced by the fact that those who conduct real-world attacks wear uniform clothing and insignia that identify them as members of a nation-state's armed forces.[289] It is also enhanced by the fact that real-world warfare involves the physical violation of territorial boundaries. As we have seen, nation-states are defined by the territory they control;[290] acts of war have, therefore, inherently involved breaching the integrity of the victim state's borders. The threat to social order consequently comes not from "insiders," as when crime and terrorism are involved, but from an "outsider"—a rival sovereign.[291]

The threat dichotomies—internal versus external threat, crime and terrorism versus war—we reviewed earlier consequently provide a stable, reliable way of parsing real-world attacks. We may be somewhat uncertain as to whether a particular event is crime or terrorism, but that is ultimately of little moment because we use the same approach for both, because both

---

286 *See*, e.g., Alan Cooperman, *Capture Focuses on Christian Terrorism*, Grand Rapids Press (June 2, 2003), 2003 WLNR 13819663 (authorities trying to determine if Olympic bomber Eric Rudolph was a "Christian terrorist" or a criminal); Patrick May & Martin Merzer, *No Place to Hide*, Miami Herald (April 21, 1995), 1995 WLNR 2638059 (authorities not sure if bombing of Oklahoma City federal building was terrorism or a crime).

287 *See*, e.g., Attack on Pearl Harbor—Wikipedia, http://en.wikipedia.org/wiki/Attack_on_Pearl_Harbor.

288 *See*, e.g., Battle of Thermopylae—Wikipedia, http://en.wikipedia.org/wiki/Battle_of_Thermopylae; Invasion of Poland (1939)—Wikipedia, http://en.wikipedia.org/wiki/Invasion_of_Poland; Six-Day War—Wikipedia, http://en.wikipedia.org/wiki/Six-Day_War.

289 Under the laws of war, military combatants must (a) have a fixed command structure; (b) wear a common uniform, emblem, or insignia; (c) carry arms openly; and (d) comply with the law and customs of war. *See* Geneva Convention Relative to the Treatment of Prisoners of War, Article 4(A)(2) (August 12, 1949), http://www.yale.edu/lawweb/avalon/lawofwar/geneva03.htm; (Hague Convention (IV): Laws and Customs of War on Land annex art. 1 (October 18, 1907), http://www.yale.edu/lawweb/avalon/lawofwar/hague04.htm [hereinafter, "Hague IV"].

290 *See* Chapter 2.

291 *See Id.*

threaten internal order.[292] And the monopolization of territory and military force by nation-states means that in the real-world, we will never be uncertain as to whether we are confronted with a threat to internal order (crime/terrorism) or a threat to our nation-state's ability to maintain external order (war).[293] As we have seen, in the real-world, only nation-states wage war.

As we will see in the next two chapters, these threat dichotomies break down when attacks are vectored through the virtual world of cyberspace. By giving nonstate actors access to a new, diffuse kind of power,[294] cyberspace erodes nation-states' monopolization of the ability to wage war and effectively levels the playing field between all actors.[295] As we will see, in the twenty-first century, states generate crime and terrorism as well as war, and individuals wage war in addition to committing crimes and carrying out acts of terrorism.

---

292 *See Id.*

293 *See Id.*

294 *See* Information Operations: Warfare and the Hard Reality of Soft Power, *supra* at 10–14.

295 *See id.* at 70 ("[T]echnology . . . has revolutionized warfare by taking the elements of power and dispersing them to the people."). *See also* John Robb, Brave New War 74–75 (John Wiley & Sons 2007).

FOUR

Attribution

Attacks

Is Estonia at war? Even the country's leaders don't seem sure.[296]

BEFORE WE EMBARK ON OUR analysis in this and later chapters, I want to note another mysterious series of cyberattacks on a nation-state, this one targeting a United States agency. We will use these attacks and the Estonian attacks to analyze how cyberspace challenges the principles underlying the threat taxonomy outlined in Chapter 3.

In the summer of 2006, the U.S. Department of Commerce's Bureau of Industry and Security (BIS) was the target of a "debilitating" series of attacks on its computer system.[297] The BIS, a "sensitive" Commerce Department agency, is responsible for controlling the export of U.S. software and technology that has commercial and military uses.[298]

The attackers sought access to BIS employee user accounts, presumably to obtain data concerning BIS operations.[299] They penetrated the BIS' system with a "rootkit" program, a type of software that lets attackers conceal their

---

296 Phillip Ball, Homunculus (May 25, 2007), http://philipball.blogspot.com/2007/05/does-this-mean-war-this-is-my-latest.html.

297 Alan Sipress, *Computer System Under Attack*, Washington Post A21 (October 6, 2006), http://www.washingtonpost.com/wp-dyn/content/article/2006/10/05/AR2006100501781.html. *Id.* In 2008, number of government systems came under cyberattacks that appeared to originate in China. *See*, e.g., Jordy Yager, *More Congressional Computers Hacked from China*, The Hill (June 21, 2008), http://thehill.com/leading-the-news/more-congressional-computers-hacked-from-china-2008-06-21.html; Peter Brooks, *Flashpoint: The Cyber Challenge*, Armed Forces Journal (March 5, 2008), http://www.armedforcesjournal.com/2008/03/3463904.

298 Sipress, *Computer System Under Attack*, *supra*. According to its Mission statement, the bureau's objective is to "[a]dvance U.S. national security, foreign policy, and economic objectives by ensuring an effective export control and treaty compliance system and promoting continued U.S. strategic technology leadership." "About BIS," Bureau of Industry and Security – U.S. Department of Commerce, http://www.bis.doc.gov/about/index.htm.

299 *See* Sipress, *Computer System Under Attack*, *supra*.

presence and gain privileged but unauthorized access to a computer system.[300] Initially, the BIS tried to defend itself against the attacks, but it ultimately decided that Internet access was a vulnerability it could neither tolerate nor mitigate.[301] The agency therefore disconnected its computers from the Internet and replaced them with new ones that were not networked; this restored security but interfered with employees' ability to perform their duties.[302] A Commerce Department representative said the attacks were traced to websites hosted by Chinese Internet Service Providers, but the attackers were never identified.[303]

Consider that for a moment: the attackers were never identified. This observation has several implications, the most obvious of which is that those who carried out the attacks were never identified. As we have already seen, that is far from remarkable; given the opportunities cyberspace creates for the remote commission of attacks and attacker anonymity, it is more common than not for cybercriminals to go unidentified and unapprehended.[304]

That statement assumes we are dealing with cybercriminals, and that assumption brings us to another implication of the observation noted above: Not only were the BIS *attackers* never identified, the *nature* of the attack itself was never identified. We will assume for the purposes of analysis that the attack came from China,[305] but what kind of attack was it? Was it cybercrime—Chinese hackers counting coup[306] on U.S. government computers? Was it cyberterrorism—an initial effort toward a takedown of U.S. government computers by terrorists (who might or might not have been Chinese) pursuing idiosyncratic ideological goals? Or was it cyberwarfare—a virtual sortie by People's Liberation Army (PLA) hackers?

---

300 *Id. See*, e.g., "Rootkit," Wikipedia, http://en.wikipedia.org/wiki/Rootkit.

301 Sipress, *Computer System Under Attack, supra* (quoting Commerce Department spokesperson).

302 *Id.* When the original computers were examined, the analysis showed they had been so compromised by the attacks they could not be salvaged. *See id.*

303 *Id.*

304 *See* Chapter 3. *See also* Chapter 5.

305 Sipress, *Computer System Under Attack, supra.* Cyberattackers can route their attacks through intermediate systems to disguise their true originating point. *See*, e.g., *Tiny Nevada Hospital Attacked by Russian Hacker*, USA Today (April 7, 2003), http://www.usatoday.com/tech/webguide/internetlife/2003-04-07-hospital-hack_x.htm (Russian hacker routed attack on Nevada hospital through Al-Jazeera's website to make it appear the attack came from Qatar).

306 Counting coup—Wikipedia, http://en.wikipedia.org/wiki/Counting_coup.

In Chapter 3, we developed a taxonomy of threats to social order—crime/cybercrime, terrorism/cyberterrorism, and warfare/cyberwarfare—and explained that nation-states must keep these threats to internal and external order under control if they are to survive and prosper.[307] Nation-states employ two related, sequential processes in their efforts to control these threats: attribution (Who launched the attack? What kind of attack is it?) and response (Who should respond to the attack: Civilian law enforcement? The military? Both? Neither?).

Like the threat categories we explored in the last chapter, these correlate processes have evolved over the centuries in an essentially ad hoc fashion and now satisfactorily control the incidence of real-world threats (crime, terrorism, and warfare). As this chapter and the next three chapters explain, however, the approaches we currently use for attribution and response are becoming increasingly unsatisfactory as cyberspace becomes a viable vector for attacks of whatever type. We must, therefore, either (a) modify the existing, general approach we use for real-world threat attribution and response so it also encompasses cyberthreats or (b) develop a supplementary, specialized approach for cyberthreat attribution and response. We will examine the need for, and relative utility of, each alternative in the next several chapters.

We begin with attribution. As I noted above, attribution encompasses two issues:[308] (a) Who carried out an attack? and (b) What kind of an attack was it? The first issue, attacker-attribution, goes to assigning responsibility for *committing* an attack. The second issue, attack-attribution, goes to assigning responsibility for *responding* to an attack. We will examine attack attribution in this chapter and attacker-attribution in Chapter 5. We then take up the problem of response in Chapters 6 and 7.

---

307 *See* Chapter 2.

308 For our purposes, attribution is a legal rather than a technical or a technical-legal concept. Our analysis will therefore focus on the information decision-makers need to determine whether an attack is a threat to internal (crime/terrorism) or external (warfare) order—i.e., whether it is a matter to be resolved by civilian law enforcement or by military personnel.

Because we are focusing on legal decision **making**, the characterization of attribution given above contains fewer elements than the characterization used by those who parse the technical aspects of attack attribution. For a technically-legally focused characterization of attribution, *see*, e.g., Dorothy E. Denning, Cyber Conflict Studies Association, Presentation—Attribution Workshop, Cyber Attack Attribution: Issues and Challenges, slide #2 (March 2005), http://www.cyberconflict.org/attributionworkshop.asp (identifying four levels of attribution: identification of attacking machines, identification of primary controlling machines, identifications of humans responsible for attack, and identification of sponsor organization).

Attack-attribution: Real-world

As a combined function of history and empirical circumstance, real-world attacks have fallen discretely into either of two categories: crime/terrorism or warfare. As we have seen, the difference lies in the distinction societies have drawn between threats to internal order and threats to external order.[309] In the real world that distinction traditionally divided attacks into two, and only two, categories: crime/terrorism (internal) and war (external).[310]
As we saw in Chapter 3, this division and the distinction it was predicated upon arose from the realities of the physical world. Until at least the last century, the limitations of travel and the nation-state monopolization of military-grade weaponry made it functionally impossible for nonstate actors to challenge a nation-state's ability to maintain its territorial integrity as a sovereign entity. Challenges to external order consequently arose only from other nation-states and were resolved through military combat.

Nonstate actors could, and did, challenge a state's ability to maintain internal order: Criminals' pursuit of self-gratification and the more doctrinaire activities of terrorists threatened to erode real-world social order in varying ways and to varying degrees. For at least the last century and a half, nation-states have employed a unique strategy—civilian law enforcement—to control internal threats.[311] As we saw in Chapter 2, this two-pronged strategy consists of (a) adopting laws that criminalize crime and terrorism and (b) using a specialized, quasi-military civilian police force to identify and apprehend those who violate the laws. Individual violators are prosecuted, convicted, and sanctioned by agents of the state, which presumptively deters them from reoffending and deters others from following their example.[312]

Attack-attribution in the real world has therefore seldom been problematic because the respective distinguishing characteristics of crime/terrorism and warfare have been apparent and unambiguous. Essentially, the attack-attribution process tacitly proceeded in two stages: the first was to determine if the attack challenged external order (attack by another nation-state); if it did not, then the attack was, by default, a challenge to internal order

---

309 *See* Chapters 2 & 3.

310 *Id.*

311 *Id.*

312 *Id.*

(crime or terrorism). We will use this two-stage process in the analysis we conduct later in this chapter.

As we saw in Chapter 3, it has not been difficult to identity acts of war in the real world. When the Japanese bombed Pearl Harbor, no one who experienced the attack could have had the slightest doubt that this was warfare—not crime and not terrorism—even though it was not preceded by a formal declaration of war.[313] The assault was a military attack, which inferentially, and essentially conclusively, established that a nation-state, not private individuals, carried it out; attackers who wore uniform clothing with insignia identifying them as members of the Japanese military carried it out on U.S. territory; the equipment the attackers used bore such insignia and was of a type (e.g., ships, airplanes) that would not have been available to individual criminals or terrorists, but only to a nation-state. And, finally, the sheer scale of the attack strongly indicated a nation-state carried it out; only a nation-state could mass the military force and materiel necessary for such an assault.

Until the twenty-first century, when our lives began migrating into cyberspace, acts of war had an inherent clarity. Often, they were preceded by a formal declaration of war.[314] And when they were not, the initial assault—always on the victim state's own territory—brought its own clarity. When the German army invaded Poland on September 1, 1939, the Polish leaders and citizens knew they were at war;[315] the same has been true throughout history, from the Achaeans' attack on Troy in the eleventh century BC down to and including the United States' March 20, 2003, invasion of Iraq.[316] And as these examples illustrate, "warfare" has always been synonymous with the use of armed, military force by a nation-state.[317]

This brings us to the default options: crime and terrorism. Because both constitute a challenge to a nation-state's ability to maintain internal order,

---

313 *See* Chapter 3. *See also* "Attack on Pearl Harbor," Wikipedia, http://en.wikipedia.org/wiki/Attack_on_Pearl_Harbor (no preceding declaration of war).

314 *See* Chapter 3.

315 *See* "Invasion of Poland (1939)," Wikipedia, http://en.wikipedia.org/wiki/Invasion_of_Poland_%281939%29.

316 *See* "Trojan War," Wikipedia, http://en.wikipedia.org/wiki/Trojan_War; "2003 Invasion of Iraq," Wikipedia, http://en.wikipedia.org/wiki/2003_invasion_of_Iraq.

317 *See, e.g.*, Kenneth B. Moss, *Information Warfare and War Powers: Keeping the Constitutional Balance*, 26 Fletcher Forum of World Affairs 239, 242 (Summer/Fall 2002) ("throughout history there has been a common understanding that war involves the use of armed force"). *See also* Chapter 3.

each necessarily inflicts "harm" within the territory of a nation-state.[318] And because crime and terrorism as we have defined them are purely real-world phenomena, this has historically meant that the entire "harmful" transaction—crime or terrorist act—occurs within the territory of a specific nation-state.[319]

Crime is easy to identify for several reasons. One is that it involves the purely civilian-on-civilian infliction of familiar categories of harm, for example, theft, robbery, rape, murder, fraud, and arson.[320] Another reason is that crime is predictably routine because it is the product of familiar, individual motives, for example, self-aggrandizement, vengeance, sexual urges, and so forth. Finally, as we saw in Chapter 2, crime, unlike warfare, is limited in scale because of the constraints physical reality imposes on action in the real-world; one consequence of this is that the default model of real-world crime is individual-on-individual victimization (versus the mass struggles in warfare). The far more unusual many-to-one pattern of victimization increases the scale on which "harm" is committed somewhat but still pales next to warfare.[321] The one-to-many victimization typical of cybercrime vastly increases the scale on which "harm" is inflicted but still pales next to warfare and is essentially unknown in the real world because of the physical constraints noted above.[322]

Real-world terrorism is also usually easy to identify even though it involves predicate activity that is defined as crime (as well as a constituent element of terrorism), that is, killing/injuring people and damaging/destroying property.[323] Even if those responsible for a terrorist event do not publicly claim responsibility for what they have done, terrorism can usually be distinguished from crime at least for two reasons: One is that the infliction of "harm" is peculiar in that we cannot attribute it to a routine criminal motivation, such as self-enrichment. The other reason is that the scale on which "harm" is

---

318 *See* Chapters 2 & 3.

319 *Id.* Prior to the twentieth century, when communication technologies such as the telephone made it possible to commit certain crimes remotely, the rare exceptions to this principle in U.S. law involved cases in which John Doe, standing in Texas, shoots across the state line into Arkansas, killing Robert Brown. *See*, e.g., State v. Hall, 114 N.C. 909, 19 S.E. 602 (N.C. 1894). When this happened, courts usually found that a crime had been committed in Arkansas, on the premise that the "harm" was inflicted there. *See* State v. Chapin, 17 Ark. 561, 1856 WL 622 (Ark. 1856) ("the crime is . . . committed where the shot . . . takes effect").

320 *See* Chapters 2 & 3.

321 *See id.*

322 *See id.*

323 *See* Chapter 3.

inflicted often vastly exceeds the "harm" typical of crime. The 1993 and 2001 attacks on the World Trade Center illustrate how these factors are used in the inferential terrorism identification process: As to motive, the attacks did not conform to what we expect of crime because neither was the type of activity that is calculated to yield financial or other personal gain and neither apparently redressed personal grievances. And the scale on which "harm" was inflicted in both instances far, far exceeded that which we would expect if the incidents had been merely criminal activity. In these and similar instances (such as the 1995 bombing of the Oklahoma City Federal Building), the sheer magnitude of the "harm" inferentially indicates that we have entered the realm of violence and destruction as theater—of terrorism. [324]

I am not suggesting that we can simply equate the apparent irrationality of inflicting "harm" with terrorism because there are irrational crimes. In the United States, for example, we can read almost daily about murders that were committed for no rational reason, such as killing one's family, shooting customers and staff at a McDonald's, or randomly killing students at a high school. Most of these ostensibly irrational crimes are "personal"; that is, they grow out of workplace or family relationships. These crimes involve people who know each other, at least to some extent, and a perpetrator who is the victim of real or imagined grievances.[325] In these crimes, there is a link, a factual nexus between the perpetrator and the victims; this nexus allows us to reconcile the "harm" inflicted with what we know of crime and criminal motives.

We do on occasion have what seem to be purely irrational incidents of mass murder and serial killings but here, too, it is usually obvious we are dealing with crime, not terrorism. As one author explains, the motive for mass murders is "likely to be personal," while serial killings often have a "sexual component."[326] And though the underlying (if distorted) logic of these crimes

---

324 *Id.*

325 *See* James Alan Fox & Jack Levin, Overkill: Mass Murder and Serial Killing Exposed 221 (NY: Plenum Press 1994):

> Most mass killers target people they know—family members, friends, or coworkers—in order to settle a score, to get even with . . . individuals whom they hold accountable for their problems. Others seek revenge against a . . . category of people . . . suspected of receiving an unfair advantage. But a few . . . mass murders stem from the killers' paranoid view of society . . . [as] a wide-ranging conspiracy in which large numbers of people . . . are out to do them harm.

See also id. at 201–236. *See*, e.g., Gary M. Lavergne, A Sniper in the Tower: The Charles Whitman Murders 90–99 (University of North Texas Press 1997).

326 Sara Knox, *A Gruesome Accounting: Mass, Serial and Spree Killing in the Mediated Public Sphere* 1 Journal for Crime, Conflict and the Media 1 (2004). *See also* James Alan Fox & Jack Levin, Overkill, *supra* at 15–23.

may not be immediately apparent, we usually derive from the circumstances of their commission that they are, in fact, "crimes." We realize, for example, that the mass murderer who climbs a tower and shoots anyone unlucky enough to come within range is engaged in a one step removed "personal" assault, that is, is taking revenge for real or wholly perceived wrongs randomly and indiscriminately.[327] We also, eventually, realize that the otherwise incomprehensible activities of serial killers are the product of what seems to be a distinct psychopathology that drives them to "hunt humans." [328]

As we saw in Chapter 3, terrorism is fundamentally different. Here, the infliction of "harm" derives neither from "personal" concerns nor skewed psyches, but from ideology. In terrorism, the link between conduct and result is actually but not ostensibly rational; those of us who are not involved in the terrorist event cannot understand the logic linking the conduct and the resultant "harm." Why, for instance, would people intentionally fly an airplane into the World Trade Center? Such an act is irrational when assessed against our understanding of criminal motives for inflicting "harm"; the perpetrators appear to have gained nothing while losing their own lives. The act becomes quite rational, of course, once we realize that we have entered the theater of terrorism—once we recognize the ideological premises that inspired the Al Qaeda members who flew the planes.

As I noted above, another factor we use to differentiate crime and terrorism is the scale on which "harm" is inflicted. As we saw in Chapter 3, because crime is the product of individual motivations and is subject to physical constraints that govern activity in the real-world, it tends to be committed on a limited scale. A mugger robs one victim, a rapist attacks another, a murderer kills yet another; in each instance, as with most crime, we have one-to-one victimization.[329] The scale on which "harm" is inflicted in terrorism tends to be more inexact. Consider, for example, a suicide bomber. He wants to inflict as much death, injury, and destruction as possible, and so takes that into account in selecting the area where he will detonate his bomb.[330] The actual scale of "harm" he achieves will remain uncertain until

---

327 *See*, e.g., Lavergne, A Sniper in the Tower, *supra* at 90–99. *See also* Fox and Levin, Overkill, *supra* at 221.

328 *See*, e.g., Knox, *A Gruesome Accounting, supra*; Fox and Levin, Overkill, *supra* at 15–23.

329 *See* Chapter 3.

330 *See*, e.g., "Suicide attack," Wikipedia, http://en.wikipedia.org/wiki/Suicide_attack. *See also* Mohammed M. Hafez, Suicide Bombers in Iraq 91–92, 102–106 (United States Institute of Peace 2007).

the attack has been consummated because the context in which suicide bombers operate (fortuitous attacks, usually in public places) is fluid.[331] The "harm" the suicide bomber inflicts will, though, almost certainly exceed that attributable to most, if not all, crimes, because where crime inflicts a focused "harm" (e.g., profit, revenge, sexual control), a terrorist attack inflicts generalized, variegated "harms." And this, aside from anything else, usually serves to distinguish terrorism from criminal violence.[332]

Overall, the internal-external threat dichotomy and resulting allocation of responsibility for responding to threats to social order has proven quite satisfactory in the real-world. There can be initial definitional ambiguity between crimes and acts of terrorism; but because both have represented purely internal threats to social order, and because internal threats are addressed by civilian law enforcement, ambiguity as to whether an attack was crime or terrorism has little, if any, impact on the civilian response process. And for our purposes, the residual category of war has had no corresponding definitional ambiguity[333] and no operational uncertainty; once an attack is identified as war, the military is exclusively responsible for responding to it.

---

331 In the 1993 World Trade Center bombing, for example, the resulting "harm" was much less than the perpetrators had expected. *See*, e.g., Dennis Piszkiewicz, Terrorism's War with America 86–87 (Praeger 2003).

332 Within at least twenty-four hours of the 1995 bombing of the Oklahoma City Federal Building, investigators had concluded it was a terrorist attack, based on the nature and scale of the event. *See*, e.g., *Wide Range of Suspects in Terrorism Scenarios*, San Jose Mercury News (April 20, 1995); *Experts See Resemblance to Middle East Attacks*, Wichita Eagle (April 20, 1995).

333 The concept of warfare is not the unitary construct it once was. The beginning of the twenty-first century has seen "a decrease in conventional warfare with large armies and an increase in conflicts characterized as Military Operations Other Than War (MOOTW)." Eugene B. Smith, *The New Condottieri and US Policy: The Privatization of Conflict and Its Implications* 17 Parameters 104 (2002), http://www.carlisle.army.mil/usawc/Parameters/02winter/smith.htm. The U.S. Department of Defense, however, seems to apply the laws of war to all such conflicts. *See* U.S. Department of Defense, Directive No. 2311.01E § 4.2, Department of Defense Law of War Program (May 9, 2006), http://www.fas.org/irp/doddir/dod/d2311_01e.pdf (Department of Defense personnel must "comply with the law of war during all armed conflicts, however such conflicts are characterized, and in all other military operations).

The dissociation of warfare into discrete modes of conflict is relevant in analyzing how societies should adapt their military posture to this evolving military threat matrix, but is not relevant to the analysis we are pursuing here. Our focus is on how the use of computer technology impacts on attribution and response to two generic categories of attacks: military-conducted attacks and civilian-conducted attacks.

Attack-attribution: Virtual World

We will analyze the BIS attacks and then the Estonian attacks[334] in considering how the use of cyberspace impacts on the process of attack-attribution. We will do essentially the same thing in the next chapter, for attacker-attribution.

BIS Attacks

The first step is parsing out what we know of these attacks, which are described at the beginning of this chapter:

i. They were deliberate, orchestrated attacks, not computer malfunctions.
ii. They targeted computers used by a United States government agency.
iii. The attackers sought access to "sensitive" data held in those computers and may or may not have succeeded.
iv. The attacks originated extraterritorially, in China.
v. For the purposes of analysis, we will assume they were traced to Guangdong, a province reputationally associated both with China's efforts to develop cyberwarfare capability and with the activities of independent hackers from Foshan University.[335]
vi. We do not know precisely what the attackers sought to accomplish, but we do know that an agency of the U.S. government was attacked on U.S. soil.

Because we—the United States—were attacked somehow and by someone, we will at some point have to decide if, and how, we should respond.

---

334 *See supra* Chapter 1 (Estonian Attacks). The BIS attacks are described at the beginning of this chapter.

335 The BIS attacks apparently originated in China, but Guangdong was not identified as a source of the attacks.

As to Guangdong province's reputation for being involved in China's cyberwarfare efforts, *see*, e.g., *Hacker Attacks in U.S. Linked to Chinese Military*, Breitbart.com (December 12, 2005), http://www.breitbart.com/news/2005/12/12/051212224756.jwmkvntb.html; Robert Vamosi, *Is China's Guangdong Province Ground Zero for Hackers?*, ZD Net (August 30, 2001), http://techupdate.zdnet.com/techupdate/stories/main/0,14179,2808609,00.html. The "Titan Rain" cyberassaults on "secure computer networks at the [U.S.'] most sensitive military bases, defense contractors and aerospace companies" came from Guangdong province. *See*, e.g., Nathan Thornburgh, *The Invasion of the Chinese Cyberspies*, Time (August 29, 2005), http://www.time.com/time/magazine/article/0,9171,1098961-1,00.html.

If we decide the BIS attacks were cybercrime or cyberterrorism, a response will definitely be in order and will come from civilian law enforcement.[336] If we decide the attacks were a foray into cyberwarfare, the response, if any, will come from the U.S. military. In that instance, we may or may not want to respond; if we decide the attacks came from PLA military hackers who were testing our online defenses, we may not want to respond actively, at least not immediately. We may want to respond passively, observing the activity to see what we can learn about our adversary's identity, strategy, and tactics. We explore these and other, related issues in Chapter 6. Our concern here is with the first step in the attribution-response process: Identifying the nature of the attacks.

The first possibility we must consider is that the attacks constitute warfare (a challenge to external order); as we saw earlier, this is the first stage in the attack-attribution process. If we decide the attacks are warfare, we respond accordingly. If we decide they are something other than war, that is, a challenge to internal order, we consider the default options: crime and terrorism.

The basic circumstances of the attacks (extraterritorial origin + origin site associated with cyberwarfare + government computers as targets) inferentially suggest they were cyberwarfare sorties launched by the Chinese government. Historically, an attack originating from the territory of one nation-state, terminating on the sovereign territory of another nation-state, and targeting property owned and used by that nation-state has presumptively constituted an act of war.[337] But while the BIS attacks have certainly characteristics that suggest they were acts of war, the nature of the attacks is ultimately ambiguous, for several reasons.

One is that the fundamental tenor of the attacks is ambiguous in ways it would not be if they had been real-world assaults. All we really know about them is that "someone" (we deal with this issue in a moment) located in Guangdong, China, attempted to gain access to data in U.S. government computers and may have succeeded. We know "something" happened that should not have, but the base circumstances are inferentially inadequate to

---

336 *See* Brenner, *Toward a Criminal Law for Cyberspace, supra* at 49–65. Because the BIS attacks targeted federally owned and used computers, federal law enforcement agents will respond. If the attacks had targeted computers used by state or local government, then state and/or local law enforcement officers would have responded, probably with the assistance of federal law enforcement agents. If the attacks had targeted computers used by a large corporation, the response could come from federal law enforcement agents, from state law enforcement officers, or, more likely, from a combination of both.

337 *See* Chapters 2 & 3.

sustain a conclusion that this was cyberwarfare. The ambiguous tenor of the virtual attacks perhaps becomes more apparent if we analogize them to real-world activity. Using the assumptions noted above, we come up with this: Chinese military personnel surreptitiously travel to the United States, where they secretly, and repeatedly, break into the BIS offices, where they rummage through secure information held in the office and steal (attempt to steal) some quantity of that data.

Here, the physical circumstances of the attack (Chinese military's entry not only onto U.S. soil but into U.S. government agency, combined with their forcible entry into the premises and search and seizure of protected U.S. government property) resolve most, if not all, of the ambiguity we encounter in the circumstances in the actual, virtual attacks. But even considered in this light, the BIS attacks look more like espionage,[338] or sabotage,[339] than an act analogous to massed military forces' crossing a national border and using sophisticated weaponry to inflict injury, death, and destruction.[340] The latter is an unambiguous, traditional act of war; the former is . . . something else, something unascertainable based on the limited facts at hand.

Another circumstance contributing to the definitional ambiguity of the BIS attacks is that while the fact they originated from the territory of another nation-state is a legitimate, indeed, significant, factor in attack-attribution, it carries much less weight here than in the real-world. In our real-world version of the BIS attacks, Chinese military personnel enter the United States by crossing a territorial border (Canada, say) or coming ashore after being transported on military personnel carriers. In that version, their entry onto U.S. soil is unambiguously and inferentially hostile; China could not deny that its military intentionally entered the sovereign territory of the United States without being authorized to do so. And as we saw earlier, the invasion of Nation-State B's territory by Nation-state A's military forces is synonymous with an act of war, indeed, with the declaration of war.

---

338 *See*, e.g., "Espionage," Wikipedia, http://en.wikipedia.org/wiki/Espionage (defining espionage as "state spying on potential . . . enemies, primarily for military purposes").

339 *See*, e.g., "Sabotage," Wikipedia, http://en.wikipedia.org/wiki/Espionage (defining sabotage as a foreign agent's destroying or damaging property used by an enemy state in its defense).

340 We return to that issue later in this chapter. For another example of this type of attack, *see* Demetri Sevastopulo, *Chinese Military Hacked into Pentagon*, Financial Times (September3,2007),http://www.ft.com/cms/s/0/9dba9ba2-5a3b-11dc-9bcd-0000779fd2ac.html ("The Chinese military hacked into a Pentagon computer network in June in the most successful cyber attack on the US defence department, say American officials").

The problem, of course, is that the BIS attackers (whoever they were) never physically entered U.S. territory; cyberspace makes it possible for anyone with an Internet connection and basic computer skills to remotely attack a computer in another country.[341] Virtual transnational operations are not the exclusive province of nation-states, which means the transnational aspect of the BIS attacks *may*, or it may *not*, be significant in attack-attribution. The transnational aspect of virtual attacks is merely one factor we must consider in the attack-attribution process, a factor that may very well be inconclusive.

We arrive at the same conclusion in analyzing the remaining circumstances involved in the BIS attacks: (a) their originating in Guangdong province and (b) their targeting computers used by a "sensitive" U.S. government agency. Guangdong's university has been producing hackers for years,[342] and for years sport hackers from various countries have explored U.S. government computers to test their skills.[343] It is therefore equally possible either that the BIS attacks came from (a) civilian hackers in Guangdong[344] or (b) Chinese government hackers.

The equivalence of these possibilities does not change if we factor our final probative circumstance into the attribution analysis: the attackers' attempting to access "sensitive" data in the computers. This could indicate either that they were PLA hackers conducting cyberwarfare sorties on behalf of the Chinese government or that they were simply sport hackers. As I noted above, the latter's accessing U.S. government computers is far from being unknown. According to the U.S. government, for example, in 2001 and 2002 a British hacker named Gary McKinnon repeatedly, and without authorization, accessed "dozens of US Army, Navy, Air Force, and Department of Defense

---

341 *See* Chapters 2 & 3.

342 *See* Vamosi, *Is China's Guangdong Province Ground Zero for Hackers?, supra. See also* Miro Cernetig, *Chinese Hackers Happy to Take Credit for Code Red*, Open Flows (August 3, 2001), http://www.openflows.org/article.pl?sid=01/08/05/1722210&mode=thread&tid=8.

343 *See*, e.g., Colin Barker, *The NASA Hacker: Scapegoat or Public Enemy?*, ZD NET, July 13, 2005, http://news.zdnet.co.uk/security/0,1000000189,39208862,00.htm.

344 *See* Vamosi, *Is China's Guangdong Province Ground Zero for Hackers?, supra.* "Sport hackers" are civilians who break into computers for fun, "for a prize or glory." Happy hacker, Everything$_2$ (April 9, 2001), http://everything2.com/index.pl?node_id=134013. *See also* The Computer Police (March 29, 2006), http://piotech.wsd.wednet.edu/techtwounits/02ComputerEthics/Task2/13computerpolice/13computerpolice.html (federal investigators often "don't know if the attacker is a kid sport-hacking from his basement or if there is a serious threat to national security from a foreign power").

computers, as well as 16 Nasa computers."[345] The U.S. Department of Justice prosecuted McKinnon for his activities, claiming, among other things, that he sought to "access classified information."[346] McKinnon says he is merely a civilian hacker who relied on commercially available software in using U.S. government computers to research "Unidentified Flying Objects."[347] He is fighting extradition to the United States at this writing, so the case has not been resolved; it is clear, though, that whatever McKinnon was doing, he was doing it on his own—so he was a cybercriminal, at most. The same could be true of the BIS attackers . . . or not.[348]

We simply cannot ascertain the nature of the BIS attacks with the available information. The transnational aspect of the attacks, the choice of federal computers as the attack target and the attempt to access "sensitive" data are all ambiguous when it comes to determining whether the attacks were cybercrime or cyberwarfare.[349] They were *at least* cybercrime, but they *could* have been cyberwarfare.[350] Our inability to resolve this issue has a profound and detrimental impact on our ability to respond to attacks such as these, an issue we examine in more detail in Chapter 6.

We have not considered whether the BIS could be cyberterrorism for several reasons. One is that the attacks themselves bear no indicia of terrorism—the entries into the computers were surreptitious and do not seem to have been designed to destroy data or to interrupt the functioning of the BIS network.[351] It is still conceivable that they came from cyberterrorists in

---

345 *See* Clark Boyd, Profile: Gary McKinnon, BBC News (April 3, 2007), http://news.bbc.co.uk/2/hi/technology/4715612.stm.

346 *Id.* U.S. prosecutors also accused him of altering and deleting data held in U.S. military computers and of attempting "to influence and affect the US government by intimidation and coercion." *Id.*

347 *See id.* McKinnon said he believed the U.S. government had reverse-engineered an antigravity propulsion system from crashed UFO's and was keeping it secret; his "humanitarian" goal was to uncover the secret and share the technology with the world. *See id.*

348 Some see the focus on government computers and sensitive government data as a factor weighing very heavily in favor of concluding that attacks such as these are launched at the behest of the Chinese government. *See*, e.g., Peter Warren, *Chinese Hackers Attack UK Houses of Parliament*, Future Intelligence (January 20, 2006), http://www.futureintelligence.co.uk/content/view/85/63/.

349 *See*, e.g., Larry Greenemeier, *Kremlin Critics Say Cyberspace Alive with DoS Attacks*, Information Week (July 3, 2007), http://www.informationweek.com/story/showArticle.jhtml?articleID=200900283 (computer security expert warns against "jumping to conclusions" about a nation-state's authorizing and carrying out cyberattacks).

350 *See* Chapters 2 & 3.

351 *See* Chapter 3.

Guangdong, perhaps as preliminary steps ultimately intended to result in a cyberterrorist event, but because we have no facts pointing to terrorists operating out of Guangdong, and no indicia of terrorist intent, we cannot logically derive that conclusion. It is alternatively conceivable that the attacks constituted state-sponsored cyberterrorism, a variation on cyberwarfare we examine in Chapter 5, but, again, the facts we have provide absolutely no basis for drawing that conclusion.

Estonian Attacks

These attacks provide a useful analytical counterpoint to the BIS attacks: As we saw in Chapter 1, various components of Estonia's critical infrastructure came under a two-week, sustained series of DDoS attacks in April of 2007. Unlike the BIS attacks, which apparently only sought access to data, the Estonian attacks were purely destructive, in a virtual sense. As we saw, they shut down websites and networks belonging to government agencies, Internet service providers, financial institutions, phone systems, police forces, newspapers, television stations, schools, and businesses.[352] If Estonian authorities had not received prior notice of the attacks, and if the country had not been adept at defending itself against this type of attack, the effects would have been far more devastating than they were.[353]

During and for some time after the attacks, Estonian authorities believed they were cyberwarfare launched by the Russian government.[354] They based this belief on a number of compelling circumstances: The attacks inferentially appeared to be launched in retaliation for Estonian authorities' removing a statute of a Russian solider from a park; Russian-language websites were used in planning the attacks; and Estonian authorities traced Internet addresses used in the attacks to Russian government agencies.[355] They ultimately concluded that the attacks were not, in fact, Russian cyberwarfare, but we will put that aside and conduct our own analysis of the attacks.

---

352 *See* Chapter 1.

353 *Id.*

354 *Id.*

355 *Id.* They also claimed to have identified a member of the Russian security service as a "mastermind" of the attacks. *See id.*

Let us begin, as we did with the BIS attacks, by parsing out what we know about the 2007 Estonian attacks:

i. They were transnational, originating in Russia and targeting websites and networks in Estonia.
ii. They only targeted websites and networks in Estonia.
iii. They were deliberate, intentional assaults.
iv. The DDoS attacks were massive in scale, both in terms of the data load and the size of the botnets used in them.[356]
v. The attackers in part used botnets rented from cybercriminals, from which some inferred that they were backed by a government with significant assets.[357]
vi. Those who planned the attacks were fluent in the Russian language.
vii. The attacks followed action by Estonian authorities that insulted Russian citizens and Estonian citizens of Russian descent.
viii. They targeted critical infrastructure components, not for exploration, theft, or extortion, but specifically to cause damage in the form of disrupted and denied services.
ix. Internet addresses belonging to Russian government agencies were used in the attacks.
x. The Russian government publicly and repeatedly denied involvement in the attacks.
xi. The sequence of attacks included less sophisticated activity, such as putting a mustache on an online photo of the Estonian Prime Minister.[358]
xii. Sophisticated computer expertise is no longer a precondition for launching DDoS attacks, even sophisticated attacks; "commercial" tools are available online that make it relatively easy to assemble botnets and engage in other malicious activity.[359]

---

356 *See* Mark Lander and John Markoff, *Digital Fears Emerge after Data Siege in Estonia*, New York Times (May 29, 2007), http://www.nytimes.com/2007/05/29/technology/29estonia.html?ex=1182484800&en=ac3eadbe88fdb21c&ei=5070 (largest attacks "blasted streams of 90 megabits of data a second at Estonia's networks," the equivalent of downloading the Windows XP operating system every six seconds for 10 hours).

357 *See* Iain Thomson, *Russia "Hired Botnets" for Estonia Cyber-war*, IT News (June 1, 2007), http://www.itnews.com.au/News/NewsStory.aspx?story=53322. *See also* Lander and Markoff, *Digital Fears Emerge after Data Siege in Estonia*, *supra*.

358 *See* Peter Finn, *Cyber Assaults on Estonia Typify a New Battle Tactic*, Washington Post (May 19, 2007), http://www.washingtonpost.com/wp-dyn/content/article/2007/05/18/AR2007051802122_pf.html (a "Hitler" mustache).

359 *See, e.g.*, Joel Hruska, *DIY Trojan-building Tools for Sale on the Internet*, ars technical (July 25, 2007), http://arstechnica.com/news.ars/post/20070725-diy-trojan-tools-for-sale-on-the-internet.html.

As with the BIS attacks, we will begin our attack-attribution analysis with the first stage of the process: whether the Estonian attacks were cyberwarfare. If we conclude that they were not cyberwarfare or that we cannot determine whether they were cyberwarfare, we will proceed to the second stage in the analysis: whether the attacks were cybercrime or cyberterrorism.

Here, as with the BIS attacks, the most compelling evidence of cyberwarfare is the fact that the attacks originated in one nation-state and targeted the assets of another. The attacks differ in that the BIS attacks only targeted U.S. government computers, but the Estonian attacks targeted both government and private computer systems. That difference, in and of itself, does not strongly militate against a finding that the Estonian attacks were cyberwarfare; as we saw at the beginning of this chapter, cyberwarfare will no doubt target the privately held components of a nation-state's critical infrastructure, as well as government computer systems.

What does militate against a finding that the Estonian attacks were cyberwarfare is the lack of clarity as to whether the attacks did, indeed, originate with the Russian government. We will assume for the purposes of analysis that they did originate in Russia because there is at least some reliable evidence to that effect. Their originating in Russia is also inferentially supported by certain other factors, including the use of Russian-language websites in planning the attacks and their apparently being, at least in part, retaliation for Estonia's removing the statute of the Russian soldier.[360] We therefore assume that the attacks originated in Russia, but we cannot, without more, conclude they were launched at the behest of the Russian government.

For one thing, we lack circumstantial evidence comparable to the evidence that the BIS attacks came not merely from China, but from Guangdong, the center of the PLA's cyberwarfare effort. The Estonians claimed their experts traced some of the attacks to Internet addresses used by Russian government officials, including an address used by an official who worked with President Putin,[361] but that evidence, which we assume is reliable, is not as compelling as the Guangdong-as-point-of-attack-origin data. First of all, we have no indication that the Russian Internet addresses were associated with agencies that are, or are likely to be, involved in the initiation of cyberwarfare.

Inferentially we may have just the opposite: It seems unlikely that administrators such as the one who works for Putin would play a hands-on role in

---

360 *See* Chapter 1.

361 *Id.*

launching cyberwarfare; they might very well order cyberwarfare attacks, just as they might well order real-world military attacks, but the realities of bureaucracy and a probable lack of technical expertise on the part of those employed in these agencies suggests they were not perpetrators of the Estonian attacks. This conclusion is supported by research conducted after the attacks: A DDoS expert who conducted a post mortem analysis of the attacks found that the evidence did not establish " 'a clear line from Moscow to Talinn.' "[362] As those who understand cyberspace realize, the Internet protocol addresses networked computers use can easily be spoofed, or faked, so an attack that seems to come from one computer actually comes from another, in a different location.[363] It is also possible that some of the Estonian attacks came from Russian government computers; we cannot eliminate that possibility, however unlikely it seems. But that is beside the point: Because we cannot link the Estonian attacks to the Russian government with any level of confidence, we cannot use this circumstance as data supporting a finding that the attacks were cyberwarfare.

The other notable circumstances the Estonians cited in concluding that the attacks were Russian cyberwarfare were that they were planned on Russian language websites and seemed to have been retaliation for Estonia's removing the statue of the Russian solider at a time when Russians and Estonians of Russian descent were preparing to honor Russia's fallen soldiers.[364] We will assume that both of these facts are true, and we assumed earlier that the attacks came from Russia. Even if we consider these facts in combination, however, they cannot conclusively establish that the attacks were Russian cyberwarfare. They inferentially support that possibility, but they also support the inference that the attacks were "hacktivism"—an online protest by Russians and Russian sympathizers who were offended by the removal of the statue.

That is the conclusion foreign experts from the private and governmental sectors arrived at after conducting their own post mortem analyses

---

362 Jeremy Kirk, *Russian Gov't Not Behind Estonia DDoS Attacks*, PC World (June 1, 2007), http://www.pcworld.com/article/id,132469-pg,1/article.html.

363 "In computer networking, . . . IP address spoofing refers to the creation of IP packets with a forged (spoofed) source IP address with the purpose of concealing the identity of the sender or impersonating another computing system. IP stands for Internet Protocol." "IP Address Spoofing," Wikipedia, http://en.wikipedia.org/wiki/IP_address_spoofing. *See*, e.g., Cyrus Farivar, *Cyberwar I: What the Attacks on Estonia Have Taught Us about Online Combat*, Slate (May 22, 2007), http://www.slate.com/id/2166749/.

364 *See* Chapter 1.

of the attacks. They pointed out that while the attacks seemed sophisticated to the Estonians, they really were not: A "former senior U.S. cybersecurity official" described them as " 'brute force, crude' " attacks lacking in the " 'elegance and precision' " associated with the sophisticated cyberwarfare capabilities of "major powers" like the United States and China.[365] The former chief scientist at the Pentagon's Defense Advanced Research Projects Agency concurred, noting that the Estonian attacks were " 'more like a cyber riot than a military attack.' "[366] The conclusion these and other experts drew was that " 'no state actor' " was responsible for the attacks.[367] They decided the attacks were most likely hacktivism—loosely coordinated online activity undertaken by individuals who believed Estonia needed to be "punished" for disrespecting Russian soldiers and the contributions they made during World War II.[368] Aside from anything else, this conclusion is consistent with certain aspects of the attacks, such as someone's adding a "Hitler mustache" to a photograph of the Estonian Prime Minister that was posted on his political party's website.[369]

Before we consider where we are in our own analysis, I want to note an implication of the experts' assessment of the technical crudity of the Estonian attacks: That crudity may, as seems likely, truly establish that they were not cyberwarfare but were, instead, an online political protest. But think about the experts' reasoning for a moment: They seem to be assuming that online warfare is no different from real-world warfare in one important respect—it presents itself as what it is, "warfare." In the real-world, countries use offensive covert and semicovert sorties in an effort to gain tactical advantages over rival countries; the use of deception is not unknown among nation-states. They have not, however, been able to dissemble in launching acts of war; countries have fabricated justifications for attacking other countries

---

365 Shaun Waterman, *Analysis: Who Cyber Smacked Estonia?*, United Press International (June 11, 2007) (quoting Bruce Brody), http://www.upi.com/Security_Terrorism/Analysis/2007/06/11/analysis_who_cyber_smacked_estonia/2683/.

366 *Id.*

367 *Id.* Other experts noted that launching the Estonian attacks would have created "real risks" for Russia. *See* John Schwartz, *When Computers Attack*, New York Times (June 24, 2007), http://www.nytimes.com/2007/06/24/weekinreview/24schwartz.html?ex=1340337600&en=6e966bd8b6fe8d8d&ei=5088.If Russia were responsible for the attacks, and if that came to light, then Russia could very well face military retaliation from Estonia or, perhaps, NATO. *See id.*

368 *See* Waterman, *Analysis:*, *supra*. We return to hacktivism in our consideration of the attacks as cyberterrorism, *infra*.

369 *See* Chapter 1.

(e.g., *Lebensraum* or having weapons of mass destruction), but they could not disguise the attacks as something other than warfare. Now, perhaps they can: In addition to the other factors we examined in Chapter 3—the anonymity and surreptitiousness cyberspace provides for those who engage in aggressive or illegal activity—nation-states may be able to camouflage online acts of war by avoiding the technical "elegance and precision" experts associate with cyberwarfare. We have, it seems, yet another factor that can complicate the attack-attribution process.

As far as our application of that process is concerned, I reach the same conclusion as the experts cited above. In parsing what we know about the Estonian attacks I noted that none of the technical aspects of the attacks—the use of DDoS attacks, the use of botnets, the scale of the attacks, their transnational aspect or even the use of rented botnets for part of the attacks—conclusively or even strongly indicates they were cyberwarfare. As I also noted, anyone with a not terribly sophisticated level of computer expertise can purchase programs that are for sale online and use them to assemble botnets and launch DDoS attacks. It would be simple for any half-way-adept hacker, or group of hackers to do this. As we saw in Chapter 1, cybercriminals routinely use DDoS attacks in committing extortion and other crimes.

The attackers' use of Russian-language websites to coordinate the attacks indicates they were civilian, not military attacks: Why would Russian military hackers (assuming they exist)[370] need to rely on publicly accessible websites to coordinate cyberwarfare? And why would a government that is assumed, correctly or incorrectly, to have the technical capacity to launch cyberwarfare attacks need hired botnets to carry out a portion of those attacks? The use of rented botnets indicates that these were civilian-launched, not government-launched, attacks. The same, finally, is true for the scale of the attacks; as the experts cited above concluded, the scale may have seemed overwhelming to the Estonians, but it is not what we expect to see in cyberwarfare. As we saw in Chapter 1, experts estimated that the botnets used in the Estonian attacks may have encompassed a million compromised computers, or zombies; but as we also saw, cybercriminals had by then already used larger botnets in committing routine crimes.

---

370 There have, for years, been indications that Russia, like other countries, is trying to develop a cyberwarfare capacity. *See*, e.g., Stephen Green, *Pentagon Giving Cyberwarfare High Priority*, Copley News Service (December 21, 1999), http://www.fas.org/irp/news/1999/12/991221-cyber.htm. *See generally* Steven A. Hildreth, Cyberwarfare, Congressional Research Service (June 19, 2001), http://www.fas.org/irp/crs/RL30735.pdf.

We cannot conclusively determine either that the 2007 Estonian attacks (a) were *not* cyberwarfare or (b) *were* cyberwarfare. That means we must proceed to the second stage in the attack-attribution analysis: determining whether they were cybercrime or cyberterrorism.

We have already concluded, with what I believe to be a fair degree of confidence, that the attacks were not launched either by the Russian government or by any other nation-state (barring the dissembled cyberwarfare possibility I noted above). This means the attackers were civilians, and that, in turns, means the attacks were cybercrime or cyberterrorism. Civilians do not, as we saw earlier, wage war, at least not as we currently understand warfare.[371]

Having concluded, if only for the sake of analysis, that the attacks were cybercrime or cyberterrorism, we must decide which alternative is correct. Determining whether the attacks were cybercrime or cyberterrorism is not of particularly pressing import for the response process we examine in Chapter 6 because civilian law enforcement responds to crime and to terrorism.[372] It can, though, be useful for law enforcement officers to know what they are dealing with, as that can influence the course of the investigation. The profile and tactics of cybercriminals can differ markedly from those of cyberterrorists, as we saw in Chapter 3: The hypothetical cyberterrorists in the Cyber Storm scenario had very different motives than the adventitious Berlin hackers who exploited the scenario for their own purposes. Their tactics can differ, as well: Cyberterrorists are likely to pursue destruction and disruption for their own sake, while cybercriminals see them as a means to an end, which is usually enriching a set of cybercriminals.

We will return to these issues in the next chapter. We still have two tasks to complete in this chapter, the first of which is to finish our attack-attribution analysis of the Estonian attacks.

As we saw above, the attacks inferentially seem to have been an instance of hacktivism, which we need to define. The definition we will use is adapted from a dissertation by Alexandra Samuel: Hacktivism is the use of illegal or legally ambiguous digital tools in pursuit of political ends.[373] My definition differs from the one Ms. Samuel proposed in this respect: She includes a requirement that the use of digital tools be for nonviolent ends, a device she

---

371 *See* Chapter 3.

372 *Id.*

373 *See* Alexandra Whitney Samuel, Hacktivism and the Future of Political Participation, Chapter 1, page 2 (Harvard University Department of Government, September 2004), http://www.alexandrasamuel.com/dissertation/pdfs/index.html.

uses to differentiate hacktivism from cyberterrorism.[374] I do not think including that requirement produces an accurate definition because, as we have seen in the real-world, political protests can include violence.[375] The violence can be accidental or intentional; when people strongly believe in their cause, they may also believe some resort to violence is justified in the pursuit of a higher purpose.

I am not equating hacktivism with cyberterrorism. Hacktivism, I submit, can constitute either cybercrime or cyberterrorism, depending on the nature and extent of the "harm" it inflicts: If hacktivists deface a website in pursuit of their political goals, that is cybercrime; if they shut down computer systems and leave people in the upper latitudes without heat in December, that, I submit, is cyberterrorism, because here they create a situation that is almost certain to result in death, injury, and property destruction.

As far as our analysis is concerned, it seems the Estonian attacks were hacktivism. They were not "regular" cybercrime, at least not in its currently most common incarnations, because the attackers made no attempt to enrich themselves by, say, siphoning funds from the financial institutions that came under attack. Their goals instead seem to have been purely destructive, in the cyber-sense of denying access to websites and web services, and that indicates we are dealing either with simple hacktivism (hacktivism as cybercrime) or with aggravated hacktivism (hacktivism as cyberterrorism).[376]

As to which the attacks were, well, this is one of those cases in which there is no simple, straightforward answer. We can argue that the Estonian attacks should be regarded as simple hacktivism because there is no indication the

---

374 *See id.* at 2-6. Ms. Samuels seems to define "violence" as encompassing death or injury to human beings and the destruction of or damage to property. *See id.* at 3.

375 *See, e.g., Americas Summit Protest Turns Violent*, CNN (November 5, 2005), http://www.cnn.com/2005/WORLD/americas/11/05/bush.summit/index.html.

376 As we saw in Chapter 3, cybercriminals have launched purely destructive DDoS attacks, though these are becoming less common. In Chapter 3, I mentioned Mafiaboy, the teenager who in 2000 launched a series of DDoS attacks that shut down various websites, including Amazon, Yahoo! and CNN. Mafiaboy's goal seems to have been destruction for destruction's sake. He, unlike modern cybercriminals, did not approach DDoS attacks as a tool he could use to enrich himself. Instead, he seems to have been engaged in the cyber-analogue of vandalism, a course of conduct for which he was prosecuted and sanctioned. As we saw in Chapter 3, "pure" DDoS attacks such as those launched by Mafiaboy are criminalized in the United States and elsewhere. Under the distinction I draw in the text above, Mafiaboy's attacks constituted hacktivism as cybercrime—instead of cyberterrorism—because of the relatively minor level of "harm" he ultimately inflicted.

attackers meant to cause death, injury, or tangible property destruction, or did so;[377] unlike the Cyber Storm terrorists, they did not turn off heat, shut down commuter trains or crash air traffic control systems, nor did they try to do so. They did disrupt the functioning of important systems, but the systems they targeted were primarily government, financial, and communications systems; inferentially, their conduct resembles a street protest more than the terrorist attacks we examined in Chapter 3.

On the other hand, we can also argue that the Estonian attacks were cyberterrorism because while the attackers did not directly seek to inflict injury, death, or property damage, their shutting down certain systems—such as communication and police systems—could (and may) have had that result. Someone might have needed to access Estonia's 911 system because they were having a heart attack or because their home was on fire but could not because the attackers had shut down the system. Even if the attackers did not actually cause death, injury, or property destruction, and even if they never intended for that to happen, their conduct could still constitute cyberterrorism. U.S. law, for example, defines cyberterrorism broadly enough that it encompasses recklessly or negligently inflicting death or injury or creating "a threat to public health or safety."[378]

Ultimately, the decision as to whether the Estonian attacks were simple hacktivism or cyberterrorism will lie with a prosecutor, in the unlikely event the attackers are ever identified and apprehended. We take up the challenges

---

377 We can argue about whether they actually caused intangible property damage. I would argue that they did because they denied thousands of people access to various types of web-based services, some of which (banking, communications, police) are essential for everyday life in the modern world. An advocate for the Estonian attackers would presumably argue that "services" are not "property," so we cannot legitimately hold them liable either for destroying or stealing property.

My first response to that argument would be to point out that criminal statutes, in the United States and elsewhere (I do not have access to Estonian law), define "services" as a type of property. *See*, e.g., Alabama Code § 13A-7-24. I would then argue that if one is denied access to services-as-property, that is the functional equivalent of destroying or stealing that property because it results in the property's not being available to those who should legitimately be able to access it when they desire to do so. My second, contingent response would depend on whether Estonia has outlawed "pure" DDoS attacks; if it has, I would argue that the criminalization of these attacks implicitly recognizes that they "harm" citizens by denying them what is in effect a commodity, a service upon which they rely in order to function in modern society. Finally, my third and final response to the advocate for the attackers would be that by denying Estonian citizens access to their financial institutions, the attackers effectively deprived them of their property, i.e., effectively deprived them of the ability to use funds being held in the financial institutions.

378 *See* 18 U.S. Code § 1030(a)(5)(B).

that presents in the next chapter, but before we turn to that issue, I need to note two final areas of concern in attack-attribution.

Mixed Metaphors

We take up the issue of attack response in Chapter 6, but I want to say a few words here about how the interaction between attack-attribution and response processes can impact on our ability to deal with cyberthreats. As we saw earlier, the response strategies nation-states rely on are predicated on the premise that we know, or can quickly determine, what ***kind*** of an attack occurred, and only need to identify and neutralize the attacker(s). As we saw in Chapter 3, legal systems incorporate that premise by allocating responsibility for crime/terrorism to civilian law enforcement and for war to the military. As we will see in Chapter 6, the allocation is scrupulously partitioned; civilian law enforcement does not respond to war and the military does not respond to crime.

As may be apparent, this rigidly partitioned response authority creates several concerns when the attacks come from cyberspace. One is that the response process will be delayed while decision-makers try to determine the nature of an attack. Another is that the decision-makers will misunderstand the nature of an attack. What if a BIS-style attack were to target a corporate computer system? The nature of the attack target inferentially supports the conclusion that this is cybercrime because we assume criminals target civilians.[379] That conclusion would further be reinforced if the attackers' actions conformed to what we expect of cybercriminals, that is, if their efforts consisted of trying to extract funds from corporate accounts or personal information from customer databases. Because we routinely equate an

---

379 This assumption derives from the internal-external threat dichotomy. We know civilians, and civilian entities, suffer "harm" of varying types and degrees in warfare but this is a collateral consequence of war. *See* Chapter 3. We have therefore equated direct, intentional attacks on civilians with crime and, to a lesser degree, with terrorism.

The assumption is also based in the limitations of physical reality: In the real-world, it is not possible for Nation-State A to launch a physical attack on the headquarters of Corporation XY, which is in Nation-State B, without starting a war. If Nation-State A used long-range missiles to damage Corporation XY's headquarters, that would in a sense constitute the commission of a crime against Corporation XY, but the criminal offense would be subsumed in the larger consequences of the attack. Nation-State B would construe the attack as war because while it inflicts "harm" only, and directly, on a purely civilian entity, it violates Nation-State B's sovereign territorial integrity and, in so doing, threatens Nation-State B's ability to maintain external order. Nation-State B would therefore treat the attack as a declaration of war and respond accordingly. *See* Chapter 3.

attack on civilians with crime, not with war, an attack such as this would be construed as cybercrime and law enforcement would respond.

The problem is that relying on this assumption (i.e., that civilian attacks constitute crime) could be a mistake. The attack on our fictive corporate entity could be cyberwarfare, instead of cybercrime. As we saw in Chapter 3, China's focus on cyberwarfare includes attacks on civilian entities, especially financial and infrastructure entities. A 2007 Congressional report warned that "U.S. federal agencies should now be aware that in cyberspace some countries consider that no boundaries exist between military and civilian targets."[380] If our default approach to attacks continues to be based on the civilian-attacks-are-crime assumption, we will certainly have a situation (or many situations) in which an act of cyberwarfare is construed, and responded to, as if it were merely cybercrime. The consequent, possibly lengthy, delay in realizing what we truly confront could have serious consequences for our financial system or other aspects of our infrastructure, especially if we misinterpreted a *series* of cyberattacks.

An analogous, but perhaps less serious, problem arises if the attack on our hypothetical corporate entity is cyberterrorism. Cyberterrorist attacks are unlikely to be isolated incidents; it is far more likely that a cyberterrorist event will be part of a sequence of attacks, which may be separated spatially and temporally, that is, have different points and times of origin. Because the corporate attack we hypothesized above seems to be "mere" cybercrime, it would be dealt with by civilian law enforcement. Except for serial killers and the odd career robber or serial arsonist, civilian law enforcement is not accustomed to approaching an attack—a crime—as part of a sequence; the officers who respond to our fictive attack are consequently not likely to consider it as part of a much larger attack sequence.[381] This means that the law enforcement response to a coordinated, sequenced cyberterrorism attack would probably be discrete and isolated; officers in different locations would respond to incidents without realizing that they were, in fact, part of a single, larger attack. And as we saw above, the same could be true for cyberwarfare.

---

380 John Rollins and Clay Wilson, Terrorist Capabilities for Cyberattack: Overview and Policy Issues 17, Congressional Research Service (Library of Congress, 2007).

381 The likelihood that they will consider this possibility will diminish, accordingly, if the attacks constituting the larger attack sequence (a) occur discretely at locations in different states or in widely separated parts of a single state; (b) occur over a week, two weeks, a month, or longer; and (c) display different attack signatures. As to (c), discrete attacks in a sequenced effort could, say, target one or more financial institutions, then commuter trail transport, then a power grid, and, finally, communication systems. The spatial, temporal, and target differentiation in the attacks would mask the fact that they are components of a single event.

This problem arises both because of our partitioned responsibility for responding to crime/terrorism versus warfare and because we tend to assume that crime, of whatever type, is a localized phenomenon. A subsidiary factor contributing to the problem is that the markers we rely upon to differentiate crime/terrorism from warfare in the real-world are absent or unreliable when it comes to virtual attacks. In the real-world, we rely on three markers—or indicia—in determining the nature of an attack: (a) point of attack origin, (b) point of attack occurrence, and (c) motive for the attack.

As we have seen, the utility of the first two markers erodes as attacks migrate online. The same is also true, but in a different way and for different reasons, of the third factor—motive. Technology enhances our ability to inflict harm, but does not alter the human psyche; unless and until technology transforms us into cyborgs or some other variety of posthuman life,[382] it is reasonable to assume the motives that have historically driven us to inflict "harm" will continue to account for our doing so, on- or off-line. Motive, therefore, will continue to be a valid differentiating factor for cyberattacks: profit drives most cybercrime, ideology drives cyberterrorism, and nation-state rivalries (will) drive cyberwarfare. The difficulty here arises not with our ability to rely upon established motivations as a "marker" that inferentially indicates the nature of an attack. It arises, instead, with our ability to ascertain the motive behind an attack.

We saw this with the BIS attacks: We know *what* the BIS attackers did, but we cannot ascertain *why* they did it. The same has been true of other highly publicized U.S. government and/or corporate attacks, including "Titan Rain" and "Moonlight Maze."[383] It is likely to be true of future attacks, as well; though the motive behind what are almost certainly routine cybercrimes is usually apparent (greed, revenge, power over another), that may not always be true. Terrorists, for example, are using cybercrime to finance their real-world efforts,[384] and that creates a mixed-motive scenario: The motive for committing the cybercrimes is profit, which is a criminal motive; but the motive for obtaining the profit is to facilitate acts of terrorism, a different kind of motive. This particular mixed message scenario, however, should have few operational implications for the response process because civilian law enforcement responds both to crime and terrorism.

---

382 *See*, e.g., Transhumanism—Wikipedia, http://en.wikipedia.org/wiki/Transhumanism.

383 *See*, e.g., Moonlight Maze—Wikipedia, http://en.wikipedia.org/wiki/Moonlight_Maze; Titan Rain—Wikipedia, http://en.wikipedia.org/wiki/Titan_Rain.

384 *See*, e.g., Rollins and Wilson, Terrorist Capabilities for Cyberattack, *supra* at 16–21.

A variation of this "mixed message" scenario can have serious implications for a state's ability to respond to attacks. When what is ostensibly cybercrime is state-sponsored—as is increasingly true of economic espionage—the efficacy of the civilian law enforcement response process breaks down.[385] The sponsoring state will almost certainly refuse to cooperate with the investigative efforts of the victim state's law enforcement officers, and thereby thwart the crime response process.[386] The same result will ensue when a nation-state sponsors cyberterrorism. In neither instance does the sponsoring state have any incentive to cooperate with those who are attempting to bring its agents to justice in the state they victimized; indeed, the opposite is true.[387] The sponsoring state has every incentive *not* to cooperate with law enforcement from the victim state, especially if the state-sponsored cybercrime/cyberterrorism is an instance of war-but-not-war, that is, a covert campaign aimed at undermining the economic or structural stability of the victim nation-state.[388] This is a class of attacks with which we must be especially concerned, because the "harm" sought to be inflicted here is systemic, not individual, "harm."

Another class of attacks with which we must be especially concerned is the BIS-style attack—attacks in which no apparent motive exists. The scenario in which we cannot ascertain whether attacks are cybercrime, cyberterrorism, or cyberwarfare creates the greatest challenges for our current response model, and consequently creates the greatest risks for the victim state. It creates marked risks for countries such as the United States, which rigidly partition response authority between civilian and military agents. In these systems, if potential responders cannot ascertain what kind of an attack occurred, they can neither assume nor assign responsibility for responding to it—which creates the possibility that no appropriate response will ensue.[389]

This, in turn, creates the possibility that a country like the United States could be the target of cyberwarfare and not realize it until the attacker had inflicted substantial systemic damage; this would be particularly true if a dispersed attack seemed to represent cybercrime. Local authorities would deal discretely with

---

385 *See*, e.g., Susan W. Brenner and Anthony C. Crescenzi, State-Sponsored Crime: The Futility of the Economic Espionage Act, 27 Houston Journal of International Law 389 (2006).

386 *See id.*

387 *See id.*

388 *See id.*

389 *See*, e.g., Testimony Before the S. Comm. on Governmental Affairs, 108th Cong. (March 2, 2000) (testimony of James Adams, Chief Executive Office, Infrastructure Defense, Inc.) http://www.senate.gov/~gov_affairs/030200_adams.htm (lack of response to Moonlight Maze).

each node of the attack, not realizing they were responding to part of a greater whole. It also creates the possibility that a concerted attack—cybercrime or cyberterrorism—by an organized group of non-nation-state actors could inflict comparable, though perhaps not as substantial, systemic harm on the victim nation-state.[390] This latter scenario raises yet another possibility—that of sporadic, concerted attacks by one or more organized groups of non-nation-state actors.[391] While the "harm" these attacks inflicted would not be the immediately devastating kind of "harm" associated with real-world warfare or real-world terrorism on a 9/11 scale, it would be damaging, particularly if it were repeated.

This last possibility highlights another difficulty with our current segmented response processes: Civilian law enforcement and military personnel have a very limited ability to join forces in combating attacks.[392] This is due to the persistence of the internal-external threat dichotomy. Civilian law enforcement and military personnel have an even more limited ability to join forces with each other and with civilians in responding to attacks.[393] In the modern nation-state model, the state monopolizes the response processes; civilians can join the military or become law enforcement officers, but aside from that, they have no legitimate role in responding to threats. This is also due to the internal-external threat dichotomy. Nation-states assume (a) they can maintain internal order exclusively with law enforcement personnel, and external order exclusively with military personnel; and (b) that by doing this they establish a secure enclave in which civilians need have no concern with, and no responsibility for, internal or external attacks. As we have already seen, and as will see in more detail later, cyberspace erodes the validity of both assumptions.

Eroding Construct(s)

Who has seen tomorrow's war? No one.[394]

We have so far assumed the continuing validity of the threat taxonomy (crime, terrorism, and warfare) that is the conceptual predicate of the rules

---

390 *See*, e.g., *id.*

391 *See*, e.g., *id.*

392 *See* Chapters 6 & 7.

393 *See id.*

394 Qiao Liang and Wang Xiangsui, Unrestricted Warfare 124 (Beijing: People's Liberation Army Literature and Arts Publishing House 1999).

and orchestrations nation-states rely on to maintain order.[395] Its validity, though, seems to be eroding as cyberspace undermines the empirical assumptions that shaped the taxonomy's threat categorizations; it is becoming evident that attackers' use of cyberspace can erode distinctions among the threat categories.

Aspects of that erosion account for the difficulties we encountered in our analysis of the Estonian attacks, particularly our analysis of whether they were cyberwarfare. In Chapter 3, we reviewed the modern conception of warfare that assumes war is a zero-sum exercise in which one "side" wins and the other loses.[396] This is the conception with which we are familiar: In World War II, the Allies won, and the Axis powers lost; in World War I, the Entente Powers won and the Central Powers lost,[397] and in the U.S. Civil War, the North defeated the South.

As I noted in the previous section, cyberwarfare almost certainly will not conform to this zero-sum conception of warfare, at least not exclusively. There is, as others have pointed out, no logical reason why war has to be zero-sum; our modern conception of warfare evolved from historical practice, in which more or less formal aggregations of combatants ("armies") came together at a particular place and struggled until one of them gained a decisive advantage. A "war" could be decided by a single battle or, as has been typical in modern history, by a series of battles fought over a protracted period and a wide geographical area. In the real-world, we are still fighting wars that conform to the model Napoleon perfected in the late eighteenth and early nineteenth centuries.[398] We use the same dynamic; only the tactics and weapons have changed. Others have demonstrated that this conception of war is ill-suited to a world in which real-world military conflicts take on

---

395 *See* Chapter 3.

396 *See* Carl von Clausewitz, On War 101 (J.J. Graham, trans.) (Penguin 1982) ("War is nothing but a duel on an extensive scale. . . . [It] *is an act of violence intended to compel our opponent to fulfill our will*) (emphasis in original). *See*, e.g., "Game theory," Wikipedia, http://en.wikipedia.org/wiki/Game_theory (zero-sum exercises are those in which one player benefits only at the expense of another). *See also* General Rupert Smith, The Utility of Force 58–65 (Alfred A. Knopf, 2007). We return to this issue later in the chapter.

397 *See*, e.g., "World War I," Wikipedia, http://en.wikipedia.org/wiki/World_War_I.

398 *See* General Rupert Smith, The Utility of Force, *supra* at 31 ("Our understanding of . . . military operations and wars stems from the nineteenth century, when the paradigm of interstate industrial war was forged. . . . The Napoleonic Wars were its starting point.")

unconventional forms, such as the insurgency/terrorist activity/guerrilla tactics the United States is confronting in Iraq.[399]

Our concern, of course, is not with real-world military conflicts, but with online conflict. In cyberspace, not only are there no compelling reasons for war to be zero-sum, there are distinct advantages to pursuing non-zero-sum conflicts. So, for example, rather than launching a zero-sum war against Country B to gain tactical ascendance over it, Country A uses cybertactics to destabilize Country B's currency and/or its economy.[400] The attacking nation-state in scenarios such as this might ultimately want the victimized nation-state to collapse so that it can assume control over the territory the victim controls . . . or it might not. The situation the United States is facing in Iraq as I write this and the post-World War II reconstruction of Europe both illustrate the downside of being the ultimate victor in a zero-sum, territorially based war.[401]

Instead of pursuing the latter, vanquishing its enemy in real-world combat and then having to assume at least some responsibility for the defeated populace, aggressive states like Country A might prefer simply to eliminate another nation-state's capacity as a serious competitor.[402] In addition to avoiding the physical costs (injury, loss of life, and equipment) of traditional warfare, this option gives Country A plausible deniability if it is accused of having been the aggressor—the instrument of Country B's decline. Cyberspace therefore creates the potential for what may be a new kind of warfare: non-zero-sum, surreptitious warfare.[403] The twenty-first century's version of the Vietnam era slogan, "what if they gave a war and no one came?," may be "what if we were at war and didn't know it?"[404]

---

399 *See id.* at 153–224.

400 *See*, e.g., Walter Gary Sharp, Sr., CyberSpace and the Use of Force 91-93 (Aegis Research Corporation 1999) (similar cyberwarfare scenario). *See also* Brenner & Crescenzi, State-Sponsored Crime, *supra* at 449–451.

401 *See*, e.g., "Marshall Plan," Wikipedia, http://en.wikipedia.org/wiki/Marshall_Plan.

402 This, too, could be accomplished virtually. *See* Sharp, CyberSpace and the Use of Force, *supra* at 92–93 (electronic blockade launched by State B brings State A "to its knees" by shutting down all telecommunications into and out of State A).

403 We return to this issue in discussing state-sponsored terrorism later in this chapter.

404 *See*, e.g., Valentinas Mite, *Estonian Attacks Seen as "Cyberwar,"* ISN (May 31, 2007), http://www.isn.ethz.ch/news/sw/details.cfm?id=17677:

> Tuuli Aug . . . of . . . the Estonian daily newspaper "Eesti Paevaleht," says that she felt the country was under attack by an invisible enemy.
>
> "It was extremely frightening . . . because we are used to having internet all the time and then suddenly it wasn't around anymore,' she says. `You couldn't get information;

This brings me to a related issue: the difficulty of knowing whether or not you are, in fact, at war. Even though the Estonian attacks ultimately proved not to be cyberwarfare, and even though that fact should probably have been apparent to the Estonian authorities sooner than it was, those attacks still illustrate the potential ambiguity of cyberwarfare. I say "potential ambiguity" because I think cyberwarfare will come in different varieties, some of which will more closely resemble traditional zero-sum warfare; and the extent to which ambiguity will inhere in a particular cyberwarfare incident will be a function of how closely that incident conforms to what we expect of traditional, real-world conflicts.

One author gives this example of an essentially zero-sum incident of cyberwarfare: Angry because State A has been manipulating global markets and devaluing its currency, State B launches an electronic blockade of State A, cutting off any and all electronic transmissions into or out of the country.[405] The blockade has a number of adverse consequences for State A's economy, governance, and ability to provide essential services to its citizens; State A therefore "surrender[s] to the will of State B."[406] This example is, at the least, a very close analogue of the zero-sum warfare we have seen in the real-world: State B can have no doubt as to whether it is under attack by a hostile nation-state, and its only options are to launch retaliatory attacks in an attempt to "win" the struggle, or surrender and "lose."

This level of clarity will, however, probably not be typical of all or even most instances of cyberwarfare. The ambiguity of cyberwarfare will increase as attacks move away from the zero-sum model and toward the surreptitious non-zero-sum attacks we examined above. The BIS attacks we reviewed at the beginning of this chapter could have been an instance of non-zero-sum cyberwarfare. If we assume, for the sake of analysis, that they were hostile acts directed by the Chinese government at the United States, we have the basic, definitional requirements of warfare;[407] it would be non-zero-sum warfare because the attacks were not designed to result in an ultimate "victory" for the attacker. They were instead designed to achieve a lesser goal,

---

you couldn't do your job. You couldn't reach the bank; you couldn't check the bus schedule anymore. It was just confusing and frightening, but we didn't realize it was a war because nobody had seen anything like that before." '

405 *See* Sharp, CyberSpace and the Use of Force, *supra* at 92–93.

406 *See id.*

407 As we saw earlier, to qualify as warfare, the attacks would have had to have been launched by the Chinese government as a hostile assault on the United States as a nation-state.

the ultimate effects of which would be to enhance China's strategic position at the United States' expense.[408] As this example illustrates, one advantage of non-zero-sum cyberwarfare is that it lets an aggressive nation-state engage in a surreptitious sequence of attacks the purpose of which is to incrementally improve its strategic position with regard to another nation-state. The attacks are in effect battles in a virtual war, a war the opponent may not be aware of for some time, if ever. Since non-zero-sum cyberwarfare offers this and other advantages, low-level, ambiguous attacks are likely to be far more common than their more traditional counterparts.

"Armed Attack"

Before we turn to other cyberwarfare issues, I need to note an unresolved issue and its implications. As we saw in Chapter 3, war is synonymous with the use of "armed force," and "armed force" is synonymous with traditional military operations. During the Estonian attacks, the question arose as to whether Estonia, a member of the North Atlantic Treaty Organization, could invoke the NATO Charter provision in which the member states agree to assist a NATO state that is the victim of an "armed attack" by another state.[409] No one knew

---

408 A series of attacks in 2007 seem to have had a similar goal. *See* Sevastopulo, *Chinese Military Hacked into Pentagon, supra.* In June of 2007, Pentagon computer networks suffered a series of attacks that the Pentagon said came from the Chinese military, from PLA hackers. *See id.* According to stories about this incident, China routinely probes U.S. military networks, and the Pentagon "is assumed" to do the same for China' military networks. *See id.* The June, 2007 attacks, though, raised the U.S. military's "concerns to a new level because of fears China had shown it could disrupt systems at critical times." *Id.* According to a former Pentagon official, the PLA hackers "demonstrated the ability to conduct attacks that disable our system . . . and the ability in a conflict situation to re-enter and disrupt on a very large scale." *Id.*

If the reports of these attacks are accurate, they at least arguably represent a type of non-zero-sum cyberwarfare: They seem to have been part of a sequenced campaign to compromise the U.S. military's defenses against hostile intrusions by (a) establishing a "back door" into Pentagon systems and (b) achieving the ability to disrupt those systems. The presumable purpose is to put the PLA in a position to impair the Pentagon's ability to respond a cyber- and/or real-world attack.

409 *See* The North Atlantic Treaty, Article 5 (Washington, D.C., April 4, 1949), http://www.nato.int/docu/basictxt/treaty.htm ("The Parties agree that an armed attack against one . . . of them . . . shall be considered an attack against them all and . . . that . . . each of them . . . will assist the Party . . . attacked by taking forthwith, individually and in concert with the other Parties, such action as it deems necessary, including the use of armed force, to restore and maintain the security of the North Atlantic area").

the answer,[410] but the unacknowledged consensus among NATO officials seems to have been that the provision does not encompass cyberattacks.[411]

NATO officials later said the issue "urgently" needed to be addressed,[412] and there was a development of sorts in early 2008: The leaders who participated in a meeting of the North Atlantic Council held in Bucharest on April 3, 2008, issued a declaration containing 50 separate provisions. The 47th provision addressed "cyber defence," noting that NATO "remains committed to strengthening key Alliance information systems against cyber attacks," has "adopted a Policy on Cyber Defence," and is "developing the structures . . . to carry it out."[413] According to this provision, the policy emphasizes "the need for NATO and nations to protect key information systems," "share best practices," and "provide a capability to assist Allied nations, upon request, to counter a cyber attack."[414] NATO has created a "Cyber Defence Management Authority" that will manage the defense of its communication and information systems and "could support individual Allies in defending against cyber attacks upon request."[415] In May 2008, NATO took another step toward implementing its cyber defense initiative when "top military commanders" from seven NATO nations (Estonia, Latvia, Lithuania, Germany, Italy, Spain, and Slovakia) and U.S. General James Mattis, Supreme Allied Commander of NATO Transformation[416] signed an agreement creating a Cooperative Cyber Defence Center that will be located,

---

410 *See*, e.g., Anne Applebaum, *e-Stonia under Attack*, Slate (May 22, 2007), http://www.slate.com/id/2166716/.

411 Estonian Defense Minister Jaak Avikson said, "NATO does not define cyber-attacks as military action" within the Treaty. *Russia's Cyber War with Estonia Rocks NATO*, The Inquirer (May 18, 2007), http://www.theinquirer.net/default.aspx?article=39714. In asking NATO to create a policy on the issue, he noted that "[i]f a power station is attacked and destroyed by a rocket, it is deemed a military attack, but if it is attacked over the Internet and the same result is achieved, international legislation cannot deal with it." Henry Meyer and Ott Ummelas, *Estonia Asks NATO to Help Foil "Cyber Attack" Linked to Russia*, Bloomberg.com (May 17, 2007), http://www.bloomberg.com/apps/news?pid=20601085&sid=abGseMma5MjU&refer=Europe.

412 *See*, e.g., NATO Ponders Cyberwarfare, Homunculus (June 20, 2007), http://philipball.blogspot.com/2007/06/nato-ponders-cyberwarfare-if-i-were.html.

413 *See* Bucharest Summit Declaration (April 3, 2008), North Atlantic Treaty Organization, http://www.nato.int/docu/pr/2008/p08-049e.html.

414 *See id.*

415 Defending against Cyber Attacks, NATO, http://www.nato.int/issues/cyber_defence/practice.html.

416 NATO Transformation has "the lead" strategic role for "the transformation of NATO's military structures, forces, capabilities and doctrines" in order to improve NATO's effectiveness. Supreme Allied Commander NATO, NATO, http://www.nato.int/issues/sact/index.html.

appropriately enough, in Tallinn, Estonia.[417] It will research cyberwarfare, conduct training, develop cyber defense standards and be a source of expertise.[418] All of this, though, is merely preparation for eventually grappling with the problem of cyberwarfare; none of these efforts has established that a cyberattack triggers the mutual defense provisions of the North Atlantic Treaty.[419]

The NATO Treaty comes into play only in instances when a NATO country is (really) the victim of a cyberwarfare attack and seeks assistance from other member countries. The NATO issue—whether a cyberattack constitutes an act of warfare—also arises under the laws of war, where its implications can be even more serious for the nation-state undergoing such an attack. Most scholars have concluded that a cyberattack does *not* constitute an act of warfare under the United Nations Charter or other international agreements unless it is accompanied by the use or threatened use of physical force.[420] If that conclusion

---

417 *See* Vladimir Socor, *NATO Creates Cyber Defense Center in Estonia*, Eurasia Daily (May 15, 2008), http://www.jamestown.org/edm/article.php?article_id=2373060. The Center is expected to reach full operational capacity in the latter half of 2008; at that point, it should have a staff of 30, half from the seven founding countries and the others from other NATO countries. *See id.*

418 *See id.*

419 *See* Paul Gallis, The NATO Summit at Bucharest, 2008 3, Congressional Research Service (May 5, 2008) (NATO nations are not prepared to consider a cyber-attack sufficient to trigger their obligations under Article V of the NATO Treaty). *See also* note 114, *supra.*

420 *See* Chapter 3. *See*, e.g., Vida M. Antolin-Jenkins, *Defining the Parameters of Cyberwar Operations: Looking for Law in all the Wrong Places*?, 51 Naval Law Review 132, 171–172 (2005); Michael N. Schmitt, *The Sixteenth Waldemar A. Solf Lecture in International Law*, 176 Military Law Review 364, 419 (2003). *See also* U.S. Army Judge Advocate General's Legal Center and School, Operational Law Handbook 425 (August 2006), http://www.jagcnet.army.mil; Michael N. Schmitt, *Computer Network Attack and the Use Of Force in International Law: Thoughts on a Normative Framework*, 37 Columbia Journal of Transnational Law 885, 907–909 (1999). One author has parsed specific types of cyberattacks into those that would constitute a use of armed force (attacks that intentionally cause a destructive effect within the territory of the victim nation-state) and those that would not (e.g., online espionage, interrupting communications, economic warfare). Sharp, CyberSpace and the Use of Force, *supra* at 120, 139–140.

Some scholars believe the U.N. Charter does encompass cyberwarfare. They rely on the International Court of Justice's decision in *Legality of the Threat or Use of Nuclear Weapons, Advisory Opinion*, 1996 I.C.J. 226 (July 8, 1996), in which the court found that the Charter's provisions on "armed attacks" apply to "any use of force, regardless of the weapons employed." *Id.* at 244. The court was specifically, and only, concerned with whether the use of nuclear weapons constituted an "armed attack." *See id.* at 240–244. It is not a great conceptual leap to bring nuclear weapons within traditional understandings of "armed attacks" or the use of "armed force"; nuclear weapons are merely an evolved, incredibly destructive type of bomb, and bombs have long been understood to constitute "weapons" the use of which will be considered a use of armed force. Like others, I think it is a much greater, and far more problematic, conceptual leap to bring computer operations within the traditional definitions of "weapon" or "armed force."

is correct, then a nation-state that correctly determines it is the target of cyberwarfare attacks by a hostile nation-state cannot retaliate with conventional military force; it can, presumably, launch retaliatory cyberattacks.[421] This creates the possibility that nation-states with sophisticated cyberwarfare capabilities could attack states that have few if any, cyberwarfare capabilities with essential impunity. Under the current law of war, the attacked state apparently could not respond to such an attack with military force (unless it were accompanied by the use of such force); because it lacks the capacity to retaliate effectively in cyberspace, the victim nation-state might, like the hypothetical victim of the electronic blockade I noted earlier, be forced to surrender to its virtual adversary.

Nation-states

In Chapter 3 we saw that war is, and has for centuries been, the exclusive province of nation-states. That proposition reflects empirical reality: Only nation-states can assemble the manpower, equipment, and other resources needed to wage war. War has by default been a contest between nation-states, a contest triggered by one state's attacking another. I am going to argue that cyberspace will alter that conception of war; in making that argument, I will retain the premise that "war" necessarily involves a hostile attack on a nation-state or its equivalent.

Before I proceed with my argument, I should explain why I included "or its equivalent" in that statement and what I mean by this phrase. A number of scholars contend that we will see the decline and disappearance of the nation-state, a process that will begin in the perhaps not-too-distant future.[422] Essentially, those who take this view agree with Martin van Creveld that

> government and state are emphatically not the same. The former is a person or group which makes peace, wages war, enacts laws, exercises

---

421 *See.* e.g., Eric Talbot Jensen, Computer Attacks on Critical National Infrastructure: *A Use of Force Invoking the Right of Self-Defense*, 38 Stanford Journal of International Law 207, 230–231 (2002).

422 *See*, e.g., Martin van Creveld, The Rise and Decline of the State 415–421 (Cambridge University Press 1999). *See also* John Robb, Brave New War 184–188 (John Wiley 2007); Jeffrey K. Walker, *Thomas P. Keenan Memorial Lecture: The Demise of the Nation-State, the Dawn of New Paradigm Warfare, and a Future for the Profession of Arms*, 51 Air Force Law Review 323 (2001). Some who subscribe to this view of a "post-Westphalian" world order see nation-states as surviving but ceasing to be the dominant global force. *See*, e.g., Walker, *Thomas P. Keenan Memorial Lecture*, *supra* at 325–332.

> justice, raises revenue, determines the currency, and looks after internal security.... The latter is merely one of the forms which ... the organization of government has assumed, and which, accordingly, need not be considered eternal and self-evident any more than were previous ones.[423]

Creveld says the nation-state has three essential characteristics: (a) sovereignty, which means it monopolizes the above functions; (b) territoriality, which means it exercises its powers and carries out the above functions only within its territorial borders; and (c) it is an "abstract organization."[424] He sees the third characteristic as the most important because it means nation-states have an independent persona; each is an entity distinct from those it governs.[425] Creveld believes this makes the nation-state a fungible construct: a corporate entity that can and will be replaced by other types of corporate organization.[426] He says nation-states will be superseded because (a) they can no longer carry out their sovereign functions effectively and (b) competing corporate entities are more attuned to the realities of modern life than nation-states.[427]

According to Creveld, many traditional nation-state functions have already been taken over by private corporate entities, and the rest will be also.[428] The superseding corporate entities may be territorially based, but that is not a given because they will not derive their power from controlling territory; some will be territorially based, but many will share control of a territorial area with other, similar entities.[429] He suggests that these successors of the nation-state will not be sovereign but will, instead, parse out functions among each other.[430] Essentially, Creveld sees the future as bringing a decentralization of government, a system in which the entities that carry out the functions now monopolized by nation-states will be more idiosyncratic in nature and structure and more integrated with each other than are nation-states.[431] Individuals will become clients of these new governing entities,

---

423 van Creveld, The Rise and Decline of the State, at 415.

424 *See id.* at 416.

425 *See id.*

426 *See id.*

427 *See id.* at 416–417.

428 *See id.* at 417–420.

429 *See id.*

430 *See id.*

431 *See id.*

which will assume responsibility for maintaining internal and external order. Creveld does not see crime, warfare, or terrorism disappearing with nation-states; he suggests, though, that warfare may be less common due to the disappearance of often self-aggrandizing nation-states.[432]

While others have slightly different conceptions of what the future of governance will hold, many scholars agree with Creveld that the nation-state's influence will at the very least decline. This creates the possibility—the likelihood—that governance and warfare become the province of nonstate actors: Creveld's corporate governing entities or something similar.

That is why I include "or its equivalent" in the definition of war outlined above; as we saw in Chapters 2 and 3, the distinguishing characteristic of warfare in our threat taxonomy is that it involves a threat to *external* order, that is, to a governing system's ability to resist aggressive acts by entities external to that system. For centuries, governing systems have been nation-states; if Creveld is right, that will change. Other governing systems will become responsible for resisting aggressive acts by entities external to their system; they, too, will wage war. Defining warfare as a hostile attack on a nation-state or its equivalent encompasses this possibility while retaining the essential defining characteristic of warfare—challenges to order among governing systems.

That brings me back to my point: In the near future *de facto* cyberwarfare will cease to be (if it has not already ceased to be) the exclusive province of nation-states.[433] By "*de facto*" cyberwarfare, I mean attacks that conform to the definition of cyberwarfare we developed in Chapter 3 but with the modification I noted above: Cyberattacks that (a) are launched by non-nation-state actors against a nation-state or a corporate entity that has assumed some/all of the functions of a nation-state in order (b) to undermine the victim's viability as a governing system. In each instance, the attacks challenge a system's ability to maintain external order and are therefore factually indistinguishable from traditional warfare.

While these attacks are warfare in fact, they are not "legally" warfare. Under the law of war, an attack by someone(s) or something other than a nation-state on a nation-state does not constitute "legal," *de jure* warfare because it does not result in nation-state versus nation-state conflict.[434]

---

432 *See id.*

433 *See*, e.g., "De jure," Wikipedia, http://en.wikipedia.org/wiki/De_jure (in Latin, *de jure* "means `based on law', as contrasted with *de facto*, which means `in fact' ").

434 *See id.*

Non-nation-state attacks are, essentially by default, defined as threats to internal order—cybercrime or cyberterrorism.

That outcome is unsatisfactory because it distorts the true nature of the threat and, in so doing, can impede a victim's ability to respond. As far as cyberwarfare is concerned, anyway, we need to revise the current definition of warfare so it encompasses non-nation-state attacks on nation-states or their equivalent. To understand why we need to do this, it is perhaps useful to consider a development in the real-world: the dramatic resurgence of mercenary armies or "private military companies."[435] Nation-states with substantial militaries use mercenaries on occasion to supplement their own resources; smaller, more fragile countries rely on them for protection against aggressive neighbors and other enemies.[436] So far, private combatants at least ostensibly wage war only on behalf of nation-states,[437] but some observers believe they "may be a threat to legitimate governments."[438] Many also believe they signal a resurgence of an old trend, the privatization of warfare,[439] which may spread to cyberspace.

The cyberworld is certainly a hospitable environment: Developing the capacity to wage real-world warfare takes a great investment of money and other resources, some of which can be difficult to obtain;[440] developing the capacity to wage cyberwarfare is, as we have seen, a much less onerous process.[441] I suspect that many, if not most, multinational corporations could today launch cyberwarfare attacks on hapless nation-states, if they were so inclined.[442]

---

435 *See*, e.g., Christopher H. Lytton, *Blood for Hire: How the War in Iraq Has Reinvented the World's Second Oldest Profession*, 8 Oregon Review of International Law 307, 318–319 (2006) (noting that "PMCs exist today in record numbers").

436 *See id.* at 320–330.

437 *See id. See also* Kristen Fricchione, Casualties in Evolving Warfare: Impact of Private Military Firms' Proliferation on the International Community, 23 Wisconsin International Law Journal 731 (2005). For a list of private military companies, *see* Mercenary/Private Military Companies, GlobalSecurity.org, http://www.globalsecurity.org/military/world/para/pmc-list.htm.

438 United Kingdom House of Commons Report, Private Military Companies: Options for Regulation 15 (The Stationery Office, February 12, 2002).

439 *See* Lytton, *Blood for Hire, supra* at 308 ("Mercenaries were fixtures on the international stage until the late nineteenth century, when they . . . faded into history as nation-states and their armies came to dominate the use of military force"). *See also id.* at 307–319.

440 *See generally* Mandel, Armies Without States, *supra* at 7–15.

441 Virtual warfare is also much safer, at least for the warriors: Real-world combatants risk death or injury; cybercombatants do not.

442 For such a scenario, *see* Liang and Xiangsui, Unrestricted Warfare, *supra* at 194 (an "economically powerful company" attacks "another country's economy").

I am not saying that multinational corporations will go rogue and begin attacking nation-states. I am saying that as the power to wage warfare ceases to be a nation-state monopoly, we must expect that attacks analogous to but much more sophisticated and much more devastating than those Estonia underwent in May of 2007 will be directed at nation-states. I cannot predict when these attacks will begin to occur, but I predict they will occur in the not-too-distant future. We have for the last decade or so seen the emergence of putative private-nation-state conflicts. The individuals who comprise the Al-Qaeda organization have, for instance, twice declared war on the United States, declarations the United States declines to recognize, based on the current law of war.[443] Despite the "war on terror" rubric, then, the United States' approach to Al-Qaeda has essentially been to treat them as terrorists, rather than military combatants.[444]

Al-Qaeda's actions regarding the United States illustrate how, even in the real-world, nonstate actors can aspire to and, some would say, actually engage a nation-state in warfare. Though it declines to accord Al-Qaeda members the status of military combatants under the law of war, the United States has used military force in retaliating against them. The use of military force creates questions as to what kind of struggle the United States is conducting with this non-nation-state entity: It cannot, as I explained above, be "war," at least not in the legal sense; and as we will see in Chapter 6, it cannot be conventional law enforcement because military force is not used in responding to crime or terrorism. The United States-Al-Qaeda struggle illustrates the kind of non-traditional, mixed metaphor conflicts we already see in the real-world and will eventually see in the cyberworld. Aside from anything else, the United States' experience with Al-Qaeda illustrates why we need to expand our definition of warfare to encompass attacks by non-nation-state aggressors.

As far as virtual conflicts are concerned, I do not think non-nation-state actors will launch zero-sum cyberwarfare against nation-states, at least not

---

443 *See*, e.g., Lloyd C. Anderson, The Detention Trilogy: Striking The Proper Balance Between National Security and Individual Liberty in an Era of Unconventional Warfare, 27 Whittier Law Review 217, 218–19 (2005) (Al-Qaeda declared war on the United States in 1996 and 1998).

444 *See*, e.g., Douglas A. Hass, Note, *Crafting Military Commissions Post-Hamdan: The Military Commissions Act of 2006*, 82 Indiana Law Journal 1101, 1106–1107 (2007) (Al-Qaeda's 9/11 attacks on the United States "did not create `international armed conflict'" under the laws of war because Al-Qaeda did not act on behalf of Afghanistan or any other nation-state). *See also* Bruce Ackerman, *This Is Not a War*, 113 Yale Law Journal 1871, 1873–1874 (2004).

for the far foreseeable future. I do think they will eventually begin to use non-zero-sum cyberwarfare to pursue political and economic goals.[445] An Al-Qaeda-style group might, for instance, use non-zero-sum cyberwarfare attacks in an effort to destabilize a vulnerable nation-state so a group sympathetic to its cause could seize power; it might, instead, use the attacks to coerce the victim state into making changes in its policies; or an unscrupulous multinational corporation might use similar tactics in an effort to extort the victim state's cooperation in an economic initiative.[446]

As to nation-states, they may use non-zero-sum cyberwarfare attacks in much the same way and for the many of the same reasons as non-nation-state actors. A nation-state might, for example, use such attacks to weaken another state so that it ceases to be a serious economic competitor; these attacks could extend over a very long period, to ensure permanence in the erosion of the victim state's economic viability and/or conceal the aggressor state's role in the decline. Alternatively, an aggressor state might use more debilitating attacks to reduce the victim to a client-state (thereby avoiding the need for, and complications of, physical conquest).

As these examples illustrate, non-zero-sum cyberwarfare will involve directing low-level but ultimately debilitating cyberattacks at the victim state. The attacker's goal is to erode the victim's general sovereign viability or its viability in a particular area, such as its economy. The 2007 Pentagon attacks I noted earlier may be an instance of this type of cyberwarfare.[447] They seem to have been part of a sustained effort the purpose, and result, of which was to give the PLA the ability to disrupt Pentagon networks. If that is true, these attacks qualify as non-zero-sum cyberwarfare: China used them in an effort to weaken the U.S. military's ability to respond to a direct, zero-sum assault in the real- or cyber-world.[448]

---

445 Two People's Liberation Army analysts call this "secondary warfare" or "analogous warfare." *See* Liang and Xiangsui, Unrestricted Warfare, *supra* at 116.

446 For similar scenarios, *see id.* at 47–55.

447 *See* Sevastopulo, *Chinese Military Hacked into Pentagon, supra.* As we saw earlier, in June 2007 Pentagon computer networks suffered a series of attacks the Pentagon said came from China, from PLA hackers. Though PLA probing of U.S. military systems is apparently not uncommon, these attacks raised particular concern because they revealed that the PLA had the ability to disrupt Pentagon networks "on a very large scale." *See id.*

448 As I noted earlier, attacks such these are conceptually distinct from espionage or sabotage. Sabotage consists of destroying or damaging "materials, premises or utilities used for national defense or for war." Black's Law Dictionary, "sabotage" (8th ed., Thomson West 2004). In the 2007 attacks, the presumptive PLA attackers neither "damaged" nor "destroyed" property in the traditional sense. One could argue that they "damaged"

It is also possible that nation-states (or their eventual equivalents) will use zero-sum cyberwarfare attacks to wage war entirely in cyberspace.[449] I have difficulty conceptualizing a purely virtual nation-state on nation-state conflict, perhaps because I am so accustomed to assuming that nation-state conquest will involve at least some real-world territorial incursions. Like others,[450] therefore, I tend to assume that zero-sum nation-state warfare will incorporate both cyberattacks and the use of more conventional, real-world military operations.

In the rather more distant future, we may see zero-sum virtual wars raging between the corporate entities some believe will succeed nation-states. And it is conceivable that conflicts between these entities would be purely virtual: Some of them might exist only in cyberspace; others might have a territorial base, but not a military. Entities in the latter category could use the service of mercenaries, but because territory will not be the defining characteristic

---

Pentagon software by orchestrating what I am assuming was a "back door" to the Pentagon networks, but this neither impaired ("damaged") nor eradicated ("destroyed") the software's ability to carry out its intended functions. The PLA hackers made alterations in the software and may even have made certain additions to it, but neither of those would qualify as sabotage in the traditional sense. Because this activity has a distinct, much more nuanced effect than does overtly damaging or destroying real property, it should be construed as a very different endeavor.

Espionage consists of "using spies to collect information about what another government . . . is doing." Black's Law Dictionary, "espionage" (8th ed., Thomson West 2004). The activity presumptively attributed to the PLA hackers differs from espionage in two ways: It did not involve the use of spies; instead, foreign military agents gained access to secure, sensitive U.S. government networks without being authorized to do so. That conduct is more analogous to trespass, or burglary, than it is to espionage. Also, the presumptive PLA hackers did more than simply "collect information"; they implemented measures which gave them the ability to compromise the functioning of those networks. This conduct, too, is more analogous to burglary than it is to espionage. Because the attacks on the Pentagon do not qualify either as sabotage or espionage, they must, as I noted above, be construed as a very different endeavor.

449 A 2004 study noted, for example, that nation-state launched cyberattacks targeting the

> transportation, communications, or banking sector . . . would . . . entail significant economic costs that would affect jobs and growth. Cyber attacks could . . . lead to disruptions . . . that go beyond . . . nuisance to inflict sustained uncertainty, confusion, and even chaos across significant elements of the population In the most extreme of cases, these disruptions could cause human casualties.

Charles Billo & Weston Chang, Cyber Warfare: An Analysis of the Means and Motivations of Selected Nation-States 13–14 (Institute for Security Technology Studies 2004).

450 Pentagon advisor John Arquilla has outlined a detailed scenario in which cyberwarfare attacks on the United States are combined with conventional military operations. *See* John Arquilla, *The Great Cyberwar of 2002*, Wired (February, 1998), http://www.wired.com/wired/archive/6.02/cyberwar.html?topic=&topic_set=.

of these successor governing entities, virtual conflicts might well become the default option.

Scenario

Under current law, none of the conflicts outlined above legally constitute warfare except nation-state on nation-state conflicts involving the use of military force.[451] As I said earlier, that is an unsatisfactory state of affairs, for several reasons.

One is that it impedes the victim's ability to respond. Assume a variation of the Estonian assault in which the attacks are much more sophisticated and devastating: They shut down the country's power grid and financial networks, disable air and ground traffic control systems, and impede communications in both the private and governmental sectors. Essential services, such as emergency services, are disrupted, and the country is effectively cut off from the rest of the world. If we assume the attacks occur in winter, then the combination of disrupting emergency services and shutting down the power grid could jeopardize human life.

We will assume the assault is being carried out by a group of nonstate actors who are using the attacks to coerce Estonia into withdrawing from an antiglobalization initiative.[452] Their assault qualifies as non-zero-sum cyberwarfare under our definition because nonstate actors are attacking a nation-state *qua* nation-state. As to the last issue, a United Nations resolution defines "war" as attacking the "territorial integrity or political independence

---

451 As we saw earlier in this chapter, nation-state on nation-state conflicts that involve the use of cyberspace coupled with military attacks would constitute warfare under the current law of war.

452 As to the likelihood of such a scenario's occurring, *see*, e.g., Liang and Xiangsui, Unrestricted Warfare, *supra* at 47–48:

> During the 1990's, . . . with . . . military actions launched by non-professional warriors and non-state organizations, we began to get an inkling of a non-military type of war which is prosecuted by yet another type of non-professional warrior. This person is not a hacker . . . and . . . not a member of a quasi-military organization. Perhaps he . . . is a systems analyst or a software engineer, or a financier . . . or a stock speculator. . . . [H]e . . . has a firmly held philosophy . . . and. . . . does not lack the motivation or courage to enter a fight. . . .

Precisely in the same way that modern technology is changing weapons and the battlefield, it is also at the same time blurring the concept of who the war participants are. From now on, soldiers no longer have a monopoly on war.

For a listing of the various types of warfare these nonprofessionals may engage in, *see id.* at 51–55 (e.g., financial warfare, network warfare, media warfare, psychological warfare).

of another state."[453] Under this definition, we have cyberwarfare here because the attackers' goal is *not* to enrich themselves, as it would be if this were cybercrime, or to demoralize a civilian populace in an effort to promote an ideology, as it would be if this were terrorism; their goal is, instead, to undermine Estonia's political independence by forcing its government to change policy. Our hypothesized nonstate actors are engaged in the functional equivalent of traditional warfare, albeit by nontraditional means.

Now assume Estonian computer experts trace the physical locus of the attacks to an island off the coast of Russia; the island is part of Russia's sovereign territory. Either because they assume the attacks are cybercrime or because they do not know what else to do, the Estonian authorities contact Russian authorities[454] and ask them to arrest those responsible for the attacks. The Russian authorities refuse, pointing out that the people in question are not committing crimes in Russia and that it has no treaty or other obligation to become involved in a quarrel between some of its citizens and the country of Estonia.[455]

What can Estonia do? Under the current law of war, it cannot retaliate with military force unless another nation-state attacks it with military force. Because cyberattacks do not qualify as military force, and because the attack does not appear to come from a nation-state, Estonia cannot lawfully use military force against the attackers. Its only options, therefore, are to do nothing or to return like for like, that is, to launch retaliatory cyberstrikes; we will assume the Estonian authorities find the first option unacceptable, and so elect the second, but that has its own problems.

The obvious problem is practical: If the attackers have shut down Estonia's networks and communication systems, it is in no position to launch counter cyberstrikes. In the real-world, Estonia might then resort to military force, out of sheer frustration.[456] But because our analytical purposes require that

---

453 William C. Bradford, *"The Duty to Defend Them:" A Natural Law Justification for the Bush Doctrine of Preventive War*, 79 Notre Dame Law Review 1365, 1373–1374 (2004) (citing U.N. Resolution on the Definition of Aggression, General Assembly Resolution 3314, U.N. Doc. A/9631 (1975) (defining aggressive war).

454 We will assume they were somehow able to get a communication out notwithstanding the ongoing attacks, or simply sent someone across the border.

455 For a very similar scenario involving Russian cybercriminals, *see* Ariana Eunjung Chung, *A Tempting Offer for Russian Pair*, Washington Post (May 19, 2003), http://www.washingtonpost.com/ac2/wp-dyn/A7774-2003May18?language=printer.

456 That does not seem to have been the attackers' goal here, but it could have been. As this scenario illustrates, a devious nation-state or group of nonstate actors could orchestrate a scenario such as this, hoping it drives the victim-state to attack the state that

we limit aggression in this hypothetical to the use of cyberforce, we will assume Estonia somehow summons the ability to launch retaliatory cyberstrikes.

Implementing that option creates a political problem: If Estonia were to direct retaliatory cyberstrikes at Russian networks being used by Estonia's attackers, Russia might see this as a hostile act, one that would not qualify as a traditional act of war but could still warrant a hostile response. Because Russia cannot lawfully use military force to respond to a nonmilitary attack, it might respond in kind by launching its own cyberattacks on Estonia. If we take this scenario to the next logical level, Estonia would then respond by directing retaliatory cyberattacks at Russia (again assuming it summons the capacity to do so).

The two instances in which Estonia launches cyberattacks at Russia differ in terms of the goals of the attacks and their respective targets: Estonia's target in its initial attacks is the networks the non-state actors who are attacking Estonia use; its goal is to disable those networks or otherwise prevent their being used to continue these attacks. In the later attacks, Estonia's target is, in effect, Russia, because now it is defending itself from what it must see as cyberwarfare launched by that country. The specific targets of these attacks will be whatever Russian systems Estonia thinks are appropriate, given its goal; its goal is to stop the Russian attacks either by disabling the systems Russia is using to launch them or by inflicting so much damage on Russian systems generally that the Russians decide to quit. Russia's goals and choice of targets in its attacks on Estonia will mirror those of Estonia in these later attacks; Russia will no doubt perceive the initial Estonian attacks as cyberwarfare, which means it will target whatever Estonian systems it deems appropriate given its goal of stopping the attacks.

Analysis

This rather convoluted scenario illustrates what I mean by eroding concepts. We have four discrete attack events (nonstate actors attack Estonia, Estonia attacks Russian networks, Russia attacks Estonia, Estonia attacks Russia), each of which involves a threat to external order, but none of which qualifies as war under current law.

---

either hosted the systems used in the attack or was made to appear to be the architect of the attack.

The externality of the threat is apparent in the nation-state on nation-state conflicts—Estonia's attacking the Russian networks,[457] Russia's retaliating against Estonia, and Estonia's retaliating against Russia—because in each instance one nation-state is attacked by another. As we saw in Chapter 2, an attack from another nation-state is the ultimate external threat to a state's ability to maintain social order. External threat is also present but is less apparent in the initial assault—the nonstate actors attacking the Estonian computer systems.

If we simply consider the dynamic of this assault—private actors attacking computer systems for their own purposes—it seems we have cybercrime, and cybercrime, like crime, is a threat to internal order.[458] The proposition that (cyber)crime threatens internal order is, however, predicated on the assumption that (cyber)crime is a purely internal phenomenon: domestic actors "harm" domestic victims.[459] Here, we have the basic crime dynamic (private actors "harming" victims), but these perpetrators are operating outside the sovereign system that hosts their victims; the threat they pose is consequently external in origin.[460] Now, to say this threat is external is not, necessarily, to say that it is the equivalent of an attack by a nation-state;[461] our experience of non-nation-state cyberattacks on nation-states is so limited that we cannot assess their actual or potential equivalence to real-world warfare. My goal, here, is but to demonstrate how the use of cyberspace can erode the internal-external threat dichotomy.

As to why none of the attack events constitutes war, the global answer is that each lacks an essential element: military force. As we saw in Chapter 3,

---

457 As I noted above, this is not *really* a nation-state on nation-state conflict because Estonia's goal is not to attack Russia *per se* but to stop the attacks from the nonstate actors who seem adventitiously to be in Russia's territory. I include it in the category of nation-state on nation-state attacks for two reasons: One is that the Russians will probably perceive it that way. The other reason is that while Estonia's attacks are really directed at a civilian threat—cybercrime or cyberterrorism, as we will see later in the text—they target networks that are in Russia and are probably owned by Russian nationals. Aside from anything else, those two factors, alone, establish Estonia's actions as a *de facto* threat to Russia's ability to maintain external order.

458 *See* Chapters 2 & 3.

459 *See id.*

460 *See id.*

461 Nor is it to say that it cannot be the equivalent of an attack by a nation-state. *See*, e.g., Liang and Xiangsui, Unrestricted Warfare, *supra* at 131 (noting the potentially severe threat posed by "non-state forces who do not acknowledge any rules and specialize in taking . . . national order as their goal of destruction").

the use of military force in combat is a defining component of war. It is a military struggle between two or more nation-states.

There are actually two deficiencies in the nonstate actors' assault on Estonia. One is that it was conducted in cyberspace and involved the transmission of electronic signals, instead of armed combat between national armies using guns, bombs, and other implements of war. The other deficiency is that the assault was not a struggle between nation-states. As we have seen, war is a conflict between nation-states; countries do not wage war with individuals.[462] The nonstate actors' attacks on Estonia must, therefore, be either cybercrime or cyberterrorism.

The absence of military force is ultimately the reason why the remaining events cannot qualify as warfare, but the nation-state on nation-state requirement is also arguably problematic in certain of the events. It is therefore instructive to parse out the extent to which each conforms to the traditional conception of warfare.

Estonia Attacks Russian Networks

As to its original attacks, Estonia really had no quarrel with Russia.[463] Estonia attacked the Russian networks not because it was pursuing a war of aggression with Russia but because this was the only way to defend itself against the Russian-based but nonstate actors who were crippling its computer systems. That makes the nation-state on nation-state aspect of this event conceptually ambiguous: In offensive warfare, the aggressor state's purpose is to subdue the victim state and compromise its territorial integrity or political independence; the aggressor state consequently attacks the victim as a nation-state.[464]

---

462 *See* Chapters 2 & 3.

463 Estonia may not be particularly happy with the Russians' declining to arrest the perpetrators of these attacks, but it would presumably concede that they were within their rights in doing so.

464 *See*, e.g., Jeffrey C. Tuomala, *Just Cause: The Threat that Runs So True*, 13 Dickinson Journal of International Law 1, 10 (1994):

> [T]here are three . . . lawful objectives for which nations wage war. These objectives include: (1) obtaining compensation or reparations for losses; (2) punishing offenders by reprisal for wrongs done; and (3) defending against unlawful attacks. Nations attain the first two objectives by resorting to offensive war and the third by waging defensive war.

And as we saw in Chapter 3, while the United Nations Charter outlaws self-aggrandizing aggressive war of the type Germany pursued in World War II, it does allow states to use defensive force if and when they are attacked.

In its initial assault, Estonia attacked Russia not as a nation-state but as the host of an implement of cybercrime; indeed, Estonia can argue that it did not "attack" Russia at all. It will point out that its attacks were not general assaults intended to compromise Russia's territorial integrity or political independence. Instead, they had a very narrow focus that was tailored to a specific purpose: preventing nonstate actors in Russia from inflicting further "harm" in Estonia. The Estonians will argue that such a response is permissible under the United Nations Charter; one scholar believes the Charter allows a state to use defensive force against non-state actors operating from another state's territory as long as the defensive force is not intended to change or "directly disrupt" the targeted state's territorial integrity or political independence.[465]

Others disagree,[466] and even this scholar concedes that host states, such as Russia in our scenario, are likely to regard any use of force within or directed at their territory as an act of war.[467] The United Nations has on several occasions condemned such extraterritorial armed reprisals against non-state actors as "armed aggression," which violates its Charter, as well as other principles of international law.[468]

If we adopt the United Nations' approach, then Estonia's different purpose and narrowed focus in its initial assault do not matter; its cyberstrikes had a deleterious impact on networks in Russia, on property Russian citizens owned.[469] Russia can therefore argue that this conduct is functionally indistinguishable from Hitler's invading Poland; in both instances, an incursion from another nation-state has compromised a nation-state's territorial integrity.

---

465 *See* Jordan J. Paust, *Use of Armed Force against Terrorists in Afghanistan, Iraq, and Beyond*, 35 Cornell International Law Journal 533, 536 (2002).

466 *See*, e.g., Leila Nadya Sadat, *Terrorism and the Rule of Law*, 3 Washington University Global Studies Law Review 135, 144 note 30 (2004).

467 *See id.*

468 *See* United Nations Security Council Resolution 573, U.N. SCOR, 40th Sess., 2615th mtg., U.N. Doc. S/RES/573 (1985) (Israeli reprisal against PLO Headquarters in Tunisia was an "act of armed aggression . . . against Tunisian territory in flagrant violation" of the U.N. Charter and international law). *See also* United Nations Security Council Resolution 487, U.N. SCOR, 36th Sess., 2288th mtg., at 10, U.N. Doc. S/INF/37 (1981); United Nations Security Council Resolution 188, U.N. Doc. S/5650 (1964).

469 *See*, e.g., Relcom.ru, http://www.relcom.ru/English/ (commercial Russian Internet service provider). *See also* Sadat, *Terrorism and the Rule of Law*, *supra* (directing reprisals at another state for an "attack that appears to emanate from a group found in that state can be likened to the collective punishment of the citizenry of the state in question").

The resolution of this issue is ultimately irrelevant because the decisive factor in the war-not-war analysis is the absence of military force. If Estonia had sent military airplanes to bomb the Russian island from which the nonstate actors were operating, that would have been an act of war because it used military force against another nation-state. But because it relied on cyberforce, Estonia's attacks on the Russian networks cannot qualify as warfare.

Russian Counterstrikes

Russia had a quarrel with Estonia because Estonia attacked Russia, in a sense; the networks Estonia targeted are Russian-based and Russian-owned property. The Russian counterstrikes are consequently more clearly nation-state on nation-state conflict than were the initial Estonian attacks. Russia was retaliating for what it no doubt believed was cyberwarfare launched by Estonia. And though Estonia did not send troops across the Russian border, the Russians can, as we saw above, say it achieved the same effect by sending electronic signals into Russia to damage Russian property. The Russians would argue that given this, their counterstrikes were justified self-defense.

As we saw in Chapter 3, the United Nations Charter allows one country to defend itself against an "armed attack" by another. Russia, however, cannot invoke that provision because, as we have seen, the use of cyberspace does not constitute an "armed attack"; armed attacks involve military force.[470] Russia's cyberstrikes therefore (a) cannot be defensive warfare because they were not a response to an "armed attack," and (b) cannot be aggressive warfare because the use of cyberspace does not constitute military force.

Estonian Response

Basically, the same analysis applies to Estonia's retaliatory cyberstrikes. Here, too, we have nation-state on nation-state conflict, and the Estonians will also claim that their response qualifies as defensive warfare under the United Nations Charter. They will say they justifiably responded to what they believed was aggressive cyberwarfare. Estonia's argument here is somewhat less compelling than Russia's because it in a sense attacked Russia; it can therefore be seen as the aggressor state. Estonia can try to rebut that

---

470 *See* Chapter 3.

interpretation by arguing that its initial attacks were not, and could not have been interpreted as, cyberwarfare given their narrow focus and given that Russia was aware of Estonia's reasons for launching them.

The self-defense argument fails here, too, though, because while Estonia's cyberstrikes were in fact defensive, they were not a response to an "armed attack." Like the Russian retaliation, Estonia's cyberstrikes (a) cannot be defensive warfare because they were not a response to an "armed attack," and (b) cannot be aggressive warfare because the use of cyberspace does not constitute the use of military force.

Permutations

The preceding analyses illustrate the limitations of our current threat taxonomy. The attacks by the nonstate actors cannot be warfare, but we can still fit them into the taxonomy, as cybercrime or cyberterrorism. While that may not, as we will see, be the optimum approach for attacks of this type, it means they at least come within our defined universe of threats.[471] The real problem lies with the nation-state attacks: If the Estonian and Russian cyberattacks are not warfare, what, if anything, are they?

We have one, and only one, external threat category: warfare.[472] Nation-state on nation-state conflicts are war; there is no other option. There have on occasion been ostensible efforts to articulate intermediate categories: In 1950, President Truman said the United States' activity in Korea was a "police action," but he disingenuously employed this euphemism to avoid asking Congress to declare war.[473] And at the time, Americans understood that the country was, in fact, at war.[474] In the early twenty-first century, President Bush characterized the United States' response to Al-Qaeda attacks as a "war on terror," but that, too, was a misnomer; as we saw earlier, the "war

---

471 The Russian response to the Estonian request for assistance in our scenario suggests that while the attacks come within our threat conceptualization, the issue of response is problematic. *See* Chapter 6.

472 *See* Chapter 2.

473 *See, e.g.*, Louis Fisher, *The Korean War: On What Legal Basis Did Truman Act?*, 89 American Journal of International Law 1, 25, 32–34 (1995).

474 *See, e.g.*, Weissman v. Metropolitan Life Insurance Company, 112 F. Supp. 420, 425 (S.D. Cal. 1953) ("We doubt very much if there is any question in the minds of the majority of the people of this country that the conflict now raging in Korea can be anything but war").

on terror" is properly understood as a law enforcement action (just as the "war on poverty" was property understood as a social justice initiative).[475]

At the end of the twentieth-century, the U.S. military developed the concept of "military operations other than war,"[476] or MOOTW, which might seem to be an intermediate category of nation-state conflict. According to the Joint Chiefs of Staff, MOOTW encompass "the use of military capabilities across the range of military operations short of war."[477] They are used to deter war, promote peace, and support "civil authorities" during "domestic crises."[478] They can involve "combat and noncombat operations,"[479] as well as the use of cyberspace.[480] MOOTW, though, are by definition undertaken only when a state of war does not exist or, in other words, when there is no nation-state on nation-state conflict.[481]

We consequently have no alternatives: Our only external threat category is the zero-sum conception of warfare we reviewed earlier, and it leaves our hypothetical Russian and Estonian attacks in limbo. Cyberattacks like these will be nonevents unless we modify our current threat taxonomy to expand our options. Logically, we can do this either (a) by importing attacks such as these into the internal threat category or (b) by expanding our conceptualization of external threat so it can encompass these attacks.

As to the first alternative, importing any nation-state-sponsored activity into the internal threat category may seem hopelessly illogical and impracticable

---

475 *See,* e.g., Hass, *Crafting Military Commissions Post-Hamdan, supra. See also* Kenneth Roth, *The Law of War in the War on Terror,* Foreign Affairs (January/February 2004), http://www.foreignaffairs.org/20040101facomment83101/kenneth-roth/the-law-of-war-in-the-war-on-terror.html.

476 *See,* e.g., "Military Operations Other than War," Wikipedia, http://en.wikipedia.org/wiki/Military_operations_other_than_war.

477 The Joint Chiefs of Staff, J-7, Operational Plans and the Interoperability Directorate, U.S. Joint Doctrine—Joint Force Deployment: Military Operations Other Than War 3, http://www.dtic.mil/doctrine/jrm/mootw.pdf.

478 *Id.*

479 *Id.* For the different types of MOOTW, *see* "Military operations other than war," Wikipedia, *supra. See also* Major Michael L. Smidt, *Yamashita, Medina, and Beyond: Command Responsibility in Contemporary Military Operations,* 164 Military Law Review 155, 158 (2000).

480 *See,* e.g., Commander Roger D. Scott, *Legal Aspects of Information Warfare: Military Disruption of Telecommunications,* 45 Naval Law Review 57, 59–60 (1998).

481 *See,* Major Michael L. Smidt, *Yamashita, Medina, and Beyond, supra* at 230–231 (MOOTW occur when there is no "state against state conflict" and the law of war does not apply); Liang and Xiangsui, Unrestricted Warfare, *supra* at 50 (MOOTW are "carried out when there is no state of war"). *See also* U.S. Army Judge Advocate General's Legal Center and School, Operational Law Handbook 28 (August 2006), http://www.jagcnet.army.mil (MOOTW "do not involve an international armed conflict").

given the inherent differences between the threats. While consummated nation-state threats do have effects in the victimized country, the threats themselves are external; if, for example, Country A directs hostile activity into Country B, the effects of the consummated threat occur in Country B, but the threat itself is external. It comes from an "outsider"—from a distinct, remote sovereign. Country B therefore cannot rely on its internal control mechanisms in dealing with this threat; it must instead use a specific external response strategy.[482] Internal threats, in contrast, come from domestic activity: from individuals who are citizens of and/or present in a nation-state. Internal threat responses therefore target local threat sources, which are generally easier to inhibit: These threats come from individuals rather than a rival nation-state, individuals who will have internalized many, if not all, of the domestic rules that promote order within the society.[483] This reduces the overall magnitude of the threat individual miscreants pose; it also tends to make them more vulnerable to capture and more amenable to correction, once captured. Finally, the dichotomy is driven by differential threat dynamics: External threats from a nation-state are cumulate, focused and systemic; they directly and often catastrophically challenge the sovereign viability of the victim nation-state. Internal threats are discrete, erratic and erosive; absent insurrection, they undermine a nation-state's viability only to the extent they are allowed to aggregate unchecked.

Notwithstanding the influence of cyberspace, logic and empirical reality both support our continued use of the internal-external threat dichotomy, at least for the foreseeable future. But this does not mean its categories must remain impermeable; indeed, we have at least one instance of some blurring between them. I refer to the concept of "state-sponsored terrorism."

The contemporary version of state-sponsored terrorism emerged in the 1970s.[484] Its distinguishing characteristic is that a nation-state uses individuals to carry out terrorist acts, acts which it instigates, supports, or authorizes.[485]

---

482 *See* Chapters 2 & 3. *See also* Chapter 6.

483 *See* Chapters 2 & 3. *See also* Chapter 6.

484 *See*, e.g., Bruce Hoffman, Inside Terrorism 185–186 (Columbia University Press 1998) (noting that state sponsorship of terrorism had occurred earlier, but it was not until the 1970s that states began "to embrace terrorism as a deliberate instrument of foreign policy").

485 Christopher C. Joyner & Wayne P. Rothbaum, *Libya and the Aerial Incident at Lockerbie: What Lessons for International Extradition Law?*, 14 Michigan Journal of International Law 222, 229 (1993). For the distinction between state-sponsored terrorism and "state terrorism," *see* M. Cherif Bassiouni, *Legal Control of International Terrorism: A Policy-Oriented Assessment*, 43 Harvard International Law Journal 83, 84 (2002) (state terrorism

This allows the sponsoring nation-state to pursue an "unlawful policy" while it maintains "plausible denial."[486] Another advantage for a sponsoring state is that the acts terrorists carry out on its behalf are an inexpensive and "potentially risk-free means of anonymously attacking stronger enemies and . . . avoiding . . . reprisal."[487]

As one author noted, state-sponsored terrorism "blurs the distinction between criminal activity and warfare."[488] State-sponsored terrorism, which so far seems to have manifested itself solely in real-world attacks, is often described as a type of warfare: "low-intensity" or "secret" or "undeclared" warfare.[489] It at least partially conforms to the model of non-zero-sum cyberwarfare we examined earlier in this chapter; as we saw there, a nation-state might use non-zero-sum cyberwarfare to undermine another nation-state's viability as an economic or political competitor while allowing it to survive as a weakened sovereign entity. State-sponsored terrorism, as it is currently practiced, is a version of this model of non-zero-sum warfare. It notably differs from that model in two respects: The model assumes cyberattacks (only); and it encompasses both nation-state and non-nation-state conducted non-zero-sum cyberwarfare.

Despite their similarities, we do not define state-sponsored terrorism as warfare. As we saw in Chapter 3, terrorism is generically construed as an internal threat and therefore defined as a crime.[490] International law distinguishes "armed attacks" of warfare from "'the commission of ordinary crimes including acts of terrorism,'" even state-sponsored acts of terrorism.[491] The distinction has two operational consequences: It reinforces the premise that

---

is carried out by state actors; state-sponsored terrorism is carried out by nonstate actors who are supported by a state).

486 Joyner & Rothbaum, Libya and the Aerial Incident at Lockerbie, *supra*.

487 Hoffman, Inside Terrorism, *supra* at 186.

488 Mark D. Kielsgard, *A Human Rights Approach to Counter-terrorism*, 36 California Western International Law Journal 249, 272 (2006).

489 *See*, e.g., Kenneth W. Abbott, *Economic Sanctions and International Terrorism*, 20 Vanderbilt Journal of Transnational law 289, 304 (1987); James Kraska, *Torts and Terror: Rethinking Deterrence Models and Catastrophic Terrorist Attack*, 22 American University International Law Review 361, 379 (2007); James C. Duncan, *Battling Aerial Terrorism and Compensating the Victims*, 39 Naval Law Review 241, 244 (1990).

490 International law recognizes some exceptions to this, such as when terrorism is carried out by a state. *See*, e.g., Mary Ellen O'Connell, *Enhancing the Status of Non-State Actors through a Global War on Terror?*, 43 Columbia Journal of Transnational Law 435, 444–449 (2005).

491 O'Connell, *Enhancing the Status of Non-State Actors through a Global War on Terror, supra* at 448 (quoting Marco Sassbli, Use and Abuse of the Laws of War in the "*War on Terrorism*," 22 Law & Inequality 195 (2004).

real-world terrorist attacks (anyway) are not warfare, and so do not trigger a nation-state's right to retaliate with armed force.[492] It also reinforces the proposition that the appropriate nation-state response to terrorism is to apprehend, prosecute, and punish the perpetrators for the crimes they committed.

This can be—but more often is not—an effectual response to international terrorism. In 1999, Osama bin Laden and fourteen other members of Al-Qaeda were indicted in the United States on federal terrorism charges based on their involvement in the 1998 bombing of two U.S. embassies in Africa.[493] Only four of those defendants—none of them, of course, bin Laden—were ever captured, tried, and convicted for their roles in these acts of terrorism.[494] While the indictment was pending, bin Laden and other members of Al-Qaeda carried out the 9/11 attacks on the United States, attacks that in substantial part targeted the very city where the prosecution against bin Laden was pending. As I write this, six years have passed since the 9/11 attacks, and the United States has not captured bin Laden, and may never do so.

Real-world terrorists are far more visible than their virtual counterparts;[495] logically, then, it should be that much easier to identify and apprehend them. The fact that the United States, which has put enormous resources into the effort, has so far failed to capture the most notorious terrorist in the world demonstrates how very difficult it can be to use the criminal justice process to sanction actualized terrorists and in hopes of deterring prospective terrorists. Though he has ties in various countries, bin Laden is not a state-sponsored terrorist;[496] he has eluded American authorities without official assistance from a sponsoring state. As other authors have noted, finding and apprehending true state-sponsored terrorists can be a difficult, if not impossible task. Aside from anything else, the sponsoring state may, as Libya did with the accused Lockerbie bombers, simply refuse to extradite the suspects so the victim state can try them.[497]

---

492 *See*, e.g., Gregory E. Maggs, *The Campaign to Restrict the Right to Respond to Terrorist Attacks in Self-Defense under Article 51 of the U.N. Charter and What the United States Can Do About It*, 4 Regent Journal of International Law 149, 164–167 (2006). *See also* Chapter 3.

493 *See*, e.g., United States v. Bin Laden, 58 F.Supp.2d 113, 114 (S.D.N.Y. 1999).

494 *See* United States v. Bin Laden, 397 F.Supp.2d 465, 473 (S.D.N.Y. 2005).

495 *See* Chapter 3.

496 *See*, e.g., Fareed Zakaria, *Terrorists Don't Need States*, Newsweek (April 5, 2007), http://www.msnbc.msn.com/id/4616799/.

497 *See*, e.g., Jonathan A. Frank, *A Return to Lockerbie and the Montreal Convention in the Wake of the September 11th Terrorist Attacks: Ramifications of Past Security Council and International Court of Justice Action*, 30 Denver Journal of International Law and Policy 532, 533–538 (2002).

In a law review article, I pointed out how similar problems arise with another type of state-sponsored criminal activity: economic espionage.[498] As I explain, economic espionage can be "civilian," that is, conducted by private parties for their own, personal aggrandizement, or it can be state sponsored. In recent years, we have seen a great increase in state-sponsored economic espionage: Currently, a number of nation-states are in effect engaging in economic warfare by having their agents steal proprietary information from other countries; the offending states undertake the thefts to advance their interests at the expense of the state from which the information is stolen.[499] Stealing proprietary information is a crime in many countries, including the United States.[500] Because the United States recognizes the significance, and the increasing incidence, of state-sponsored economic espionage, it makes the latter a separate crime; federal law consequently criminalizes both "civilian" and state-sponsored economic espionage.[501] The problem with this approach is, again, enforcement: Sponsoring states have absolutely no incentive to cooperate with U.S. authorities investigating and/or seeking to extradite those who stole proprietary information at the sponsoring state's behest and for its benefit.[502]

The problem of enforcement—for state-sponsored terrorism or state-sponsored crime—is only exacerbated when the underlying conduct moves from the real-world to the virtual world. We will examine response issues in more detail in Chapter 6; I raise them here simply to further illustrate the inherent futility of using a "crimes" approach for what is actually low-level warfare.

As I see it, our current threat taxonomy suffers from a fatal deficiency, one that becomes particularly pronounced when activity migrates to the virtual world but can also be apparent in real-world attacks. The deficiency lies, as explained earlier, in the rigidity and exclusiveness of the threat categories. By defining state-sponsored terrorism as "crime," we automatically and necessarily remove it from the category of nation-state on nation-state conflict; we do the same thing with state-sponsored economic espionage (and will no doubt do the same thing with other types of state-sponsored crime if and when they emerge). In doing so, we ignore an essential, critical aspect of state-sponsored activity that does not rise to the level of traditional,

---

498 *See* Brenner & Crescenzi, State-Sponsored Crime, *supra* at 392–417.

499 *See id.*

500 *See id.* at 417–440.

501 *See id.*

502 *See id.* at 417–440.

zero-sum warfare: It is *not* "crime." It may be "crime" in a literal sense (one person "harms" another), but from a broader perspective, it is something, something more: covert, sublimated nation-state on nation-state conflict. Because this is also an external threat—low-level aggression by a nation-state—it cannot effectively be addressed as if it were internal.

Rigid adherence to the internal-external dichotomy blinds us to the reality that threats are morphing. The dichotomy assumes the two threat categories are mutually exclusive, as they once were, but that is no longer true. State-sponsored terrorism and state-sponsored crime are binary events: Each is a "crime" at one level and "warfare" at another. When a state-sponsored actor steals proprietary information from, say, a British company and sends that information to the sponsoring state, the theft is simultaneously "crime" and "warfare." It is a crime because the thief directly inflicts "harm" on the corporate victim; it is warfare because his conduct inflicts a surpassing, systemic "harm" on Britain's economy. The same binary pattern manifests itself in state-sponsored terrorism. If a terrorist group acting on behalf of another nation-state bombs an Ankara hotel while it hosts a political rally, the act concurrently "harms" (a) the individuals who are killed or injured at the scene (crime) and (b) the Turkish government by eroding confidence in its ability to protect citizens from what may be a transparently external threat (warfare).

State-sponsored terrorism and state-sponsored crime demonstrate that threat categories can blur and blend. Here, nation-states replace civilians as perpetrators of crime and terrorism. If nation-states can infiltrate the internal threat category, I see no reason why civilians, including civilian entities, cannot infiltrate the external threat category. Civilians could, as we saw earlier, wage war, at least of a type.

That brings me to the final point I want to make in this chapter: We not only need to accept that threat categories can blur, and threats intermingling internal and external elements can emerge; we also need to revise our unitary conception of warfare. The internal category has never been unitary: It has always encompassed an array of discrete behaviors (crimes), each of which erodes internal order by inflicting a distinct "harm." As societies become more complex, the array increases in complexity.

The external category has heretofore been unitary for a very simple reason: The threat scenarios it encompassed were functionally indistinguishable. The internal category developed an array of proscribed behaviors out of necessity: Individuals can "harm" each other in myriad ways; nation-states must control the incidence of each of these "harmful" behaviors if they are to maintain internal order. The external category had no corresponding need for an intricate threat array because its threat was constant: an armed attack by one nation-state upon another. The weapons used to attack might vary, as might the motives behind the attacks, but the "harm" was invariable—an assault upon the viability, upon the very existence, of the victim state.

The external threat category has in effect functioned as a criminal code that contains only one crime: murder. Criminal codes proscribe the "harm"—killing a human being—but do not concern themselves either with method or motive. The same has been true of the external threat category: Once it becomes apparent a nation-state is bent on warfare ("murder"), the law of war lets the potential victim respond in self-defense (a privilege law also accords to potential murder victims).

The influence of cyberspace alters this state of affairs by opening it up to new offenders and by fracturing the unitary "harm" of warfare into a perhaps modest array of "harms." As we have seen, cyberspace makes it possible for non-state actors to in effect wage warfare. As we have also seen, cyberspace will fracture the unitary concept of warfare into an array of conduct involving the infliction of "harms" of varying types and varying degrees. If nation-states (or their successors) are to resist these threats and maintain order in the global environment, they must be able to identify the precise nature of an attack, because the nature of the attack dictates the appropriate level and type of response.

That among other things means parsing the "harm" encompassed by our external threat category into discrete "harms," the exact nature and number of which will become apparent as our experience with this new threat activity increases. This will certainly involve classifying the "harms" according to the type of injury (e.g., physical, economic, psychological) they inflict and the extent to which they successfully inflict that injury. It may also mean classifying "harms" on the basis of the threat source, that is, based on whether they originate with a nation-state or with non-nation-state actors.[503] In a sense, what we need to do with the external threat category is to replicate what we have done with our internal threat category: develop an array of discrete threats that is sufficiently nuanced to allow us to identify and respond appropriately to different types of attacks.

---

503 For why we might want to do this, *see* Liang and Xiangsui, Unrestricted Warfare, *supra* at 48:

> Non-professional warriors and non-state organizations are posing a greater and greater threat to sovereign nations, making these warriors and organizations more and more serious adversaries for every professional army. Compared to these adversaries, professional armies are like gigantic dinosaurs which lack strength commensurate to their size in this new age. Their adversaries, then, are rodents with great powers of survival, which can use their sharp teeth to torment the better part of the world.

FIVE

Attribution

Attackers

Cyberspace attackers may be governments, groups, individuals or some combination.[504]

WE EXAMINED ATTACK-ATTRIBUTION in the last chapter. In this chapter, we will examine the related issue of attacker-attribution. The need to identify those who are responsible for an attack has been, and will remain, a constant in the process of maintaining sovereign order. As we will see, identifying the attacker is an essential prerequisite to responding to an attack, an issue we will take up in the next two chapters. And as we will also see, identifying the attacker can play an integral role in ascertaining the nature of an attack.

Our examination of attacker-attribution will proceed in two stages: We will consider how attacker-attribution for the several threats in our taxonomy is approached in the real-world and then consider how it becomes increasingly problematic as attacks migrate online. We will begin our analysis in each stage with warfare, and then examine crime and terrorism, respectively.

Attacker-attribution: Real-world

Attacker-attribution has traditionally been less problematic for warfare than for crime or terrorism. There are several reasons why this was true, one of which was that until relatively recently the law of war required that states declare their warlike intentions before attacking each other.[505] In complying with this requirement, a nation-state eliminated any doubt as to which state was responsible for an attack. As we saw in Chapter 3, the United Nations

---

504 Joseph S. Nye, Jr., *U.S. Cyber Defense Needs to Surge*, Newsday (March 25, 2002).

505 *See* Chapter 3.

Charter eliminated this requirement by abolishing war "as a category of international law." Countries are therefore no longer obliged to declare war, unless their own law imposes such an obligation.[506]

Even without a formal declaration, it is generally not difficult to identify the nation-state that is responsible for an act of war in the real-world. The initial attack may be a surprise, as it was with Pearl Harbor, but attributing the attack to a specific state tends to be a relatively simple process. Military attackers will wear distinctive, uniform clothing bearing insignia that indicates their national affiliation; their equipment will display similar insignia.[507] So when the Japanese attacked Pearl Harbor, insignia on their airplanes revealed their country of origin.[508] The language the attackers use is another clue to their national affiliation, which can also be inferred from the circumstances of the attack itself (such as timing, weapons, targets, and/or tactics).[509] The location from which an attack is launched is yet another clue: If Country A is being attacked by missiles coming from Country B, Country A's decision-makers can reliably infer that either (a) Country B, or (b) a state with which Country B is affiliated (Country C, say) is responsible for the attack.

Identifying those responsible for crime is can be far more difficult. As we saw in Chapter 2, criminals have a strong incentive to avoid identification because it is usually the first step in their being apprehended, convicted, and sanctioned for their misdeeds. Aside from a few rare exceptions,[510] criminals

---

506 *See*, e.g., Curtis A. Bradley and Jack L. Goldsmith, *Congressional Authorization and the War on Terrorism*, 188 Harvard Law Review 2047, 2061–2062 (2005) (U.N. Charter explains why no state "has declared war since the late 1940s"). Commentators disagree on whether U.S. law requires a Congressional declaration of war. *Compare* Donna M. Davis, *Preemptive War and the Limits of National Security Policy*, 10 IUS Gentium 11, 19 (2004) (Constitution requires a declaration of war) with Bradley and Goldsmith, Congressional Authorization and the War on Terrorism, *supra* at 2061 (declaration of war not constitutionally required).

507 *See* Chapter 3. One reason for having military personnel wear uniforms and insignia is to make it possible to "clearly distinguish combatants" from noncombatants. "Military Uniform," Wikipedia, http://en.wikipedia.org/wiki/Military_uniform#Purpose. Another is to let members of a nation-state's military recognize each other and identify members of opposing military forces.

508 *See*, e.g., Report of Action at Pearl Harbor ¶ 4, U.S.S. Vega (December 10, 1941), http://ibiblio.org/hyperwar/USN/ships/logs/AK/ak17-Pearl.html. ("Japanese insignia on the planes was plainly visible through glasses").

509 *See*, e.g., "Spartan Army," Wikipedia, http://en.wikipedia.org/wiki/Spartan_Army#Tactics (Spartan army was known by its use of the phalanx).

510 *See*, e.g., "Bonnie and Clyde," Wikipedia, http://en.wikipedia.org/wiki/Bonnie_and_Clyde (Bonnie Parker wrote poems about the pair's exploits and sent them to newspapers, which published them).

do not intentionally identify themselves as the architects of their crimes, though an occasional publicity-seeker may do so indirectly by using a *nom de crime* (such as "the Zodiac Killer").[511] Criminals inadvertently leave clues law enforcement uses to identify them, but their conscious goal is to remain unidentified and unapprehended.

As we saw in Chapter 2, because crime control is essential for the maintenance of internal order, nation-states have developed a standardized, generally effective approach for identifying those who commit crimes in their territory. As we saw in Chapter 3, this approach assumes activity in the real world because, until recently, physical reality was the only arena of crime commission.

It consequently focuses on finding attacker-attribution evidence at a physical crime scene by locating (a) witnesses who saw the perpetrator and who can describe, perhaps even identify, him, and (b) physical evidence (e.g., DNA, fibers) that can be traced to someone whose presence at the scene is unexplained or otherwise inferentially suspicious.[512] Because it assumes conduct in the real world, this approach presumes that the perpetrator of an attack—a crime—was, and is, physically present in the local geographical area.[513] The latter assumption gives rise to the "dragnet" tactic in which officers comb that area in an attempt to locate people who have seen the perpetrator or those who know him, and his whereabouts.[514] If attacker-attribution fails for a crime, officers will assume the attacker remains in the area and will therefore be alert for the possibility that he will re-offend and be identified at that point.[515] As we saw in Chapter 3, law enforcement officers can also use identifiable geographical and offense patterns in local crime to assist them in finding perpetrators.[516]

---

511 *See*, e.g., "Zodiac Killer," Wikipedia, http://en.wikipedia.org/wiki/Zodiac_killer. The *nom de crime* tactic is not intended to reveal the perpetrator's true identity. It is, instead, a compromise: a way he can "take credit" for the crimes while retaining enough anonymity to avoid capture.

512 *See* Jon Zonderman, Beyond the Crime Lab: The New Science of Investigation 1–44 (John Wiley & Sons 1999). *See also* Barry A. J. Fisher, Techniques of Crime Scene Investigation 25–56, 93–256, 372–374 (7th ed., CRC Press 2003).

513 *See* Chapters 2 & 3.

514 *See id.*

515 *See id.*

516 *See*, e.g., Nancy G. La Vigne and Elizabeth R. Goff, *The Evolution of Crime Mapping in the United States*, in Mapping and Analysing Crime Data: Lessons from Research and Practice 203–222 (Kate Bowers & Alex Hirschfield, eds., Taylor & Francis 2001).

In terms of its complexity, attacker-attribution for terrorism occupies a middle ground between war and crime. While those who carry out a particular terrorist attack may not identify themselves personally,[517] they may identify themselves as acting on behalf of a specific terrorist group, so the group can take credit for the attack.[518] Terrorism perpetrators increasingly identify themselves as representatives of a particular terrorist group in "martyrdom messages" recorded prior to an attack, especially a suicide attack.[519] It is also increasingly common for the terrorist group sponsoring an attack to claim credit for it in a message posted online or in a videotape delivered to media outlets.[520] If the sponsoring terrorist group does not claim credit for an attack, the structure and style of the attack may inferentially identify the organization responsible for it.[521]

With terrorism, as with war, authorities can usually identify the entity responsible for an attack; even when a terrorist organization does not publicly claim responsibility, authorities can generally infer the identity of the culpable group based on the target of, and the tactics used in, an attack.[522]

---

517 Terrorists are more likely to identify themselves if they do not anticipate escaping to commit further attacks. Terrorists whose goal is to commit further attacks eschew self-identification for the same reason crime perpetrators try to avoid identifying themselves. Identification facilitates apprehension, which negates their ability to commit further acts of terrorism. *See,* e.g., Carlos the Jackal—Wikipedia, http://en.wikipedia.org/wiki/Ilich_Ram%C3%ADrez_S%C3%A1nchez.

518 Terrorist groups differ in terms of their attitude toward publicly taking credit for attacks. *See,* e.g., Kim Cragin and Sara A. Daly, The Dynamic Terrorist Threat 37–38 (Rand 2004) (Real Irish Republican Army and Hamas take credit for the attacks they sponsor, while the Revolutionary Armed Forces of Colombia and Al Qaeda generally do not).

519 *See id.* at 38. *See also Video Shows Laughing 9/11 Hijackers in Afghan Hideout,* CNN (October 1, 2006), http://edition.cnn.com/2006/WORLD/meast/10/01/hijackers.video/index.html; "Martyrdom Video," Wikipedia, http://en.wikipedia.org/wiki/Martyrdom_video.

520 *See,* e.g., "7 July 2005 London bombings," Wikipedia, http://en.wikipedia.org/wiki/7_July_2005_London_bombings#Claims_of_responsibility. The increasing tendency of groups to claim responsibility can produce conflicting claims of responsibility for a particular attack. *See,* e.g., Hugh Miles, *Terrorist Attacks Kill 90, Wound Hundreds at Sharm el-Sheikh Resort,* Militant Islam Monitor (July 23, 2005), http://www.militantislammonitor.org/article/id/853; John Ward Anderson, *Suicide Blast Kills Four in Tel Aviv,* Washington Post (February 26, 2005, at A01), http://www.washingtonpost.com/wp-dyn/articles/A55514-2005Feb26.html.

521 *See,* e.g., Scott MacLeod, *Is Al-Qaeda in Sinai?,* Time 17 (October 12, 2004) (synchronized attacks "a common Al-Qaeda tactic"); *World Nations Beef Up Security after London Bombings,* Al Jazeera, July 8, 2005, http://www.aljazeera.com/me.asp?service_ID=8870 (synchronized attacks are "classic for al Qaeda"). It is also possible to infer responsibility for an attack from the likely motive for the attack.

522 *See id.*

But because our attribution and response strategy treats terrorism as crime rather than warfare, merely identifying the responsible entity is not sufficient; authorities must identify the individuals who carried out an attack so they can be apprehended, convicted, and sanctioned for their crime.[523] The criminal investigation approach outlined above is, therefore, also used to identify and apprehend individual terrorists.[524]

Attacker-attribution: Virtual World

The BIS episode we examined at the beginning of Chapter 4 illustrates how online attacks complicate attacker-attribution across the three dimensions of crime, terrorism and war. Attacker-attribution becomes problematic at each level because, as we saw in Chapter 3, the approaches we use to identify attackers all implicitly assume activity in the real, physical world. As we also saw in Chapter 3, because cyberattacks do not take place in physical reality, the attack signatures of cybercrime, cyberterrorism, and cyberwarfare display few of the empirical characteristics common to their real-world counterparts.[525]

We will use the BIS and Estonian attacks to analyze attacker-attribution in the virtual world.[526] As we saw in the previous section, the real-world crime-terrorism and warfare attacker-attribution calculi rely heavily on the "place" where an attack occurred and/or originated from in determining attacker identity. With virtual attacks, "place" tends to be at once more ambiguous and less conclusive than in real-world analyses.

---

523 *See* Chapter 2. *See also* Peter J. Spiro, *Globalization and the (Foreign Affairs) Constitution*, 63 Ohio State Law Journal 649, 665 (2002) ("Terrorism . . . ultimately reduces to a kind of criminal activity, which can be addressed as such").

524 *See*, e.g., Dennis Piszkiewicz, Terrorism's War with America: A History 85–95 (Praeger 2003) (investigation and apprehension of 1993 World Trade Center bomber Ramzi Yousef).

525 An attack signature encompasses the essential elements of an attack. *See* "Attack signature," Symantec, http://www.symantec.com/avcenter/attack_sigs/ (attack signature "is a unique arrangement of information that can be used to identify an attacker's attempt to exploit a known operating system"). *See*, e.g., Bryan Sartin, *Tracking the Cybercrime Trail*, Security Management 95–96 (September 2004) (FBI agents examined "audit logs to find the hacker's . . . attack signature—that is, how the hacker broke in and what the hacker did once he . . . had access").

526 The Estonia attacks are described in Chapter 1, and the BIS attacks are described at the beginning of Chapter 4.

Point of Attack Origin

How does one know an attack "isn't coming from a kid in Tarzana [California] who is bouncing off a Chinese server?"[527]

"Place" is ambiguous in point of attack origin analysis for cyberattacks because, for example, the fact that attacks were routed through Internet servers located in China does not necessarily mean the attacks originated in China. Online attackers commonly use "stepping stones"—computers owned by innocent parties but controlled by the attacker—in their assaults.[528] The stepping stone computers can be anywhere in the physical world because real-space is irrelevant to activity in cyberspace. So though the use of Chinese servers in the BIS attacks *might* mean the attacks came from China, it might *not* mean that at all; the attacks might instead have come from Russia or Brazil or Peoria or Washington, D.C. [529] The attacker could have been operating from a site a few blocks from the BIS offices in Washington.

This possibility opens point of attack origin up to manipulation, as well as obfuscation: An attacker in Brazil who knew of the Unites States' concern about China's cyberwarfare initiative might use Chinese servers to mask the true source of the attack and mislead the investigators trying to identify him.[530] The attacker could do this to obfuscate, to protect himself from being identified, apprehended and sanctioned for his crimes, if he is a mere cybercriminal. He might instead be operating on behalf of a nation-state, a state with an interest in fomenting hostility between the United States and

---

527 Robert Marquand & Ben Arnoldy, *China Emerges as Leader in Cyberwarfare*, Christian Science Monitor (September 14, 2007), http://www.csmonitor.com/2007/0914/p01s01-woap.html?page=1 (quoting James Mulvenon).

528 *See*, e.g., Jiangqiang Xin, *et al.*, *A Testbed for Evaluation and Analysis of Stepping Stone Attack Attribution Techniques*, 25th IEEE International Performance Computing and Communications Conference (Phoenix 2006), http://www.public.iastate.edu/~zhanglf/doc/Testbed_TridentCom2006.pdf ("To evade detection, network attackers often launch their attacks . . . through . . . previously compromised computers, namely stepping stones").

529 Investigators determined that the BIS attacks came "through Chinese servers," but that does not necessarily mean they came from China. *See*, e.g., Gregg Keizer, *Chinese Hackers Hit Commerce Department*, Tech Web, October 6, 2006, http://www.techweb.com/showArticle.jhtml;jsessionid=OM4E5LCHY4W0WQSNDLRCKHSCJUNN2JVN?articleID=193105174.

530 *See*, e.g., Nathan Thornburgh, *The Invasion of the Chinese Cyberspies*, Time 34 (September 5, 2005), http://www.time.com/time/magazine/article/0,9171,1098961-1,00.html (China "is known for having poorly defended servers that outsiders from around the world commandeer as their unwitting launchpads").

China. In this scenario, the attacker (who could be anywhere) routes the BIS attacks through Chinese servers in an effort to convince the United States that China is attacking it; the sponsoring nation-state's goal in this scenario might be to prompt the United States to launch retaliatory attacks against China, or it might simply be to encourage an atmosphere of hostility and tension between the two countries.

This result occurred, probably inadvertently, in the May 2007 attacks on Estonia. As we saw in Chapter 1, the Estonian authorities watched as those attacks were planned on Russian-language websites and later traced at least some of the attacks to Internet protocol addresses used by computers in Russia. The combined influence of these factors—along with Estonia's expecting protests for removing a statute of a Russian soldier from a park—convinced Estonian authorities that the attacks not only came from Russian territory, they came from the Russian government itself. As we saw in Chapter 1, the Estonian authorities believed the attacks were Russian cyberwarfare and went so far as to contact NATO to see if they could obtain military assistance (of an unspecified type) in responding to them. Because it is not clear that attacks such as this trigger NATO pact obligations, Estonia's quest for NATO assistance went no further.[531] As time passed, many concluded that the attacks were not from the Russian government and were therefore not cyberwarfare. Investigators working after the fact decided the attacks were too disorganized, too haphazard to have been government sponsored; they also determined that at least some of the attacks originated outside Russia, for example, in Brazil and in Vietnam.[532] As we saw in Chapter 1, the Estonian authorities eventually asked their Russian counterparts for assistance in finding the "cybercriminals" responsible for the attack. In January of 2008, a spokesman for an Estonian prosecutor's office announced that a twenty-year-old hacker had been convicted for organizing "some of" the attacks on Estonian government sites during the May 2007 cyber assaults.[533] Dmitri Galushkevich, an ethnic Russian student living in Estonia, used his

---

531 *See* Chapter 1.

532 *See* Chapter 1. *See also* Anne Applebaum, *For Estonia and NATO, A New Kind of War*, Washington Post A15 (May 22, 2007), http://www.washingtonpost.com/wp-dyn/content/article/2007/05/21/AR2007052101436.html.

533 *See*, e.g., Robert Vamosi, *First Conviction for Estonia's "Cyberwar,"* CNet News (January 24, 2008), http://news.cnet.com/8301-10789_3-9857492-57.html.

home computer to launch at least a portion of the cyberattacks.[534] After being charged, he admitted his role in them and was sentenced to pay a fine of about $1,600.[535]

While the attacks were in progress, the Estonian authorities rationally, though perhaps erroneously, concluded they were cyberwarfare based primarily on their apparent point of attack origin. The inferential circumstances they relief upon in identifying point of attack origin were the use of Russian-language sites (presumably operating from Russia) to plan the attacks and the fact that at least a percentage of the attacks originated from computers in Russia. In hindsight, both circumstances are equally attributable to the attacks being an online protest by Russians and by Russian sympathizers, all outraged by Estonia's apparently disrespecting the Russian military's contributions during World War II.[536] Estonia's misidentifying the nature of the attack and the identity of the attackers were consequently the product of inadvertence, not intentional manipulation by the attackers; and this inadvertence probably accounts for the fact that the misidentification came to light. If a clever cyberattacker—a state-sponsored cyberattacker—wanted to make it appear that Russia was attacking Estonia (or China was attacking the U.S., or vice versa), it might be far more difficult to determine that the attacks were in fact an artifice.

As these examples demonstrate, we cannot give point of attack origin conclusive or even substantial weight in attacker-attribution for cyberattacks. We should factor point of attack origin (or, more accurately, perhaps, apparent point of attack origin) into attacker-attribution, but in so doing, we must somehow consider its inherent capacity for ambiguity. Estonian authorities based their conclusion that the May 2007 attacks were Russian cyberwarfare on the premises that the attacks were planned in and originated from Russia. Afterward, investigators concluded that they were neither cyberwarfare nor from Russia, basing their conclusion in substantial

---

534 *See id. See also Estonia Fines Man for "Cyber War,"* BBC News (January 25, 2008), http://news.bbc.co.uk/2/hi/technology/7208511.stm; Jeremy Kirk, *Student Fined for Attack against Estonian Web Site*, InfoWorld (January 24, 2008), http://www.infoworld.com/article/08/01/24/Student-fined-for-attack-against-Estonian-Web-site_1.html. Galushkevich only *seems* to have pled guilty to launching attacks that shut down the Web site for the prime minister's political party. *See* Kirk, *Student Fined for Attack against Estonian Web Site, supra.*

535 *See* Vamosi, *First Conviction for Estonia's "Cyberwar," supra.* The court apparently imposed the lenient sentence both because Galushkevich only pled guilty to attacking one target and because he had no prior convictions. *See Estonia Fines Man for "Cyber War," supra.*

536 *See* Chapter 1.

part on the fact that the attacks used zombie computers from countries other than Russia.[537]

But while that became the generally accepted conclusion, even in Estonia, there were reports that NATO believed the Estonian attacks *were* launched by Russia, which was testing "the West's preparedness for cyber-warfare."[538] If those reports are accurate, then the NATO analysts who hold this belief must not find the attacks partially originating outside Russia to be conclusive or even compelling evidence as to the identity of the attackers. They must place greater weight on other factors, such as the Russians' motive (disrespect for its military) and the facts that (a) a number of the attacks came from Russian government offices and (b) Russian authorities refused to help Estonians trace the Internet protocol addresses of other computers used in the attacks, which could have identified attackers.[539] The last two circumstances led some to suggest that the Russian government was either responsible for the Estonian attacks or was an accomplice to crimes civilians committed.[540]

As the above analysis illustrates, point of attack origin seems to play an ambiguous and uncertain role in online attacker-attribution. Arguably, though, the real problem in this analysis is that the attacks did not have a single point of origin; instead, a percentage of them came from countries other than Russia. Perhaps the difficulty, then, lies not with the utility of point of attack origin as such, but with the factual ambiguity of the scenario we chose to analyze?

We should be able to resolve that issue by analyzing a different scenario, one in which point of attack origin is less ambiguous. We will use the BIS attacks described in Chapter 4, with certain modifications. We will assume the BIS computers were the targets of a sequenced set of attacks that had the same attack signature and were repeated over a substantial period of time. We will also assume U.S. investigators determined with a high level of confidence that the attacks came directly from servers in Beijing that belong to the

537 *See id.*

538 *See* Applebaum, *For Estonia and NATO, A New Kind of War, supra.*

539 *See* Michael Weiss, *Here Come the Cyber Wars*, Reason Online (August 17, 2007), http://reason.com/news/show/121896.html (noting that the attacks peaked on May 8 and 9, the dates on which the Soviet Army defeated the Wehrmacht).

540 *See id.* An accomplice is someone who does not actually participate in the commission of a crime but contributes to its commission by providing assistance or encouragement to the actual perpetrator. *See, e.g.,* "Accomplice," Wikipedia, http://en.wikipedia.org/wiki/Accomplice.

Chinese Internet service provider CHINANET. This latter assumption seems to eliminate the possibility I noted earlier, that is, that the attacker used computers in various locations to obscure the true point of attack origin.

Can we predicate attacker-attribution for our hypothesized BIS attacks on inferences drawn from the repetitive use of what almost certainly seems to be the same point of attack origin? Can we reliably infer that these attacks are coming either from the Chinese government (cyberwarfare) or from Chinese hackers (cybercrime/cyberterrorism)? Even if we assume all of the attacks originated from the same geographical place, it would still be risky to rely too much on mere repetition.

One problem is that neither the repetition of the attacks nor their having the same point of origin can conclusively or even substantially support attacker attribution. As we saw earlier, real-world investigators can almost conclusively infer that a single local criminal is responsible when a series of identical crimes occur in the same "place;" they can draw that inference with a high degree of confidence because the attack signature indicates a single perpetrator and the constraints of the real-world indicate he will be local.[541] In our scenario, the repetition of attacks with the same attack signature can indicate that there was one attacker; but it can also indicate that there were multiple attackers, each using the same attack tools.[542]

Our concern, though, lies not with the number of attackers as such but with whether we can determine the identity of the attackers from the originating point of the attacks.[543] While the originating point is a crime scene, it differs from real-world crime scenes in certain ways, one of which is that it does not provide direct evidence such as the testimony of eyewitnesses who can identify the attackers.[544] We must instead rely on circumstantial evidence, which is based on the process of inference.[545] Here, investigators

---

541 *See* Chapter 3.

542 *See*, e.g., *Cyber Crime Tool Kits Go on Sale*, BBC News (September 4, 2007), http://news.bbc.co.uk/2/hi/technology/6976308.stm ("easy to use tools that automate attacks").

543 To simplify the analysis that follows, we will assume there was more than one attacker.

544 *See* Chapter 3. "Direct evidence is evidence which, if believed, proves the . . . fact in issue without inference . . . [I]t . . . comes from one who . . . saw or heard the factual matters which are the subject of the testimony." American Jurisprudence 2d Evidence § 4 (Thomson West 2007).

545 *See* American Jurisprudence 2d Evidence § 4, *supra* (defining circumstantial evidence as evidence of "facts other than the fact in issue, which leads to a permissible inference concerning the existence of the fact in issue"). *See also* American Jurisprudence 2d, Evidence § 313 (Thomson West 2007) (direct evidence consists of witnesses testifying about their knowledge of the facts to be proven; circumstantial evidence consists of witnesses testifying about facts from which the jury can infer the existence of the facts to be proven).

concluded that the BIS attacks came "directly" from Chinese servers; that conclusion would reasonably support our inferring that the attackers were physically in China when they launched the attacks.[546] The inference and its reasonableness are both predicated on an assumption: that the attackers had to be in China to launch their attacks from Chinese servers. As we saw earlier, that assumption is valid for real-world crime, terrorism, and warfare but is generally not valid for cyberattacks of whatever type.[547] The reasonableness of the inference therefore depends on the likelihood that the assumption applies in this instance.

To determine the likelihood of it applying, we must analyze the investigators' conclusion that the attacks came "directly" from the CHINANET servers in Beijing. Their conclusion can mean either of two things: One is that the attacks originated with the CHINANET servers, that is, came from those servers and from nowhere else. The other possibility is that the CHINANET servers were the immediate trajectory for attacks that originated elsewhere, that is, other than on those servers.

With the first possibility, we have a reliably determined point of attack origin that reasonably supports inferring that the attackers were in China when they launched their attacks. Here, the investigators' conclusion negates the possibility they were anywhere else: In finding that the attacks originated on the Beijing CHINANET servers, the investigators implicitly, and conclusively, eliminated the possibility that the attackers used servers in another location to launch the attacks; the attackers consequently had to be in a location from which they could access the Beijing servers directly, without going through intermediaries.[548] Logically, then, we can infer that they were somewhere in China when they accessed the Beijing servers. The repetition of the attacks inferentially establishes that the attackers were in China, in an area from which they could access the CHINANET servers, for the period of time required to launch the successive attacks.

What, if anything, does this tell us about the attackers? We can say with a high degree of confidence that they were in China; but that does not necessarily

546 *See* American Jurisprudence 2d, Evidence § 313, *supra* (circumstantial evidence "must lead to a reasonable inference and not a mere suspicion of the existence of the fact . . . to be proved").

547 *See* Chapter 3.

548 *See generally* Tim Wu, *The Filtered Future*, Slate (July 11, 2005), http://www.slate.com/id/2122270/ (China's development of Internet infrastructure is designed to develop "an inward-looking network that is physically disconnected from the rest of the world").

mean that they are still in China or that they are Chinese. They might still be in China, or they might not; they might be Chinese, or they might not. It is perhaps farfetched to assume that foreigners came to China, used the Beijing CHINANET servers to launch the attacks, and then either left China or have remained there, but it is not beyond the realm of possibility.

It is, though, more reasonable to assume that because the attackers were in China when they launched the attacks and because they used the domestic CHINANET servers, they are Chinese, and they are still in China. This assumption is more reasonable than the first alternative because it reflects what we know of reality, as it has existed so far. It is, as we saw in Chapter 3, common for online attackers to route attacks through servers in countries other than the one from which they are operating. It is also common for online attackers to attack targets in other countries.[549] In both instances, the attackers exploit cyberspace's capacity to support remote attacks. The attackers have no need to relocate themselves; they can attack China or India or the United States without ever leaving their home base in, say, Russia. And that is why it is more reasonable to assume that because the attackers were in China when they launched the attacks, and because they used the CHINANET servers, they are Chinese and are still in China. Territory is irrelevant for cyberattacks, but not for people; it would be time consuming, expensive, and quite unnecessary for the attackers to travel to China to launch their attacks. It could also be counterproductive; their presence there and their use of the CHINANET servers from a hotel or other facility might bring them to the attention of Chinese authorities.

It is therefore logical to assume that our hypothesized attackers are Chinese and are still in China. The repetition of the attacks also supports both inferences. If the attackers are not Chinese, then they (a) would have to have traveled to China for the express purpose of using the CHINANET servers to launch the attacks or (b) be foreigners who are, or were, there for other purposes. The repetition of the attacks undermines, but does not eliminate, the likelihood that the first alternative is correct; it is possible that the attackers did this, but the expense and time required to make the trip, obtain access to the CHINANET system, and remain in China for the period of time necessary to launch the attacks makes it more reasonable to assume that the attackers are domestic, that is, are Chinese. If they are Chinese, then it is reasonable to assume they are still in China. It is also *possible* that the

549 *See* Chapter 3.

attackers are foreigners who are living in China for other, presumably legitimate purposes, but that seems unlikely. There are foreigners living in China on an extended or even permanent basis, but the number is relatively low, and they are generally there for business or professional reasons.[550] The small number of foreign residents coupled with their apparent legitimacy inferentially indicates that they are not the source of the attacks; another factor supporting this inference is the issue noted above, that is, that Chinese authorities might track foreigners' use of the CHINANET servers. It is ultimately more reasonable to assume the attackers are, and remain, domestic.

Even if we assume they are Chinese and are in China, this tells us little, if anything, about who they are. They could be members of the People's Liberation Army, civilian sport hackers or members of a terrorist group operating with or without the approval of the Chinese government. It will be very difficult to identify individual attackers if we cannot determine the nature of the attack.[551] The mere repetition of the attacks tells us nothing about which category they fall into; repetitive attacks can be equally characteristic of cyberwarfare, cybercrime, and cyberterrorism.[552]

With the second possibility, we have one locus (the CHINANET servers) being reliably excluded, which leaves us with a residual, ultimately undetermined point of attack origin. Here, the investigators' conclusion reasonably supports two, mutually exclusive inferences: that the attacks originated from outside of Chinese territory; or that they originated from other servers in China.

---

550 *See, e.g., Rules Impact Shanghai Luxury Property*, China Economic Review (January 2007), http://www.chinaeconomicreview.com/cer/2007_01/Rules_impact_Shanghai_luxury_property.html (300,000 foreigners live in China in 2007). *See also* Work in China? Get Your Docs in Order!, http://www.associatedcontent.com/article/321577/work_in_china_get_your_docs_in_order.html (foreigners need a physical exam, work and residence permits, residence registration certificate, visa, passport, employment license, and invitation letter to work in China).

551 *See* Chapter 4 (interaction between attack-attribution and attacker-attribution).

552 Mere repetition is not enough to determine the nature of the attacks, but repetition coupled with other circumstances could allow us to do so. The final analysis of the sustained Estonian attacks described in Chapter 1 found that those attacks were not cyberwarfare because of the relatively disorganized structure of the attacks. So, repetition coupled with factors such as a high level of organization in an attack and the expenditure of massive resources in the attack could support the inference that it was cyberwarfare. That inference, coupled with a reliable inference of point of attack origin, could support attacker-attribution. So if we could reasonably infer that the attacks hypothesized above were cyberwarfare, we could reasonably infer that they came from the Chinese government, most probably from the People's Liberation Army.

If we assume the attacks originated from other servers in China, we would then have to go through a variation of our analysis of the first possibility. That is, we would have to decide whether this circumstance coupled with the repetition of the attacks reasonably supports our inferring that the attackers (a) are Chinese and are still in China or (b) are not Chinese and are, or are not, in China. Here, too, it is more reasonable to infer that they are Chinese—rather than foreigners who were in China using Chinese servers—and are still in China. But again, even if we accept this as the most reasonable inference, it tells us little, if anything, about who they are.

If we assume the attacks originated outside China, we can almost conclusively infer that the attackers were not in China when they launched the attacks. And absent evidence to the contrary, we can reasonably infer that the attackers are not Chinese. This inference is based on the premise noted above, that is, that territory is irrelevant for bits and bytes but not for people.

It is conceivable that Chinese nationals acting on their own or on their government's behalf went to, say, Pakistan to launch the attacks to conceal their connection with China. But this is an almost illogical scenario: If these Chinese nationals went to the trouble of traveling to Pakistan to make it the originating point for attacks on the United States, why would they then route the attacks through the CHINANET servers? Using those servers undermines, if it does not eliminate, the plausible deniability the attackers established by leaving their own country to initiate the attacks from Pakistan.

Logically, the only reasons to do this are either to (a) frame Pakistan as the source of the attacks or (b) make it appear that someone in Pakistan was trying to frame China as the source of the attacks. But if the Chinese nationals wanted to frame Pakistan, there is no reason to use the CHINANET servers; they would want the attacks to be directly traceable to Pakistan. The only reason for doing this is to make it appear that someone in Pakistan tried to frame China. And that is a conceivable but unlikely possibility: Chinese nationals as virtual Machiavellis who frame Pakistan as the source of attacks that are ostensibly designed to frame China. The other, more likely possibility is that another, unknown set of virtual Machiavellis really did try to "frame" China by routing structurally similar attacks through its real-space.[553] Empirically, this second scenario is more likely to be true than the first,

---

553 *See*, e.g., Jeremiah Grossman—The Devil Made Me Do It (July 18, 2006), http://jeremiahgrossman.blogspot.com/2006/07/devil-made-me-do-it.html (XSS exploitation could be used to frame someone for launching attacks on government or other sites).

doubly devious one. While the first scenario is not outside the realm of possibility, Occam's razor favors the more straightforward scenario.[554]

If we assume this is, indeed, what happened, we would know Pakistan is the point of origin of the attacks, but that again tells us little, if anything, about the identity of the attackers. They might be Pakistani; this would be a reasonable inference given the premise I noted above, i.e., that attackers tend to operate from their home territory. Even if we accept that inference, we would still know little more than that the attackers are Pakistanis; they could be Pakistani cybercriminals or Pakistani cyberterrorists or Pakistani military hackers experimenting with cyberwar. If we knew from collateral evidence[555] that a group of cyberterrorists opposed to the current regime in China was operating in Pakistan, we could reasonably infer that they launched the attacks. Absent such evidence, we cannot take that next step; neither the identified point of attack origin nor the repetitive nature of the attacks markedly advances the process of attacker-attribution beyond linking the attackers with a particular territory and, perhaps, with a particular attack signature.

Is that inevitable? That is, is it necessarily true that point of attack origin will play such a minor role in identifying online attackers? To answer those questions, we need to consider the role point of attack origin plays in attacker-attribution for real-world warfare, crime, and terrorism. As we will see, it plays a significant role in attacker-attribution in this context because real-world activity is inescapably linked with territory. As we will also see, its significance for this endeavor erodes as we move into an online environment where territoriality recedes in importance.

Warfare

Point of attack origin has historically played a critical role in attacker-attribution for acts of warfare for two reasons: War is a conflict between nation-states, which are defined by the territory they control; war has consequently

554 Occam's razor, a logic principle, holds that "all things being equal, the simplest solution tends to be correct." *See* "Occam's razor," Wikipedia, http://en.wikipedia.org/wiki/Occam's_Razor.

555 This would be evidence completely unrelated to the attacks, such as news stories about the existence and agenda of this group or about other activities it had undertaken to advance its anti-Chinese government agenda.

been a territorially based endeavor.[556] If Country A was the target of a military attack originating from the territory of Country B, Country A would infer, with a very high degree of confidence, that Country B was responsible for the attack.[557]

If we apply this logic to the modified BIS attacks we hypothesized earlier, the United States could reasonably infer that the attacks were acts of cyber-warfare launched by China. It could, in effect, construe them as the virtual equivalent of Japan's real-world assault on Pearl Harbor. Here, as with Pearl Harbor, U.S. government property located on U.S. territory is the target of attacks originating in the territory of another nation-state. The problem with drawing that derivative inference[558] of responsibility in this instance lies in the permissibility of equating an online attack inferentially launched *from* Chinese territory with an attack launched *by* China as a nation-state.

Historically, it was reasonable to equate transnational attacks with acts of war because only nation-states could launch such attacks.[559] That is still true in the real-world,[560] but as we saw earlier, cyberspace gives each nation-state an incremental, highly permeable set of "virtual" national borders.[561] Anyone with Internet access and certain skills can launch a cross-border virtual

---

556 *See* Chapters 2 & 3.

557 *See id.*

558 This inference, unlike the primary inference that the attacks came from China, is based not on ascertained facts but on inferences from the primary inference. *See,* e.g., Wabash Corp. v. Ross Elec. Corp., 187 F.2d 577, 601–63 (2d Cir. 1951) (Frank, J., concurring and dissenting).

559 *See* Chapter 3. *See,* e.g., United Nations General Assembly Resolution 3314, U.N. GAOR, 29th Sess., U.N. Doc. A/RES/3314 (December 14, 1974).

560 One might argue that it is no longer necessarily true in the real-world and cite the 9–11 attacks as supporting the argument. And the 9–11 attacks were, in a sense, transnational attacks: They targeted sites in U.S. territory, and they were launched at the behest of a terrorist group, the leadership and headquarters of which are located abroad, in the territory of another nation-state.

There are two problems with this argument: One is that the 9–11 attacks were not acts of war; they were launched by a group of individuals, not by a nation-state. As we saw in Chapters 2 and 3, war is a struggle between nation-states. What we have here is a terrorist attack on a nation-state or, perhaps more accurately, on targets in another nation-state.

The other problem with the argument is that the attacks were not really transnational in nature: In each of the 9–11 attacks, the attackers commandeered domestically owned and operating airliners and used them to destroy property in the territorial United States. The attackers were transnational only insofar as the attackers themselves traveled into the United States from abroad for the purpose of launching them. This might, technically, qualify as a transnational attack, but it certainly does not qualify as warfare.

561 *See* Chapter 3.

attack not on the territory but on the internal "machinery" of a nation-state, as the BIS and similar attacks demonstrate.[562] A virtual attack is not territorially invasive, but it produces effects in the victim-state's territory that are damaging in various ways and in varying degrees.[563] The character and extent of the damage inflicted will tend to be a function of the nature of the attack, as we will see later in this chapter.

For the warfare calculus, there are two reasons why point of attack origin plays a more problematic role in online attacks. One is that the process of identifying the point of origin for online attacks is likely to rely more on inference than on direct evidence, which is the major component of real-world attacker-attribution.[564] As we saw above, this increased reliance on inference introduces an element of ambiguity into online attacker-attribution calculus. The other reason is that an identified external point of origin (attack originated in China) may not markedly advance online attacker-attribution, even for warfare. An identified point of attack origin in another nation-state cannot routinely be construed as an attack by that nation-state because cyberspace gives essentially anyone the ability to launch transnational attacks.

Crime-terrorism

Though crime and terrorism are conceptually distinct phenomena, we will consider them jointly for at least two reasons: One is that law treats terrorism as a crime.[565] The other reason is that though war is the product of nation-state activity and threatens a state's ability to maintain external order, crime and terrorism are the product of purely individual activity and threaten a state's ability to maintain internal order.[566]

In the real-world, point of attack origin has played a much more limited role in attacker-attribution for crime and terrorism than it has for war. Though point of attack origin can inferentially indicate who was responsible for a crime

---

562 *See, e.g.*, Peter Warren, *Smash and Grab, the Hi-Tech Way*, Guardian (Manchester, U.K.) (Jan. 19, 2006), *available at* http://technology.guardian.co.uk/weekly/story/0,,1689093,00.html (virtual attack "aimed at stealing sensitive information" on computers Parliament uses).

563 *See* Chapter 3.

564 *See id.*

565 *See* Chapters 2 & 3.

566 *See* Chapter 2. This proposition becomes problematic for state-sponsored crime and state-sponsored terrorism, an issue we will consider later.

or an act of terrorism, the link between point of attack origin and attacker identity is usually far more attenuated in this context than it is for warfare.

The primary reason for this is that in the real-world point of attack origin and point of attack occurrence are often so closely related as to be indistinguishable for crime, and even for terrorism.[567] A crack dealer buys and sells crack in his neighborhood; the points of origin and occurrence of his drug crimes are functionally identical. In 1982, the Irish National Liberation Army, a terrorist group, bombed a disco in Ballykelly, Northern Ireland, frequented by British soldiers.[568] The attack killed eleven soldiers and six civilians; the INLA agents responsible for it operated out of nearby Derry.[569] The points of attack origin and occurrence for this terrorist act were separated by such a short distance that they, too, are functionally identical.

When there is little or no differentiation between point of attack origin and point of attack occurrence, identifying the point of attack origin is unlikely to markedly advance the process of attacker-attribution. Assume a woman is raped as she leaves Ladies Night at a neighborhood bar. She left at closing time and was attacked in the nearby parking lot where she left her car.[570] Police are likely to infer that the attacker is a local, on the assumption that he was familiar with the bar's closing time, with its Ladies Nights, and with patrons' use of the rather isolated parking lot. This reasonable inference[571] establishes that insofar as an attack like this has a distinct point of origin, it is in the local area. The inference would structure the police's efforts to identify the rapist by focusing their efforts on the area the bar serves.

567 As we saw in Chapter 3, spatial proximity between attacker and victim has been an inevitable element of real-world crime and, to a somewhat lesser extent, terrorism. Proximity has been unavoidable because both required direct physical action by the attacker against the victim. The development of timing devices created a limited potential for the remote commission of crime and terrorism, but physical proximity remains the norm for both real-world endeavors.

568 *See* "Droppin Well Bombing," Wikipedia, http://en.wikipedia.org/wiki/Ballykelly_disco_bombing. The U.S. Department of State has designated the INLA as a terrorist organization. *See* State Department Identifies 40 Foreign Terrorist Organizations, U.S. Department of State (April 27, 2005), http://usinfo.state.gov/xarchives/display.html?p=washfile-english&y=2005&m=April&x=20050427155413dmslahrellek0.2537195.

569 *See* "Irish National Liberation Army," Wikipedia, http://en.wikipedia.org/wiki/INLA; "Dominic McGlinchey," Wikipedia, http://en.wikipedia.org/wiki/Dominic_McGlinchey.

570 We will assume that the victim cannot identify her attacker and he left no traceable DNA.

571 It is *possible* he is instead an out-of-towner who simply happened to be driving by when the victim was walking to her car. While this inference is logically permissible, experience tells us it is less likely to be correct than the inference given above. The police will therefore base their investigation on the more likely inference.

Officers would interview people who might have seen someone in the area that night or have heard someone talking about the rape. They would check the location and alibis of locals with sex crime convictions and pursue other, similar leads.

My point is that the most logical approach here, as with most real-world crimes, is to infer that the attacker is someone who has ties to the area in which the crime occurred. The logic of this inference lies in our experience of life in the real-world; the constraints of physical reality mean that our activities routinely tend to center around a specific geographical area, even if we live in a large city. While the attacker in this hypothetical *could* be an out-of-towner who simply happened to be driving by when the victim was walking to her car and decided to exploit the situation, the likelihood of that inference's being true is much less than the likelihood that the more mundane inference outlined above is the correct interpretation of what happened.

As this hypothetical illustrates, point of attack origin tends to play a relatively minor role in the inferential process law enforcement officers use in attacker-attribution for real-world crime and terrorism. It plays a relatively minor role here because these threats to internal order come primarily, if not almost exclusively, from local actors.[572] Domestic actors are presumptively in the nation-state where the attack occurred, and as we saw earlier, investigators tend to assume the domestic actors responsible for an attack remain in the area where it occurred. Even when there is significant spatial differentiation between point of attack origin and point of attack occurrence, identifying the former serves at most as a minor clue—an inferential datum that can contribute to the identification of the attacker(s) and also, for terrorism, of the sponsoring organization.[573]

As crime and terrorism migrate online, point of attack origin can play a more important role in attacker-attribution. When crime and terrorism cease to be localized phenomena, point of attack origin can become a significant indicator of the attacker's nationality and, perhaps, of his motives. The problem lies in determining point of attack origin.

In 1994, the U.S. Air Force's Rome Air Development Center ("Rome Labs") in upstate New York discovered that its computer systems had been hacked

---

572 *See* Chapter 3.

573 *See generally, Search Continues for Witness in Clinic Bombing*, CNN.com, Jan. 31, 1998, http://www.cnn.com/US/9801/31/clinic.bombing/?related; "World Trade Center Bombing," Wikipedia, http://en.wikipedia.org/wiki/World_Trade_Center_Bombing.

by unknown persons.[574] The hackers had among other things copied data from computers that contained sensitive Air Force research and development data. Since hacking is a federal crime,[575] Air Force, Secret Service and Federal Bureau of Investigation agents immediately began investigating the incidents, hoping to identify the perpetrators. They found a complex attack signature: the attackers had routed their attacks through multiple computers in various countries. Through a complex process we will review later in this section, the U.S. investigators eventually traced the attacks to the United Kingdom where, with Scotland Yard's assistance, they identified two adolescents as the Rome Labs attackers. Both were prosecuted, with mixed results.[576]

The Rome Labs case illustrates how and why the use of cyberspace can make attacker-attribution more difficult in crime and terrorism cases. Cyberspace erodes law enforcement's ability routinely to assume, as in the rape hypothetical given above, that an attacker is parochial. The viability of that default assumption still holds for real-world crime and *can* also hold for real-world terrorism, but its applicability to online crime and terrorism is increasingly problematic.

When it comes to cybercrime and even some types of cyberterrorism, the parochial-attacker assumption is most likely to hold for "personal" attacks: crimes and acts of terrorism in which the perpetrator's motives are idiosyncratically emotional.[577] In these cases—where, say, John uses cyberspace to stalk his former girlfriend or Jane uses it to attack her employer—the perpetrator and victim are in the same geographical area, but instead of using physical activity in that real-space to conduct the attack, the perpetrator vectors it through cyberspace.[578]

---

574 *See*, e.g., Richard Power, Tangled Web 66–75 (Que 2000).

575 *See* 18 U.S. Code § 1030.

576 *See* Power, Tangled Web, *supra* at 70–75 (one plead guilty, charges were dropped against the other).

577 These cybercrimes include revenge attacks by former spouses/lovers and current or former employees, as well as more generalized cyberstalking and harassment. *See* Susan W. Brenner, *Should Criminal Liability Be Used to Control Online Speech?*, 76 Mississippi Law Journal 705 (2007).

578 *See* Paul Shukovsky, *Cyberstalker Just out of Reach of Law, But Finally, He Stops*, Seattle Post-Intelligencer (February 11, 2004), http://seattlepi.nwsource.com/local/160201_cyberstalking11.html; Press Release, Office of the U.S. Attorney for the Southern District of California (August 28, 2006), http://www.usdoj.gov/usao/cas/press/cas60828-1.pdf.

This creates an epistemological issue: When attacker and attacked are in the same real-space area throughout an attack conducted online, did that attack originate (a) only in the real-space occupied by attacker and victim, (b) only online, or (c) in both locations? For the purposes of attacker-attribution, the answer should be (c).

In "personal" attack cases, the connections between attacker and victim mean that the parochial-attacker assumption is likely to be very useful in identifying the attacker. And so far, cyber-vendettas seem primarily to originate in real-world contacts between attacker and victim.[579] This means that investigators can profitably rely on the approach they use for real-world crime and terrorism, that is, focusing on inferences derived from the real-world context from which the attack(s) emerged. The attack should, therefore, be construed as originating in the real-space occupied by attacker and victim.[580]

They should not, though, focus only on the geographical point of origin of the attack. When a "personal" attacker uses cyberspace, it, too, is a "place" of origin of the attack. Its role in the investigation of "personal" attacks is consequently analogous to the role a physical point of attack origin plays in the traditional investigative process. Cyberspace, like a real-world point of attack origin, becomes a source of inferential data that can be used to identify the attacker. If, for instance, a stalker consistently uses a specific website in tormenting his victim, that website becomes "a" point of origin of the attack and should be treated as such. Investigators may be able to use that website to identify the attacker.[581] And even if the attacker has not revealed his identity to the site operator, his use of the website may provide inferential data as to his identity.

---

579 *See* Leroy McFarlane and Paul Bocij, *An Exploration of Predatory Behaviour in Cyberspace: Towards a Typology of Cyberstalkers,* First Monday (September 2003), http://www.first-monday.org/issues/issue8_9/mcfarlane/index.html. *But see* Working to Halt Online Abuse Online Harassment Statistics, http://www.haltabuse.org/resources/stats/relation.shtml.

580 Investigators dealing with these cases are, in fact, likely to assume a real-space point of origin and proceed accordingly. *See,* e.g., People v. Vijay, No. H024123, 2003 WL 23030492 (Cal. Ct. App.December 19, 2003); State v. Hoying, No. 2004-CA-71, 2005 WL 678989 (Ohio Ct. App. Mar. 25, 2005); State v. Cline, No. 2002-CA-05, 2003 WL 22064118 (Ohio Ct. App. September 5, 2003), *rev'd,* 816 N.E.2d 1069 (Ohio 2004); State v. Askham, 86 P.3d 1224 (Wash. 2004).

581 *See,* e.g., Amanda Gillooly, *Apologies Set for Charleroi Area Officials,* Valley Independent (Monessen, PA) (July 3, 2003), 2003 WLNR 13946022 (woman charged with harassment after state police traced messages criticizing local officials that were posted on a local TV station's website to her Hotmail email account).

What about attacks in which the attacker is *not* in the same real-space as the victim? An identified point of attack origin has a very different function in these cases, for two reasons. One is that identifying point of attack origin in attacks such as these serves an initial, essentially negative function in attacker-attribution. It tells the investigators that the parochial-attacker assumption and derivative investigative approach that they use for real-world crime and terrorism will probably be of little use in identifying the attacker(s). When an attack presents functionally coterminous points of attack origin and attack occurrence, we have a localized crime scene that then becomes the focal point of the investigation. Evidence, inferences, observations of witnesses, and connections between victim and attacker all radiate from and revolve around this unitary crime scene. It creates a comprehensible focus for the investigation and, in so doing, makes the investigation a manageable task. We have seen in complex serial-killer cases how expanding the usual single real-space crime scene into a variegated network of geographically dispersed, victim-idiosyncratic crime scenes can test the limits of the traditional investigative approach.[582]

But even ambitious serial killers necessarily operate on a limited geographical scale: in the United States, they have tended to confine their activities to a smaller area within a state, sometimes to the state itself, and in unusual instances, to surrounding states.[583] The physical constraints of the real-world (the need to maintain a residence, support oneself, and generally survive, combined with the logistics involved in successful killings) limit the frequency and geographical dispersion of the attacks real-world serial killers can carry out successfully.[584] As we have seen, this limitation does not apply to other offenses once cyberspace becomes a component of criminal and/or

---

582 *See*, e.g., Ann Rule, Green River, Running Red (NY: Free Press 2004); "Andrei Chikatilo," Wikipedia, http://en.wikipedia.org/wiki/Andrei_Chikatilo. Serial killers are a useful analogue here because while their attacks take place in real-space, they are not "personal":

> Murder is usually either a crime of personal relationships . . . or an unintended consequence of other crimes. Because of this, most murders are . . . simple to solve; in most familial deaths, the murderer makes little . . . effort to conceal the crime . . . ; in other cases, the murderer is usually a local. . . . These assumptions, with which any law enforcement officer naturally approaches a single murder, are barriers to catching a serial killer.
>
> Another barrier to serial killers' early capture is their . . . choices of victim. . . . They almost never have any links to their victims—they pick by whim or impulse, seeking types or opportunity rather than any easily detectable link.

"Serial killer," Wikipedia, http://en.wikipedia.org/wiki/Serial_killer.

583 *See*, e.g., "Ted Bundy," Wikipedia, http://en.wikipedia.org/wiki/Ted_Bundy.

584 *See* Chapter 3.

terrorist activity. One can strike anonymously from any point connected to the Internet and iterate the attacks with a frequency impossible in the real-world.[585]

And that brings us to the second reason why point of attack origin has a very different function in attacker-attribution when the attacker is *not* in the same real-space as the victim: Cyberspace fractures the crime-scene into shards, the number of which depends on the particular circumstances of an attack. One constant shard is the alpha point of attack origin—the place where the attacker is physically located and from which she launches the attack. Other, variable crime scene shards (beta, gamma, etc.) are the intermediary points of transmission used in the attack; each represents the occurrence of a constituent, spatially diverse event that contributed to the success of the ultimate attack. The final constant shard—the omega shard—is the place of attack occurrence, which we examine later in this chapter.

Fracturing the crime scene into shards makes identifying the point of attack origin and linking it to the attacker much more difficult. Aside from anything else, a fractured crime scene can result in false positives, that is, in investigators' assuming an intermediary point of transmission of an attack is the originating point for the attack (when, of course, it is not).

This could have happened in the Rome Labs case. The Rome Labs investigators initially traced the intruders back to an Internet service provider (ISP) mindvox.phantom.com, in New York City.[586] The mindvox.phantom.com ISP allegedly had ties to the Legion of Doom, a hacker group several members of which had been convicted of unlawful intrusion crimes a few years earlier.[587] The investigators could have logically assumed that this ISP was the originating point for the attack on the Rome Labs computers, given (a) that it was in the state where the attacks occurred and (b) its immediate connection to the attacks plus its apparent ties to hackers. Their drawing this assumption would have been consistent with the real-world approach to criminal investigation because it implicitly incorporates the premise that point of attack origin is a binary concept. Real-world "places" are mutually exclusive. It follows that a real-world "place" is the point of attack origin or it is not; and if it is the point of attack origin, other "places" cannot be. Because the New York ISP was the immediately proximate point of origin for the Rome Labs

585 *See id.*

586 *See* Power, Tangled Web, *supra* at 70–75.

587 *See id.*

attacks, it could have been deemed to be "the" point of attack origin for those attacks.

Identifying the New York ISP as the point of origin would have been a false positive, one that could have derailed the investigation. Fortunately, the investigators did not make this error; instead, they continued to investigate and, as we saw earlier, eventually identified a complicated trail of attack increments that used many computers in eight different countries.[588] They were not, however, able to track these increments back to their true points of origin.

Ironically, perhaps, the Rome Labs investigators ultimately identified the perpetrators—two adolescents using the names "Datastream Cowboy" and "Kuji"—the old-fashioned way: by using informants.[589] The investigators learned that the attackers used these *noms de hack* by monitoring their activity during some of the attacks; so they sent informants into online chat rooms to see if either was taking credit for what they had done.[590] Datastream Cowboy not only took credit, he revealed that he was from the United Kingdom and gave an informant his home phone number.[591] It was then a simple matter to identify and apprehend him; doing the same for Kuji took longer because he was "a far more sophisticated hacker."[592]

The Rome Labs episode illustrates how cyberspace can fracture the crime scene into shards and make it much more difficult to determine the ultimate point of origin of an attack.[593] Making this determination proved impossible for the Rome Labs investigators because of the intricate paths the attackers used, a circumstance far from being unique to this case.[594]

Another issue that can complicate the process of making this determination, especially when the attack employed a series of incremental attack

---

588 For a detailed analysis of the attack and the investigation, *see* The Case Study: Rome Laboratory, Griffiss Air Force Base, NY Intrusion, Security in Cyberspace, Appendix B, U.S. Senate Permanent Subcommittee on Investigations (Minority Staff Statement) (June 5, 1996), http://www.fas.org/irp/congress/1996_hr/s960605b.htm.

589 *See* Power, Tangled Web, *supra* at 70–75.

590 *See id.*

591 *See id.*

592 The Case Study: Rome Laboratory, *supra.*

593 *See,* e.g., Daniel A. Morris, U.S. Dep't. of Justice, Tracking a Computer Hacker (2001), http://www.cybercrime.gov/usamay2001_2.htm.

594 *See,* e.g., Tom Young, *IT Industry Core to Global E-Crime Battle,* IT Week (November 9, 2006), http://www.itweek.co.uk/computing/analysis/2168266/industry-core-global-crime (FBI Special Agent "estimates that fewer than five per cent of international e-criminals are caught").

stages, is the legal process involved.[595] Incremental attack stages almost always involve the use of computers in different countries.[596] To gain access to the data they need to trace an attack back through those computers, law enforcement officers must obtain assistance from government and civilian entities in each of the countries where computers were used.[597] That can be a difficult and time consuming, if not impossible, process. The formal methods investigators have traditionally used to obtain such assistance can take months or even years; because digital evidence is fragile, it can disappear by the time the investigators finally obtain the assistance they need.[598] And because some countries have not criminalized hacking or other computer crimes, it may be impossible for investigators to obtain assistance from local authorities.[599]

Even if the investigators obtain the assistance they need and are confident they have traced an attack back to its true point of origin, this may not markedly advance their effort to identify the attacker. The BIS attacks are instructive in this regard. Investigators in that case accurately ascertained that the attacks came from servers in China. But this information could neither directly nor inferentially establish who was responsible for the attacks or, indeed, what *kind* of attacks they were.

In some instances, identifying the ultimate extraterritorial point of attack origin can serve the same function an identified domestic point of origin serves in investigating real-world crime: It becomes an inferential datum that can contribute to identifying the attacker. Assume the FBI obtains information indicating that a specific Romanian gang is engaged in phishing.[600] If the FBI subsequently traces a phishing attack to Romania, it would be reasonable for the agents to infer that the attack came from the gang under suspicion.[601] The inference would be strengthened if, say, the attack were

---

595 *See*, e.g., Morris, Tracking a Computer Hacker, *supra*.

596 *See*, e.g., Young, *IT Industry Core to Global E-Crime Battle, supra* (FBI Special Agent said international cybercriminals "are specialists in . . . covering their tracks").

597 *See*, e.g., Susan W. Brenner & Joseph J. Schwerha IV, *Transnational Evidence-Gathering and Local Prosecution of International Cybercrime*, 20 John Marshall Journal of Computer & Information Law J347, 354–88 (2002).

598 *See id.*

599 *See id.*

600 *See* Rene Millman, *Half of All Phishes from Romanian Cyber Gang*, PC Pro (December 18, 2006), http://www.pcpro.co.uk/news/100351/half-of-all-phishes-from-romanian-cyber-gang.html.

601 This example suggests a longitudinal way in which point of attack origin can contribute to the identification or an attacker or attackers. If investigators can establish point of attack origin with a high level of confidence for successive attacks, then they should be able to

traced to the city from which the gang is known to operate or if the attack signature displayed elements peculiar to this gang's operations.

In sum, while point of attack origin can play a role in identifying the attacker(s) in a cybercrime or cyberterrorism event, its function tends to be limited and will no doubt become even more limited as cyberattackers become more sophisticated about hiding their tracks.[602]

Hybrid crime/terrorism

In this and preceding chapters, we have assumed there is a distinct conceptual divide between (a) war, which is conducted by nation-states and challenges external order; and (b) crime and terrorism, which are carried out by individuals and challenge internal order. Though this distinction is still useful for analyzing attacker-attribution in real-world and online attacks, it is not as stable as it once was.

As we saw in Chapter 4, over the last several decades, the hybrid phenomena of state-sponsored terrorism and state-sponsored crime have emerged as increasingly serious threats. Both present distinct legal issues, most notably regarding the efficacy of attempting to use criminal sanctions to deter an activity sponsored by a nation-state. Aside from anything else, a sponsoring state may not cooperate in the investigation, apprehension, and extradition of those who acted on its behalf in committing criminal or terrorist acts.[603]

Notwithstanding the complexities that can be involved bringing the offenders to justice, the victim-state will want to identify the person(s) and/or nation-state responsible for an attack. Doing this is essential because any attack represents a threat to that state's ability to maintain a baseline of order, whether internal or external. And identifying the attackers is an important step toward being able to respond to that threat. Attacker-attribution analysis will therefore be of at least equal importance in the hybrid context of state-sponsored cybercrime and cyberterrorism. Here, though, the analysis can be more complex because the related but conceptually distinct

---

use the repeated occurrence of attacks emanating from this same point of origin to infer some consistency in the identity of the person or persons responsible for those attacks.

602 *See generally* Brian Krebs, *Cyber Crime Hits the Big Time in 2006*, Washington Post (December 28, 2006), http://www.washingtonpost.com/wp-dyn/content/article/2006/12/22/AR2006122200367_pf.html.

603 *See* Chapter 4.

processes of attacker-attribution and attack-attribution can be inextricably intertwined.

To understand why that is so, consider the BIS attack scenario we analyzed earlier. In that scenario, we know they came from CHINANET servers in Beijing and targeted computer systems used by a sensitive United States government agency in Washington. We analyzed how the attacker-attribution calculus would proceed if the attacks were (a) cyberwarfare or (b) "personal" cybercrime or cyberterrorism. Inherent in this analysis was the need to differentiate the two categories. We differentiated them because for one of the categories (warfare) a single factor establishes both the identity of the attacker and the nature of the attack. That factor is nation-state command and control, which we have so far assumed to be a binary phenomenon; that is, we have assumed cyberattacks are necessarily either war or not-war. If the attacker-attribution calculus indicated that an attack "came from" a nation-state, we concluded it was war; otherwise, it fell into the residual category of cybercrime or cyberterrorism.

One problem with this analysis is that determining whether a cyberattack "comes from" a nation-state can be difficult because, as we saw earlier, territorial point of attack origin is often ambiguous in this context. An attack from the CHINANET computers might "come from" China itself or it might "come from" sport hackers who are adventitiously in Beijing. Point of attack origin's utility in attacker-attribution has to this point been limited to negating the proposition that an attack is an instance of cyberwarfare. If we concluded with some confidence that an attack did not "come from" a nation-state, we inferentially assigned it to the cybercrime/cyberterrorism category and embarked upon determining precisely what it was and who was responsible for it.

The other problem with the dichotomized war-crime/terrorism analysis is that nation-state "involvement" in an attack is no longer synonymous with war. In the real-world, we have intermediate categories of nation-state involvement that have, among other things, given us state-sponsored crime and terrorism.[604] State-sponsored crime has migrated online, and

---

604 *See id.*

In addition to state-sponsored crime and terrorism, we also have a proliferation of military-style conflicts that do not rise to the level of warfare. *See* Chapter 4. *See, e.g.*, Carl Bruch, *Closing Remarks* in *Symposium: The International Responses to the Environmental Impacts of War*, 17 Georgetown International Environmental Law Review 565, 648 (2005) ("you just don't have just peace and war. There are tensions; there are disturbances; there is low level conflict; there are police actions; there is peacekeeping"). And we have what

state-sponsored terrorism will certainly follow. Since state-sponsored cybercrime and state-sponsored cyberterrorism each represent a novel, hybrid threat to internal *and* external order, we need to incorporate this new category into our threat analyses. The dichotomy consequently becomes a trichotomy: warfare/cyberwarfare;[605] state-sponsored cybercrime or cyberterrorism; and "mere" cybercrime or cyberterrorism.

Empirically, it will be difficult to determine whether an attack should be assigned to the second or third category. State sponsorship necessarily involves a level of state participation in a cyberattack, but identifying a nation-state's involvement in a less-than-cyberwarfare attack will almost certainly be problematic. Point of attack origin is unlikely to be helpful in this effort for two reasons. One is that for now and for the foreseeable future, all attacks will almost certainly come from the territory of *some* nation-state. Unless and until we develop the ability to originate (or ostensibly originate) attacks from satellites or from points in or on the high seas, all human activity will be territorially based; and most of the earth's territory is under the sovereign control of a nation-state. This means that the fact an attack originates in the territory of a nation-state, even one that is known to be inclined to sponsor criminal or terrorist activity, is inconclusive. Attack origination on its territory *might* mean the state is involved in the attack, but it might not; territorial origination is inferentially even less significant here than it is for cyberwarfare. The other reason is that the fact an attack originates *outside* the territory of a particular nation-state does not necessarily mean the state is not sponsoring the attack. As we have learned in the real-world, state sponsorship of crime and terrorism can take many forms, such as providing attackers with "funding, weapons, training and sanctuary."[606]

A devious state could fund and/or otherwise support terrorists or criminals who launch cyberattacks from outside its territory on a country the sponsoring state wants "harmed," for example, economically undermined, harassed, embarrassed, made vulnerable to intimidation in the real-world. The point of attack origin might be traced and used to identify individual attackers, but it should reveal nothing about the sponsoring state's complicity

---

might be called "aspirational war": situations, such as the one that exists between Al Qaeda and the United States, in which a group of non-state actors declare war on a state, but the state declines to recognize their encounters as such.

605 If it becomes necessary to incorporate online manifestations of the aspirational war scenario noted above, we would then have a quadrachotomy, if such a term exists.

606 State Sponsors: Iran, Council on Foreign Relations, http://www.cfr.org/publication/9362/.

in the attacks; indeed, because the attacks originated outside the physical "presence" of the sponsoring state, it should be able to plausibly deny any association with them.[607]

The same result can ensue if the attacks are launched from within the territory of the sponsoring state. Physical attacks necessarily involve staging efforts, which can be difficult to conceal; that, in turn, can make it difficult for a state to disavow knowledge of (and complicity in) criminal or terrorist activity occurring inside its borders.[608] Unlike physical attacks, cyberattacks are clandestine in staging and, to a great extent, in execution. A sovereign sponsor of internally launched cyberattacks could therefore credibly deny knowledge of and involvement with them; the sponsoring state would presumably claim that it, like most other nation-states, has great difficulty in identifying cybercriminals or cyberterrorists before or after they attack.[609] A truly devious state might encourage civilian cybercriminals and/or cyberterrorists to conduct their operations from its territory and use the fog of their conventional cyberattacks to obscure the purpose and origins of its state-sponsored attacks.

Point of attack origin is unlikely to be particularly helpful in attributing state responsibility for sponsored cyberattacks because here we are dealing with tiered responsibility: Primary responsibility for an attack rests with the individuals who carry it out; secondary responsibility for the attack rests with the nation-state that sponsors their efforts. An identified point of attack origin can play a role in primary attacker-attribution for cybercrime and cyberterrorism; that role diminishes, if it does not disappear, for secondary attacker-attribution because of the sponsor's indirect participation in the attack.

Point of Attack Occurrence

As we saw above, point of attack occurrence plays a pivotal role in real-world attacker-attribution. We see below that its role diminishes as attacks move online.

---

607 This becomes easier as the state's level of sponsorship diminishes. *See*, e.g., Paul R. Pillar, Terrorism and U.S. Foreign Policy 13–14 (Brookings Institution Press, 2004) (distinguishing state-sponsors of terrorism, state-enablers of terrorism, and state-cooperators in terrorism).

608 *See*, e.g., Michael Elliott, *They Had a Plan*, Time (August 2, 2002), http://www.time.com/time/covers/1101020812/story.html (Al-Qaeda in Afghanistan prior to 9/11).

609 *See* Chapter 3.

Warfare

For real-world warfare, point of attack occurrence is the essential complement to point of attack origin—its inevitable counterpoint in the attack dynamic. Point of attack origin tells us which country has initiated war; point of attack occurrence tells us which is the "victim" of war. The points of attack origin and occurrence will consequently be in different countries when the attack constitutes an act of war.

As we saw earlier, this calculus is unambiguous in the real-world because "place" is unambiguous in the real-world. When Germany invaded Poland, it clearly initiated war. The calculus becomes ambiguous when warfare migrates online. We saw earlier how defining point of attack origin becomes problematic in this context. Even if we can ascertain with the requisite level of confidence that an online act of war "came from" a particular nation-state, we cannot reflexively attribute that attack to the nation-state from which it came.

Similar problems arise as to point of attack occurrence. We return, again, to the BIS attacks. They occurred in the United States. What, if anything, can that tell us about who is responsible for the attacks?

We will assume that the attacks originated on the CHINANET servers in Beijing. Can we reasonably infer that cyberattacks that originate in China and inflict damage in the United States constitute acts of war attributable to the Chinese government? We do not have the presence of enemy personnel and armament on U.S. soil. We have the virtual "presence" of signals—bits and bytes—that traveled through cyberspace by routine means, by the same means civilian and government traffic use every second of every day.[610] The signals bear neither state insignia nor other markers of military allegiance or intent. Our only bases for possibly concluding they constitute components of an act of war by the Chinese government are (a) their point of origin, (b) their geographic destination, and (c) the nature of the harm they inflict (damage to U.S. government computers). We will defer our consideration of the third factor until later, because it goes to the nature of the attack, and analyze only the first two factors now.

We have already analyzed the inherent ambiguity of the point of origin of this attack, that is, the difficulty we encounter in determining its point of origin. Here, the point of attack occurrence is not ambiguous in and of itself;

---

610 We are assuming effects triggered by bits and bytes can constitute acts of war. *See* Chapters 3 & 4.

we know the attack manifested its effects in the United States. The ambiguity here lies in the implications of that point of attack occurrence.

In the real-world, the occurrence of an act of war on Country A's territory is equivalent to a declaration of war by the nation responsible for that attack. As we saw earlier, this is because war has historically been about territory; the violation of one nation-state's territorial integrity by agents of another nation-state is a challenge to the victim state's ability to maintain external order, that is, to sustain its existence as an autonomous entity.[611]

In the real-world, then, the singular inference to be drawn from an attack originating in the territory of one nation-state and occurring inside the territory of another is war; real-world transborder attacks have been equated with warfare because only nation-states could (and did) launch such attacks.[612] If we were dealing with real-world attacks, therefore, the rational (if not exclusive) inference would be that the BIS attacks were acts of war launched by China.

But we are not in the real-world. We are in the cyberworld, where transborder attacks are not the exclusive province of nation-states. We consequently cannot infer from the mere fact that the attacks targeted computers on U.S. territory that they are equivalent to Hitler's invading Poland. Sport hackers in Beijing, whose goal was exploring U.S government computers, could have launched them. So could professional cybercriminals or a group of cyberterrorists.

In utilizing point of attack occurrence as a factor in attacker-attribution, we must modify the assumption that equates transborder attacks with war so that it incorporates a basic reality of the online environment: Anyone with access to the Internet and certain basic skills can launch transborder attacks on computer systems. U.S. government and civilian computers are now encompassed by an analogue of the Willie Sutton rule; they are attacked because they are attractive targets for cybercriminals, cyberterrorists and nation-states bent on cyberwarfare.[613] Because U.S. computers are attractive targets for all three categories of attackers, and because actors from any of the categories can launch transborder attacks, the fact that an externally launched attack occurs "in" the United States cannot sustain the conclusion

---

611 *See id.*

612 We are assuming effects triggered by bits and bytes can constitute acts of war. *See* Chapters 3 & 4.

613 Willie Sutton allegedly said he robbed banks "because that's where the money is." "Willie Sutton," Wikipedia, http://en.wikipedia.org/wiki/Willie_Sutton.

that the attack was an act of war on the part of the nation-state from whose territory it originated.

Point of attack occurrence can be used in the cyberwarfare attacker-attribution calculus, but its inferential weight will be less than for attacks in the real world. In the real world, point of attack occurrence is one of two interacting and essentially determinative factors in war attacker-attribution. (The other is point of attack origin.) Their coalescence in the real world can be enough to establish the fact of war and identify the state responsible for its initiation. This is a viable calculus because real-world acts of war emanate from and violate the integrity of nation-states' territory; the territorial locus is in each instance unambiguous.[614]

Because cyberspace erodes the significance of sovereign territorial boundaries, the fact that an attack crosses national borders is merely one circumstance that must be considered in determining responsibility for it. The same is true for point of attack occurrence.

Those charged with determining responsibility for a transborder computer attack must first consider whether its target was (a) the nation-state on whose territory the attack occurred or (b) a civilian entity which happens to be in that territory. In conducting this analysis, they should consider the extent to which the country on whose territory the attack occurred is the object of rivalry or hostility on the part of the state from whose territory the attack was launched. If the victim country is the object of such rancor, this may indicate that the antagonistic nation-state was responsible for the attack, or it may not. There are various reasons why an attack could originate from the antagonistic state but not amount to its initiating cyberwarfare. For instance, a third party—perhaps a third nation-state or terrorists—could be trying to exploit the tensions that exist between these state by launching an attack and gambling that the victim state will attribute it to the state from whose territory it came. Alternatively, the rancor between the two states might be irrelevant; an attack launched by cybercriminals could coincidentally originate from the antagonistic nation-state. The latter scenario is similar to what occurred with the Estonian attacks described in Chapter 1.

For cyberwarfare, determining attacker identity is often associated with establishing the nature of an attack. Those charged with carrying out the process of attacker-attribution will therefore have to determine if an attack

614 *See* Chapters 3 & 4.

is actually cyberwarfare before they can begin assigning blame to sovereign entities.[615]

Crime-terrorism

Point of attack occurrence is an integral component of attacker-attribution for crimes and acts of terrorism. Real-world investigations concentrate on the scene of the crime or terrorist event, on the place where the attack occurred. As we saw earlier, this investigative model is based on the assumption that the players in the attack dynamic (criminals/terrorists and victims) occupied shared real-space; this assumption derives from the inescapable fact that physical proximity is an essential prerequisite for the commission of real-world crime or terrorism.

The point—the place—of attack occurrence consequently plays a central role in the investigation of these real-world events. It is the most likely source of physical evidence and eyewitness testimony that can be used to identify an attacker and link him to the crime/act of terrorism. The larger spatial context in which the immediate crime scene resides provides a potential source of further testimony and data that can become the basis of inferential linkages between victim and attacker. And the place where an attack occurs can itself become a source of inference as to the likely identity of an attacker. If someone is murdered in a home with an armed alarm system, this suggests the attacker knew the victim; and if jewelry disappears from a locked safe in a jewelry store, this suggests the thief was an insider who had access to the safe's combination.[616]

Here, again, the importance of point of attack occurrence diminishes as attacks move online. A real-space attacker's gaining entry to a home that has an armed alarm system suggests the attacker knew the victim, but a cyberspace attacker's gaining entry to a home computer hooked to a cable modem does not. A hacker's transferring funds from online bank accounts might be, but very well may not be, an inside job. Although the bank presumably had measures in place that were intended to limit virtual access to the accounts, the compromise of those measures, unlike the compromise of the jewelry

---

615 *See* Chapter 4.

616 *See, e.g., Sex, Lies and the Doctor's Wife*, CBS News (June 6, 2006), http://www.cbsnews.com/stories/2005/11/09/48hours/main1028132.shtml. *See also* Ali Winston, *$13G "Inside Job" Jewelry Theft*, Jersey Journal (December 8, 2006), http://www.nj.com/news/jjournal/index.ssf?/base/news-3/116556105967210.xml&coll=3.

store safe in the example cited above, did not necessarily involve privileged physical access to the accounts or to "inside" information needed to access them.

Investigators can infer with a high degree of confidence that the compromise of the jewelry store safe came from (a) an employee/former employee who was given the combination as part of his employment or (b) someone with whom an employee/former employee shared that information. The physical constraints that govern action in the real world make it eminently reasonable to draw certain inferences from the place where an attack occurred; the absence of those constraints make it problematic, if not impossible, to predicate similar inferences on the place where a virtual attack occurred. As we saw in Chapter 3, cyberspace eliminates the influence of the spatial dimensions and physical forces that constrain action in the real world and, in so doing, erodes the significance of place in attacker-attribution.

The point of attack occurrence will still play some role in attacker-attribution for online crimes and acts of terrorism because it is literally the place where an attack occurred. More precisely, it is the place where the virtual attack was consummated. Real-world attacks are initiated and consummated in a single physical place, which then becomes the crime scene. As we saw earlier, the use of cyberspace breaks the crime scene into shards. The signals that will culminate in an attack originate at one or more physical places, depending on whether the attack comes from a single source or coordinated sources; once launched, the signals will travel divers distances along varying paths before they reach their target and consummate the attack. Here, the place where the attack actually occurs—where the harm is inflicted on the victim—is part of a larger crime scene. Like a real-world crime scene, it will contain evidence that can be used in an attempt to track the person(s) responsible for the attack. Unlike a real-world crime scene, however, it is not self-contained; the evidence found at this virtual crime scene is part of a sequence of digital evidence that is strewn around cyberspace and stored on the computer(s) that the perpetrator(s) used in the attack. Because the ultimate crime scene accounts for only part of the available evidence, its role in the inferential process of identifying the attacker is accordingly reduced.

Hybrid Crime/Terrorism

The role of point of attack occurrence in attacker-attribution for state-sponsored cybercrime and cyberterrorism is functionally indistinguishable from the role it plays in assigning individual responsibility for online crime and acts of terrorism. Here, too, it is simply part of the total crime scene: the point at which an attack is consummated. Digital evidence retrieved from the point of attack occurrence can be used in efforts to backtrack the attack to its source and can be the basis for an inference as to primary and secondary responsibility for an attack, once investigators determine that it was state sponsored. In making this determination, investigators should factor the analogue of the Willie Sutton rule into the calculus because, as we saw earlier, point of attack occurrence cannot itself sustain a finding of nation-state responsibility. And here, as with cyberwarfare and "mere" cyber-crime and cyberterrorism, determining the identity of the attacker will often be bound up with determining the nature of an attack.

This concludes our consideration of attack- and attacker-attribution. In the next two chapters, we take up what is no doubt the most important issue of all: how nation-states can respond appropriately and effectively to cyberattacks.

SIX

Response

Where We Are

Our enemies . . . do not recognize the artificial construct between law enforcement and national defense.[617]

THE LAST THREE CHAPTERS ANALYZED the unique characteristics of cyberattacks. These chapters demonstrated that cyberattacks of whatever type—crime, terrorism, or warfare—do not conform to the empirical assumptions that shaped the response strategies nation-states use to control chaos and maintain order.

This chapter and the next chapter deal with the process of responding to cyberattacks. As we saw in Chapter 2, nation-states must respond effectively to crime, terrorism, and warfare if they are to maintain the internal and external order necessary for their survival as sovereign entities. As we saw in the last several chapters, the migration of crime, terrorism, and warfare into cyberspace erodes the efficacy of the strategies states have relied upon to control the real-world manifestations of these threats.

In those chapters we analyzed the difficulties nation-states encounter in ascertaining the nature of an online attack (crime, terrorism, or warfare?) and the identity of the online attackers (criminals, terrorists, or nation-state?). In this chapter, we examine the partitioned response model that is essentially the norm in modern nation-states. As we shall see, this model divides official responsibility for responding to crime, terrorism, and warfare between two governmental institutions: civilian law enforcement and the military. While we focus almost exclusively on the laws implementing this model in the United States, comparable provisions, and a similar allocation of responsibility, are in effect in many other countries, as well.[618]

---

617 Gary Felicetti and John Luce, *The Posse Comitatus Act: Setting the Record Straight on 124 Years of Mischief and Misunderstanding Before Any More Damage Is Done*, 175 Military Law Review 87 (2003).

618 The partition between civilian law enforcement and the military is not as defined, or as rigid, in some countries as it is in the United States. *See*, e.g., Donald E. Schulz,

In the United States, response authority is scrupulously bifurcated between the military, which responds to external threats (acts of war), and civilian law enforcement, which responds to internal threats (crime and terrorism). Civilian law enforcement is made up of civilians who have been recruited and specially trained for the purpose of responding to crime and terrorism, just as the military is made up of former civilians who have been recruited and specially trained for the purpose of responding to external threats. "Pure" civilians have no role in either process, as we will see.

The chapter is divided into two sections: The first examines the military-law enforcement bifurcation, that is, the institutional division of responsibility for responding to (a) warfare (military) or (b) crime and terrorism (law enforcement). The second section examines the nonrole "pure" civilians have in the processes of responding to internal or external threats.

In Chapter 7, we consider the continuing viability of this approach. More precisely, we consider whether we must modify, or even abandon, the bifurcated response model if we are to improve nation-states' ability to respond to cyberattacks of whatever type.

Military–Law Enforcement Bifurcation

The sections below examine the compartmentalization of responsibility for dealing with internal and external threats. The first section examines the military's exclusion from civilian law enforcement; the second examines law enforcement's *de facto*, but perhaps eroding, exclusion from the process of responding to external threats.

Military Excluded from Law Enforcement

The United States' commitment to bifurcated response authority derives from English common law and the American colonists' experience with the

---

The United States and Latin America: Shaping an Elusive Future 37 (2000); Michael C. Desch, Civilian Control of the Military (Johns Hopkins University Press 1999). *But see* Daniella Ashkenazy, The Military in the Service of Society and Democracy: The Challenge of the Dual-Role Military 5 (1994) (in democratic countries, military is not responsible for domestic law enforcement except in "extreme circumstances," such as insurrection or the collapse of domestic public order).

occupying English military.[619] It has, as we shall see, been incorporated into a federal statute.

The Militia and the military

In England, the Crown's excessive use of martial law in the early seventeenth century prompted Parliament to issue the Bill of Rights of 1689, which among other things declared that a sovereign's using the military "to enforce [internal] order is not due process of law."[620] The Riot Act of 1714 later reinforced this prohibition on using the army for civilian law enforcement.[621]

These measures and the general public sentiment that resulted in their adoption are traceable to the ancient principle of posse Comitatus. The phrase translates as "power of the country,"[622] and the principle evolved from the Roman practice of letting civilians escort Roman proconsuls as they traveled through the Empire.[623] In the twelfth and thirteenth centuries, the evolved principle of posse Comitatus emerged as the premise upon which English law allocated responsibility for civilian law enforcement to local sheriffs and to a "pool of free men upon whom they relied for help."[624] It remained a basic tenet of British law for at least seven hundred years.[625]

The English colonists brought this mistrust of the military and preference for posse Comitatus with them when they came to the American continent.[626] And for many years, they followed English practice in enforcing

---

619 *See*, e.g., Nathan Canestaro, *Homeland Defense: Another Nail in the Coffin for Posse Comitatus*, 12 Washington University Journal of Law & Policy 99, 101–10 (2003).

620 *Id.* at 103.

621 *See id.* at 104.

622 *See* Black's Law Dictionary (8th ed., Thomson West 2004).

623 *See* Roger Blake Hohnsbeen, *Fourth Amendment and Posse Comitatus Restrictions on Military Involvement in Civil Law Enforcement*, 54 George Washington Law Review 404, 406 (1986).

624 *See id. See also* Canestaro, Homeland Defense, *supra* at 102.

625 *See*, e.g., Regina v. Secretary of State for the Home Department, Ex parte Northumbria Police Authority, [1989] Q.B. 26 (Court of Appeal):

> The posse comitatus was a civilian body, consisting . . . of all the able-bodied male inhabitants of the county. . . . [A]lthough the posse comitatus might be called out by a justice of the peace, it was generally done by the sheriff. This duty was given statutory recognition by the Sheriffs Act 1887, . . . which imposed . . . a fine on any who failed to respond to the sheriff's call. Although that . . . provision has now been repealed (see section 10(2) of the Criminal Law Act 1967), the sheriff's duties in keeping the peace . . .were preserved. . . .

626 *See* Canestaro, Homeland Defense, *supra* at 105.

civilian law; because there were very few British soldiers on American soil until 1763, the propriety of using the military for law enforcement did not arise until shortly before the Revolution.[627] By 1763, England had embarked upon an effort to exert more control over the colonies, which were becoming restive.[628] Over the next few years, the Crown sent thousands of British soldiers to the American colonies, where they soon took over the task of enforcing civilian law.[629] This outraged the colonists, who saw this as a blatant violation of the Bill of Rights and other provisions of English law.[630] Their outrage crystallized around the "Boston Massacre" of 1770, in which British soldiers fired into a crowd that was pelting them with rocks and snowballs; five colonists were killed and others were wounded.[631] The Boston Massacre became a symbol of British abuse of the military and influenced the drafting of the Declaration of Independence.[632]

The Declaration of Independence outlines a series of abuses by the British military and then asserts that the Crown's actions had "render[ed] the military independent of and superior to the Civil Powers" in violation of English law.[633] According to one scholar, the Declaration of Independence's "'repudiation of military intervention in domestic law enforcement,' which the founders viewed as an offense against civil liberties, became 'the bedrock of due process on which the American government was built.'"[634]

The concern with barring military involvement in civil law enforcement carried over to the drafting of the Constitution. As the Supreme Court explained in *Pepich v. Department of Defense*, 496 U.S. 334 (1990), delegates to the Constitutional Convention struggled with two "conflicting themes": "[T]here was a widespread fear that a national standing Army posed an intolerable threat to individual liberty and to the sovereignty of the separate States, while, on the other hand, there was a recognition of the danger of relying on inadequately trained soldiers as the primary means of providing

627 *See id.*

628 *See id.*

629 *See id.* at 105–106.

630 *See id.* at 106–107.

631 *See id.* at 107.

632 *See id.*

633 The Declaration of Independence para. 14 (U.S. 1776). *See also id.* at para. 13, 16, & 27.

634 Canestaro, Homeland Defense, *supra* at 108 (quoting David E. Engdahl, *Foundations for Military Intervention in the United States*, 7 University of Puget Sound Law Review 1, 7 (1983)).

for the common defense."[635] The solution the delegates to the Constitutional Convention arrived at essentially perpetuated those themes.

They ultimately accepted the need for a standing military force but were careful to ensure that the Constitution would keep the military under civilian control.[636] The Constitution therefore gives Congress, a purely civilian entity, the power to declare war, to "raise and support armies," and to "provide and maintain a navy."[637] As a way of limiting the autonomy of the military, it provides that "no appropriation of money" to support an army "shall be for a longer term than two years."[638] And, perhaps most significant, the Constitution makes the president, a civilian, commander-in-chief, to whom all the branches of the military are subordinate.[639]

The delegates to the Constitutional Convention also retained "the Militia," which was the country's "chief means of defense" until the Constitution was ratified.[640] The militia dates back to Anglo-Saxon times; it was the common law correlate of the posse Comitatus.[641] Essentially, all able-bodied adult English males were subject to being required to serve either in the posse Comitatus (to maintain internal order) or the militia (to repel external threats).[642] Because it was an "important institution" when the American colonies were settled, the colonists brought the militia with them,[643] and colonial militias played a "significant role" in the Revolutionary War.[644] After

---

635 Pepich v. Department of Defense, 249 U.S. 334, 341 (1990).

636 *See* Canestaro, Homeland Defense, *supra* at 109. *See also* Pepich v. Department of Defense, 496 U.S. 334, 341 (1990).

637 *See* U.S. Constitution article I § 8 clauses 11, 12 & 13.

638 *See* U.S. Constitution article I § 8 clause 12. According to one scholar, the U.S. Constitution institutionalizes the division of authority over the military that existed in England and in the American colonies in the mid-eighteenth century. *See* Samuel P. Huntington, The Soldier and the State 177 (Belknap Press of Harvard University Press 1957).

639 *See* U.S. Constitution article II § 1. According to Professor Huntington, the Commander in Chief clause of the Constitution is ambiguous: The clause creates the office of "Commander in Chief of the Army and Navy of the United States" but does not explain what, if any, functions this office entails. *See* Huntington, The Soldier and the State, *supra* at 178. As he notes, the functions could range from the broad power to conduct warfare to the much narrower power to command military forces in the field. *See id.*

640 *See* Joseph G. Sullivan, Note, *Who Controls the National Guard? Congress and Governors Stake Their Constitutional Claims*, 67 U. Det. L. Rev. 443, 447 (1990).

641 *See*, e.g., "Militia," Wikipedia, http://en.wikipedia.org/wiki/Militia#United_Kingdom.

642 *See id.*

643 *See id.*

644 *See* Sullivan, Note, *Who Controls the National Guard?*, *supra* at 447.

the War, the Articles of Confederation required that "every State . . . always keep up a well-regulated and disciplined militia, sufficiently armed and accoutered."[645]

By the time the Constitution was being drafted, it had become apparent militias were not adequate to ensure the new nation's defense.[646] The delegates to the Constitutional Convention therefore gave Congress the power to raise and maintain armies, but they did not abandon the militia. The Constitution gives Congress the power to call "forth the Militia to execute the Laws of the Union, suppress Insurrections and repel Invasions."[647] It also gives Congress the power to organize, arm and discipline "the Militia" and govern "such Part of them as may be employed in the Service of the United States, reserving to the States . . . the Appointment of the Officers, and the Authority of training the Militia."[648] The importance of the militia was reemphasized in 1791, when the Second Amendment was ratified.[649] Its primary purpose was to guarantee the right to bear arms in a militia acting under federal or state authority; a subsidiary purpose seems to have been to ensure the militia could serve as an alternative to a standing army.[650]

The drafters of the Constitution assumed that the security of the country would lie in the hands of a small standing army and a militia that could be

---

645 U.S. Articles of Confederation, article VI.

646 Alexander Hamilton, for example, argued that militias were inferior to professional armies:

> I expect we shall be told that the militia of the country . . . would be at all times equal to the national defense. This doctrine . . . had like to have lost us our independence. . . . The facts . . . from our own experience . . . too recent to permit us to be the dupes of such a suggestion. The steady operations of war against a regular and disciplined army can only be successfully conducted by a force of the same kind. . . . The American militia . . . have . . . erected eternal monuments to their fame; but . . . know that the liberty of their country could not have been established by their efforts alone. . . . War . . . is a science to be acquired and perfected by diligence, by perseverance, by time, and by practice.

The Federalist Papers 166 (James Madison & John Jay, eds., New American Library 1961). George Washington was also vehemently opposed to relying on militias, based on his experience with them during the Revolutionary War. *See* H. Richard Uviller and William G. Merkel, *The Second Amendment in Context: The Case of the Vanishing Predicate*, 76 Chi-Kent L. Rev. 403, 467 (2000).

647 U.S. Constitution article I § I clause 15.

648 U.S. Constitution article I § I clause 16.

649 *See* U.S. Constitution amendment II ("A well regulated Militia, being necessary to the security of a free State, the right of the people to keep and bear Arms, shall not be infringed"). *See also* Uviller & Merkel, *The Second Amendment in Context, supra* at 500–511.

650 *Id.* at 500–504.

called into the "service of the United States."[651] These military institutions would be controlled by civilian authority, which brings us to the military-law enforcement bifurcation. The Constitution neither outlaws nor restricts the use of the military in domestic law enforcement,[652] and for nearly a century, no statute addressed the issue.[653] To understand why the issue did not arise earlier, and why it was resolved the way it was, we need to review the evolution of the U.S. military and the decline of the militia.

Although the new Congress was authorized both to raise and support an army and to organize "the Militia," it at first did neither.[654] The United States inherited a small army from the Confederation, but it was quite inadequate to deal with threats from Native Americans and from the British in Canada.[655] After years of debate, Congress adopted the Uniform Militia Act of 1792, which created a national militia responsible for fending off threats of whatever type.[656] And while the inadvisability of relying on a militia became apparent in the War of 1812, the militia remained the nation's dominant source of military force until the Civil War.[657]

That is not to say there was no regular army. While its concern about the threat posed by a standing army led Congress to essentially disband the small force it had inherited from the Confederation, it soon realized the militia was not capable of dealing with the Native American tribes on the nation's frontiers.[658] In 1792, after tribes soundly defeated militia forces in several battles, Congress created an army unit called "the Legion of the United States."[659] Under the command of General Anthony Wayne, the Legion

---

651 *See*, e.g., *id.* at 512.

652 *See* Canestaro, Homeland Defense, *supra* at 109. In *The Federalist Papers*, Alexander Hamilton suggested the army could assist local magistrates under certain circumstances, such as suppressing "a small faction, or an occasional mob." The Federalist Papers, *supra* at 69.

653 *See* Canestaro, Homeland Defense, *supra* at 109–110. *See also* John R Longley. III, *Military Purpose Act*, 49 Arizona Law Review 717, 718 (2007).

654 *See* Perpich v. Department of Defense, 496 U.S. 334, 341 (1990).

655 *See*, e.g., Uviller & Merkel, *The Second Amendment in Context, supra* at 513.

656 *See id.* at 514–518.

657 *See id.* at 517–526. Among other things, during the War of 1812, militia members refused to obey orders and their incompetence is credited with allowing the British to invade the Capitol.

658 *See*, e.g., Fred J. Chiaventone, "Fallen Timbers," in The Historical Dictionary of the U.S. Army 165–166 (Jerold E. Brown, ed., Greenwood 2001).

659 *See id.*

won several strategic battles and ended the ongoing warfare with several tribes.[660] With the Native American threat resolved, Congress abolished the Legion in 1796 and replaced with a smaller force.[661] This force fought, far more effectively than the militias, in the War of 1812. To prevent repetition of the problems that arose in that war, in 1815 President Madison created a small regular army; the assumption was that the combined efforts of the professional soldiers and militia members would adequately protect the nation from external threats.[662]

When the Civil War began, President Lincoln initially relied on militia, summoning 75,000 militiamen who could serve for up to three months.[663] As the magnitude of the struggle facing the nation became apparent, Lincoln realized he needed to rely on an army, not a militia. In July of 1864, he therefore began creating a large, essentially modern army.[664] Like the militias, this army was composed of citizen volunteers; unlike militiamen, soldiers served under the command of U.S. Army officers, wore standard U.S. Army uniforms, and received Army pay.[665] They were also subject to Army discipline, which was much more stringent than that exercised in militias. After the North prevailed, it rapidly demobilized most of its citizen-soldiers, leaving a much smaller regular army to deal with Reconstruction in the South, emerging Native American threats in the West, and a threat from France in Mexico.[666] That army evolved over the next half-century into a modern, though still relatively modest, force.[667]

This brings us back to the militia. The Uniform Militia Act of 1792 was still in effect, but militias had fallen into disrepute, due to their poor performance in the War of 1812 and other exercises. This was of little significance until the

---

660 *See id.*

661 *See* Richard W. Stewart, The United States Army and the Forging of a Nation, 1175–1917 119 (U.S. Army 2005). The act creating the Legion stated it would be dissolved once the country was at peace "with the Indian tribes." *See* An Act for Making Further and More Effectual provision for the Protection of the Frontiers of the United States, 1 Stat. 241 (1792).

662 *See* Stewart, The United States Army and the Forging of a Nation, *supra* at 159–160.

663 *See id.* at 201.

664 *See id.* at 202.

665 *See*, e.g., Uviller and Merkel, *The Second Amendment in Contexte, supra* at 530.

666 *See id.* at 303–305.

667 *See id.* at 303–385. The U.S. Navy and U.S. Marines evolved on parallel tracks during this same approximate period; I focus only on the army to limit the scope of the descriptive material above. Its history suffices to illustrate how the law enforcement-military bifurcation developed.

last quarter of the nineteenth century, when the nation's Centennial and other factors resulted in renewed patriotism and a revived interest in volunteer soldiering.[668] To accommodate both, state legislators encouraged the creation of new "guard" or "national guard" units.[669] The units were so popular that in 1878 they resulted in the formation of the National Guard Association, which worked to obtain funding and recognition for the guard units in the various states.[670] As the century ended, Congress appropriated additional funds for the emerging National Guard units.[671]

The next step came in 1901, when President Roosevelt declared the country's militia law "'obsolete and worthless'" and called for legislation that would modernize the organization of the "National Guard of the several States."[672] Congress responded in 1903 by repealing the Uniform Militia Act and substituting the Dick Act.[673] The Dick Act was the first time Congress exercised its constitutional power to organize the militia; as such it heralded the end of the state-controlled militias.[674] It renamed the militia the "National Guard," authorized federal funds to equip and train Guard members, and authorized the use of federal troops to train members of the Guard.[675] In 1916, Congress federalized the National Guard by making Guard units "in

---

668 *See,* e.g., Uviller and Merkel, *The Second Amendment in Context, supra* at 532.

669 *See id.* at 532.

670 *See id..*

671 *See id.* at 532–534.

672 Pepich v. Department of Defense, 496 U.S. 334, 341 (1990).

673 *See* Act of January 21, 1903, 32 Stat. 775 (the Dick Act, which also repealed the 1791 Act).

674 *See,* e.g., Patrick Todd Mullins, Note, *The Militia Clauses, The National Guard, and Federalism: A Constitutional Tug of War,* 57 Geo. Wash. L. Rev, 328, 333 (1988).

675 *See id. See also* Act of January 21, 1903, § 3, 32 Stat. at 775. The Dick Act actually divided the militia into two categories: "the organized militia, which consists of the National Guard" and "the unorganized militia, which consists of the members of the militia who are not members of the National Guard." *See* 10 U.S. Code § 311(b) (current version of the Dick Act). Today, the organized militia also includes the Naval Militia. *See* 10 U.S. Code § 311(b). Only a few states have Naval Militias. *See* "Naval militia," Wikipedia, http://en.wikipedia.org/wiki/Naval_Militia.

A number of states have relied on Congress' recognizing the "unorganized militia" to create their own state guards, who essentially train themselves, serve without pay except when called to duty by the governor, and usually act as community servants instead of soldiers. *See,* e.g., Brian C. Brook, Note, *Federalizing the First Responders to Acts of Terrorism Via the Militia Clauses,* 54 Duke L.J. 999, 1017 (2005). *See also* 32 U.S. Code § 109 (states and the District of Columbia can "organize and maintain defense forces" in "addition to" their National Guard). To make things even more complicated, some states have a National Guard, a State Guard and an unorganized militia. *See,* e.g., Miss. Code Ann. § 33-5-1.

the service of the United States" part of the U.S. Army.[676] The adoption of this provision was a response to a 1912 opinion by the Attorney General Wickersham, in which he concluded that the militia clauses of the Constitution prohibited the Guard being used outside the territorial United States.[677]

In World War I, the President used this provision to draft Guard members into the Army; that "virtually destroyed the Guard" because "the statute did not provide for a restoration of their . . . status as members of the Guard when they were mustered out."[678] Congress addressed this problem in 1933 by amending the 1916 statute.[679] The amendments created two "'overlapping but distinct organizations'—the National Guard of the various States and the National Guard of the United States."[680] Since 1933, those who enlist in a State National Guard simultaneously enlist in the National Guard of the United States. "In the latter capacity they [become] a part of the Enlisted Reserve Corps of the Army, but unless . . . ordered to active duty in the Army, they [retain] their status as members of a separate State Guard unit."[681] A National Guard member ordered to active duty in the Army is "relieved of his . . . status in the State Guard for the entire period of federal service"; once the person is relieved from active duty in the Army, he regains his National Guard status.[682]

Essentially, then, the "military" that *could* participate in civilian law enforcement consists of the regular military organizations—the Army, Marines, Navy, and Air Force—and the National Guard, the modern militia. Before we consider the historical role the military has played in law enforcement, we should briefly survey the evolution of the U.S. Marines, Air Force, and Navy.

The Marine Corps' history is similar to that of the U.S. Army; the Corps originated with the Continental Marines organized during the Revolutionary War and disbanded afterward.[683] Congress revived the Marines in 1798

---

676 *See* National Defense Act, ch. 134, 39 Stat. 166 (1916).

677 *See* Perpich v. Department of Defense, 496 U.S. 334, 344–345 (1990). The Attorney General actually found that the Guard—as a militia—could only be used outside the United States in three situations: "to suppress insurrections, repel invasions, or to execute the laws of the United States." *Id.* at 344 note 13 (quoting 29 Op.Atty.Gen. 322,323–324 (1912)).

678 *See id.* at 345.

679 *Id.*

680 *Id.*

681 *Id.*

682 *Id.*

683 *See* "United States Marine Corps," Wikipedia, http://en.wikipedia.org/wiki/U.S._Marine_Corps.

because of concern about threats from other nations.[684] The Marine Corps existed but fell into something of a decline in the nineteenth century; it began to come into its own at the end of that century, and in the next, established itself as an essential component of the U.S. military.[685] The Air Force began in 1918 as the Air Service, a unit of the military forces the United States sent to Europe after entering World War I.[686] In 1920, it became part of the Army; it remained part of the Army until 1947, when the National Security Act made the Air Force a separate branch of the U.S. military.[687]

The Navy traces its origins to the Continental Navy created in 1775.[688] Congress was initially opposed to creating a navy because it might provoke retaliation from the British Navy.[689] It relented when it became apparent a navy was needed to ensure the flow of supplies to the Continental Army.[690] After the war, a financially strapped Congress abolished the Continental Navy and auctioned off its ships.[691] Although authorized to do so, the Articles of Confederation Congress did not create a navy in the seven years it existed.[692] The Constitution gives Congress the power to "provide and maintain a Navy,"[693] but it did not exercise that power until 1794, when it created the U.S. Navy.[694] Unlike the Army and Marines, the Navy did not languish during the nineteenth century; it played a major role in a number of

---

684 *See id.*

685 *See id.*

686 *See* "History of the United States Air Force," Wikipedia, http://en.wikipedia.org/wiki/History_of_the_United_States_Air_Force.

687 *See id.* There is an Air National Guard. It traces its origins to a New York Air National Guard unit federalized in 1916. *See* "Air National Guard," Wikipedia, http://en.wikipedia.org/wiki/Air_National_Guard. The current version of the Dick Act defines the Air National Guard as part of the "organized militia" encompassed by federal "laws relating to the militia." *See* 32 U.S. Code § 101(1).

688 *See* "Continental Navy," Wikipedia, http://en.wikipedia.org/wiki/Continental_Navy.

689 *See id.*

690 *See id.*

691 *See id.*

692 The Confederation Congress existed between March 1, 1781, and June 21, 1788. *See* "Articles of Confederation," Wikipedia, http://en.wikipedia.org/wiki/Articles_of_confederation. *See also* Articles of Confederation, Article IX, http://www.yale.edu/lawweb/avalon/artconf.htm
(Congress authorized "to build and equip a navy").

693 U.S. Constitution Article I § 8 clause 13.

694 *See* "United States Navy," Wikipedia, http://en.wikipedia.org/wiki/United_States_Navy. *See also* Act of March 27, 1794, ch. 12, 1 Stat. 350.

conflicts, and by the twentieth century, had become one of the top-ranked navies in the world.[695]

There is no naval counterpart to the National Guard,[696] even though the current version of the Dick Act defines the "organized militia" as including the "Naval Militia."[697] In 1891, Congress authorized the formation of state naval militias composed of civilian sailors (professional and recreational) and former Navy men.[698] In 1916, Congress adopted legislation that in effect federalized the Naval Militia by letting militia members transfer to the new Naval Reserve.[699] After World War I, Congress repealed the laws applying to Naval Militias and authorized the President to transfer militia members to the Naval Reserve.[700] Today only a handful of states have Naval Militias.[701]

Law Enforcement

In the early years of the Republic, militias played a significant role in law enforcement. The Judiciary Act of 1789 authorized federal marshals to "command all necessary assistance" in executing their duty, which was understood as authorizing them to summon the assistance of the military and state militias in enforcing federal law.[702] Three years later, the Militia Act of

---

695 *See* "United States Navy," Wikipedia, *supra.*

696 There is, as I noted above, an Air National Guard. *See* "Air National Guard," Wikipedia, *supra* note 71. There is no Marine National Guard.

697 *See* 10 U.S. Code § 311(b).

698 *See* Donald W. Mitchell, History of the American Navy: From 1883 through Pearl Harbor 32 (Knopf 1946). *See also* Jon B. Silvis, Naval Militia: A Historical Perspective, State Guard Association of the United States, Inc., http://www.sgaus.org/hist_nm.htm.

699 *See* Silvis, Naval Militia, *supra. See also* Naval Appropriations Act for Fiscal Year 1916, ch. 83, 38 Stat. 928 (1915). *See,* e.g., Fickett v. United States, 149 Ct. Cl. 697 (Ct. Cl. 1960).

700 *See* Silvis, Naval Militia, *supra.* The Naval Reserve was renamed the United States Navy Reserve in 2005. *See* "United States Navy Reserve," Wikipedia, http://en.wikipedia.org/wiki/United_States_Navy_Reserve.
The difference between the Reserve units—Army, Air Force, and Navy—and the Army and Air National Guards is that the Reserve forces are exclusively federal entities. As such, they are always under federal command. *See,* e.g., Jason A. Coats, *Base Closure and Realignment: Federal Control over the National Guard,* 75 U. Cin. L. Rev. 343, 344 (2006). Guard units, as we saw earlier, are under the control of state authorities unless called into federal service.

701 *See* "Naval Militia," Wikipedia, http://en.wikipedia.org/wiki/Naval_Militia.

702 *See* Canestaro, Homeland Defense, *supra* at 110.

1792—the second Militia Act adopted that year[703]—specifically authorized federal marshals and the president to summon the assistance of militias in law enforcement.[704] The implicit justification for these measures was the premise that members of the militia (and the military) were summoned as private citizens, not as soldiers.[705]

Two years later, President Washington invoked the provisions of the second Militia Act and took approximately 13,000 militiamen from four states into Pennsylvania to suppress the Whiskey Rebellion, an uprising by farmers refusing to pay a federal tax on distilled spirits.[706] It is the only time an American president has led forces—of whatever type—into the field.[707] While some believe Washington could have treated the Rebellion as an armed insurrection, that is, as a military matter, he chose to approach it as a law enforcement issue.[708] His choice triggered a set of procedural requirements that would not have applied had he treated the Rebellion as a military matter.[709] Some think Washington chose this more restrictive alternative deliberately, to show the public that the president would not abuse the

---

703 Congress actually adopted them in the same week: The Uniform Militia Act we examined in the previous section—the one that organized state militias—was adopted on May 8, 1792. *See* Act of May 8, 1792, ch. 33, 1 Stat. 271. The Militia Act, which allowed federal marshals to call upon state militias for assistance in enforcing the law, was adopted on May 1, 1792. *See* Act of May 2, 1792, ch. 28, 1 Stat. 264.

704 *See* Canestaro, Homeland Defense, *supra* at 110. *See also* Calling Forth Act of 1792, ch. 28, 1 Stat. 264, 264. The Militia Act had two clauses: The first allowed the president to summon state militias when the United States had been, or was in "imminent danger" of being, invaded by a foreign power or an Indian tribe. *See* Calling Forth Act of 1792 § 1, 1 Stat. at 264. The second clause allowed the president to summon the militias when the enforcement of the laws of the United States was being opposed or obstructed by "combinations too powerful to be suppressed by the ordinary course of judicial proceedings, or by the powers vested in the marshals." *See* Calling Forth Act of 1792 § 1, 1 Stat. at 264.

705 *See* Canestaro, Homeland Defense, *supra* at 110. The justification noted above was based on the Mansfield Doctrine, a principle Lord Mansfield articulated in response to the Crown's using the military to break up the Gordon Riots in 1780. *See id.* Under the Mansfield Doctrine, soldiers could legitimately be used in domestic law enforcement as long as they were acting as civilian members of the posse Comitatus, rather than as members of the military. *See id.* at 104–105.

706 *See* Christopher A. Abel, Note, Not Fit For Sea Duty: The Posse Comitatus Act, the United States Navy, and Federal Law Enforcement at Sea, 31 William & Mary Law Review 445, 451 n. 36 (1990).

707 Fredrick B. Wiener, *The Militia Clause of the Constitution*, 54 Harvard Law Review 181, 187–188 (1940).

708 *See*, e.g., Brian C. Brook, Note, *Federalizing the First Responders to Acts of Terrorism via the Militia Clauses*, 54 Duke Law Journal 999, 1018 (2005).

709 *See* Stephen I. Vladek, Note, *Emergency Power and the Militia Acts*, 114 Yale Law Journal 149, 160–161 (2004).

power of his office.[710] Whatever his reason were, before he summoned militia members Washington scrupulously complied with the Act's procedural requirements by having Supreme Court Justice Wilson certify that they were needed to "execute the laws" of the United States.[711]

Washington's use of the Militia Act was not an anomaly: In the years leading to the Civil War, federal marshals' use of army troops to enforce federal law "became commonplace," and in 1854, the U.S. Attorney General issued an opinion upholding the legality of the practice.[712] In that opinion, Attorney General Caleb Cushing found that a United States Marshal charged with enforcing federal law could, if "opposed in the execution of his duty, by unlawful combinations, . . . summon the entire able-bodied force of his precinct, as a posse comitatus." [713] According to Cushing, the Marshal's authority to summon assistance encompassed "not only bystanders and other citizens generally, but any and all organized armed force, whether militia of the State, or officers, soldiers, sailors, and marines of the United States." [714] The latter aspect of Cushing's opinion was patently inconsistent with the traditional distinction between the military and civilian law enforcement.[715]

That aspect of the opinion—which became known as the Cushing Doctrine—was used in enforcing the Fugitive Slave Acts but otherwise lay dormant until after the Civil War.[716] In 1868, Attorney General William Evarts extended the Cushing Doctrine to let local authorities in the post-Civil War South use the military to "suppress domestic violence."[717] Abuses resulting from the

---

710 *See id.*

711 *See id.*

712 Canestaro, Homeland Defense, *supra* at 110 (citing 6 Opinion of the Attorney General 466, 473 (1854)).

713 6 Opinion of the Attorney General 466, 466 (1854).

714 *Id.*

715 In 1860, Attorney General Jeremiah Black published an interpretation of the opinion that purported to resolve the apparent inconsistency: He found that soldiers who participated in a posse Comitatus did so not as members of the military but as private citizens who were acting "in strict subordination to the civil authority." *See* 9 Op.Att'y. Gen. 516, 522–23 (1860).

716 *See* Michael Noone, *Posse Comitatus: Preparing for the Hearings*, 4 Chicago Journal of International Law 193, 194–195 (2003).

717 *See id.* Congress also adopted a series of acts—the Reconstruction Acts—which gave federal military commanders in the former Confederacy the authority to substitute military commissions for civilian courts, essentially at their discretion. *See*, e.g., Anthony F. Renzo,

use of the military during Reconstruction eventually brought calls for a change.[718]

The Posse Comitatus Act

In 1878, a Representative from Kentucky proposed an amendment to an appropriations bill that would reject the Cushing Doctrine and restore the distinction between the military and civilian law enforcement.[719] After extensive debate, Congress enacted the measure, which became known as the Posse Comitatus Act.[720] A version of the Posse Comitatus Act has been in effect ever since.[721] In its current incarnation, the Act states that "[w]hoever, except in cases and under circumstances expressly authorized by the Constitution or Act of Congress, willfully uses any part of the Army or the Air

---

Making a Burlesque of the Constitution: Military Trials of Civilians in the War against Terrorism, 31 Vermont Law Review 447, 487–489 (2007).

718 *See*, e.g., Comment, *The Posse Comitatus Act Applied to the Prosecution of Civilians*, 53 University of Kansas Law 767, 771 (2005) (military has never before or since exercised the police function on the scale it did in the eleven former Confederate states between 1865 and 1877). *See also* Robert W. Coakley, The Role of Federal Military Forces in Domestic Disorders 1789–1878 268 (U.S. Army Center of Military History 1988).

719 *See* Noone, *Posse Comitatus, supra* at 195. *See also* Wrynn v. United States, 200 F. Supp. 457, 464 (D.C.N.Y. 1961).

720 Comment, *The Posse Comitatus Act, supra* at 772 (citing Army Appropriations Act, ch. 263, § 15, 20 Stat. 145, 152 (1878)). The measure became known as the Posse Comitatus Act because it prohibits using the military "as a posse comitatus or otherwise to execute the laws." *See* 18 U.S. Code § 1385.

For Senate debate on the measure, *see* Wrynn v. United States, 200 F. Supp. 457, 463–464 (D. N.Y. 1961).

Once enacted, the Posse Comitatus Act was almost never used. A search of cases decided between 1878 and 1945 yields only one decision in which the Act is even mentioned. *See* Ex parte Mason, 256 F. 384 (C.C.N.Y. 1882). In 1948, a federal court accurately described it as "obscure and all-but-forgotten." Chandler v. United States, 171 F.2d 921, 936 (1st Cir. 1948). *See also* Sean J. Kealy, *Re-examining the Posse Comitatus Act: Toward a Right to Civil Law Enforcement*, 21 Yale Law & Policy Review 383, 398 (2003) (quoting Chandler v. United States, 171 F.2d 921, 936 (1st Cir. 1948)). Its then-obscurity was no doubt because after Reconstruction, no efforts were made to incorporate military forces into law enforcement efforts.

A fall 2007 search of cases decided after 1945 yielded 172 citations. Although the invocation of the Posse Comitatus Act borders on the frivolous in some of these cases, the dramatic increase in its being invoked suggests that more efforts are being made to incorporate military forces into civilian law enforcement.

721 It has been slightly modified on several occasions. These modifications increased the fine for violating the Act, applied it to Alaska, and conformed its text to the style and terminology used in the modern version of Title 18 of the U.S. Code. *See* 18 U.S. Code § 1385 annotations.

Force as a posse comitatus or to execute the laws shall be fined . . . or imprisoned . . . or both."[722]

Although the current version of the Posse Comitatus Act specifically applies only to the Army and Air Force, Department of Defense regulations also extend its restrictions to the Navy and Marines.[723] It has not been applied to the Coast Guard because the Coast Guard has traditionally functioned more as a law enforcement agency than as a military entity.[724] And it has been "uniformly interpreted to apply to National Guard members only when they are in federal service and not when they are in service to their states."[725]

---

722 18 U.S.C. § 1835 (2000). It originally provided that "it shall not be lawful to employ any part of the Army of the United States, as a posse comitatus, or otherwise, for the purpose of executing the laws, except in such cases and under such circumstances as such employment of said force may be expressly authorized by the Constitution or by act of Congress." The Posse Comitatus Act, ch 263, § 15, 20 Stat 145, 152 (1878).

The reference to posse comitatus is not meant to limit the scope of the provision: "Senate debate indicated . . . that the section was not limited by the expression 'as a posse comitatus or otherwise' but was to operate as if the prohibition ran—simpliciter—against the use of the Army to execute the laws, without reference to whether employed as a posse comitatus or as a portion of the Army." Wrynn v. United States, 200 F. Supp. 457, 464–465 (D. N.Y. 1961). The reference was probably included to emphasize Congress' rejection of the Cushing Doctrine and other interpretations upholding the use of the military as a posse comitatus. *See id.* at 463.

723 *See* Comment, *The Posse Comitatus Act, supra* at 772–73 (citing U.S. Department of Defense, Directive No. 5525.5, DoD Cooperation with Civilian Law Enforcement Officials, encl. 4 at 4.3 (Jan. 15, 1986)). A federal statute requires the Secretary of Defense to establish regulations that ensure law enforcement activity "does not include or permit direct participation by a member of the Army, Navy, Air Force or Marine Corps." 10 U.S.C. § 375. Prior to the enactment of the Department of Defense regulations, the Fourth Circuit had held that the Act applies all branches of the armed services. *See* United States v. Walden, 490 F.2d 372, 375 (4th Cir. 1974).

724 *See,* e.g., Jackson v. State, 572 P.2d 87, 93 (Alaska 1977) ("the law enforcement role established for the Coast Guard by Congress indicates that Congress did not intend to make the Posse Comitatus Act applicable to the United States Coast Guard"). *See also* Comment, *The Posse Comitatus Act, supra* at 773; United States v. Chaparro-Almeida, 679 F.2d 423, 425–26 (5th Cir. 1982). Statutorily, the Coast Guard is simultaneously a law enforcement agency and a member of the U.S. armed forces. *See,* e.g., Phil DeCaro, *Safety Among Dragons: East Asia and Maritime Security*, 33 Transp. L.J. 227, 230 (2005–2006). *See also* 14 U.S. Code §§ 1, 89.

At least two circuits have held that the Posse Comitatus Act does not apply to the Navy when it is under the control of or supporting the Coast Guard. *See* United States v. Klimavicius-Viloria, 144 F.3d 1249, 1259 (9th Cir. 1998); United States v. Kahn, 35 F.3d 426, 432 (9th Cir. 1994); United States v. Mendoza-Cecelia, 963 F.2d 1467, 1477–78 (11th Cir. 1992).

725 Clark v. United States, 322 F.3d 1358, 1367 (Fed. Cir. 2003). *See also* United States v. Hutchings, 127 F.3d 1255, 1257–1258 (10th Cir. 1997); State v. Spicer, 2008 WL 948309 *2 (Del. Super. 2008). As the Supreme Court noted in *Perpich v. Department of Defense*, 496 U.S. 334, 348 (1990), members of the National Guard "must keep three hats in their closets-a civilian hat, a state militia hat, and an army hat-only one of which is worn at any particular time."

The restrictions the Act imposes were eroded in the latter part of the twentieth century. Courts narrowed its scope, and in the 1980s Congress exempted certain military actions from the Act as part of the war on drugs.[726] In 1981, for example, it passed the Military Cooperation with Law Enforcement Officials Act.[727] This Act was intended to outline the types of assistance

> the military can provide to civilian law enforcement agencies without running afoul of the Posse Comitatus Act.... The legislation attempted to "maximize the degree of cooperation between the military and civilian law enforcement" in dealing with drug trafficking and smuggling while "maintain[ing] the traditional balance of authority between civilians and the military."... The Act permits the Secretary of Defense to "make available any equipment..., base facility, or research facility of the Department or Defense to any Federal, State, or local civilian law enforcement official for law enforcement purposes."[728]

The Military Cooperation with Law Enforcement Officials Act consequently draws a clear, inviolate distinction between the military's providing civilian law enforcement with equipment and facilities and its becoming involved in the process of law enforcement. In *United States v. Johnson*, for example, the Fourth Circuit Court of Appeals held that while the Act authorized law enforcement's using "the *equipment* and *facilities* of the Armed Forces Institute of Pathology to perform blood tests for civilian law enforcement agencies, it did not authorize "military *personnel* to perform" these tests.[729] The Fourth Circuit held that the act of performing the tests is a purely law enforcement activity and, as such, must be conducted by civilians.[730]

---

726 *See*, e.g., Kealy, *Re-examining the Posse Comitatus Act supra* at 398. *See also* Laird v. Tatum, 408 U.S. 1, 19 (1972).

727 *See* 10 U.S. Code § 372.

728 United States v. Johnson, 410 F.3d 137, 147 (4th Cir. 1005) (quoting H.R.Rep. No. 97–71, pt. 2, at 3 (1981), reprinted in 1981 U.S.C.C.A.N. 1785, 1785, and 10 U.S.C. § 372(a)).

729 United States v. Johnson, 410 F.3d 137, 147 (4th Cir. 1005).

730 It also rejected the government's argument that "the performance of ordinary DUI blood tests by military personnel for civilian law enforcement" was authorized as the "permissible training of civilians." *See id.* The Fourth Circuit noted that in enacting the Military Cooperation with Law Enforcement Officials Act Congress emphasized that its authorization of the military's training law enforcement personnel "would not alter the traditional separation of the military from civilian law enforcement." *Id.* (quoting House of Representatives Report No. 97–71, pt. 2, at 3 (1981), reprinted in 1981 U.S.C.C.A.N. 1785, 1785).

After 9/11, some called for the repeal of the Posse Comitatus Act.[731] They wanted the military to play a larger role in homeland defense, but as the influence of 9/11 has waned, the general sentiment seems to be that the Act should neither be repealed nor further eroded.[732] And while the Posse Comitatus Act is the primary law prohibiting military participation in civilian law enforcement, other federal statutes and regulations also contribute to the bifurcation.[733]

Law Enforcement Excluded from Military Operations

The correlate aspect of the bifurcation—law enforcement exclusion from the conduct of military operations—derives from the law of war.[734] Under Article 48 of the Protocol Additional to the Geneva Conventions, warring countries must "at all times distinguish between the civilian population and combatants."[735] Article 50 of the Protocol defines a "civilian" as those who are not defined as a "combatant" by the Protocol and the Third Geneva Convention.[736]

---

731 *See* Kealy, Re-examining the Posse Comitatus Act, *supra* at 424. *See*, e.g., Stewart M. Powell, Bush Considers Changes to Posse Comitatus Act, Houston Chronicle (October 2, 2005). *See also* Longley, Military Purpose Act *supra* at 740.
Additional calls for the repeal of the Act came after the Hurricane Katrina disaster. Some politicians, including President Bush, argued that the military should be able to play a "broader role" in restoring order and conducting humanitarian operations after a natural disaster or, by extension, after a large-scale attack on the civilian populace. *See* Joshua M. Samek, *The Federal Response to Hurricane Katrina*, 61 University of Miami Law Review 441, 443 (2007).

732 *See*, e.g., Dan Bennett, Comment, *The Domestic Role of the Military in America: Why Modifying or Repealing the Posse Comitatus Act Would Be a Mistake*, 10 Lewis & Clark Law Review 935 (2006). *See also* Michael T. Cunningham, *The Military's Involvement in Law Enforcement: The Threat Is Not What You Think*, 26 Seattle University Law Review 699 (2003) (arguing that utilizing the military in domestic law enforcement would enhance the military's ability to "project effective, overwhelming force" in the interests of national defense). *But see* Longley, *Military Purpose Act supra* at 739–742 (arguing for repeal of the Posse Comitatus Act).

733 *See*, e.g., Adam Burton, *Fixing FISA for Long War: Regulating Warrantless Surveillance in the Age of Terrorism*, 4 Pierce Law Review 381, 389 (2006) (Foreign Intelligence Surveillance Act creates "a 'wall' of separation between agencies responsible for law enforcement and those responsible for military and foreign intelligence").

734 *See*, e.g., Hague Convention (IV): Laws and Customs of War on Land annex art. 1 (October 18, 1907), http://www.yale.edu/lawweb/avalon/lawofwar/hague04.htm. *See also* Chapter 3.

735 *See* Protocol Additional to the Geneva Conventions of 12 August 1949, and Relating to the Protection of Victims of International Armed Conflicts (Protocol I), art. 48, June 8, 1977, 1125 U.N.T.S. 3, http://www.unhchr.ch/html/menu3/b/93.htm.

736 *See* Protocol Additional to the Geneva Conventions of 12 August 1949, *supra* at Article 50(1).

The Protocol and the Third Geneva Convention both define "combatants" as members of the armed forces of a nation-state engaged in a military conflict with another state.[737] "Combatants" includes members of "all organized armed forces, groups and units which are under a command responsible to that [nation-state] for the conduct of its subordinates."[738] Only combatants are authorized to engage in military combat.[739]

Because civilian law enforcement personnel are not members of the armed forces of a nation-state under the definitions given above, they are "civilians."[740] And because they are civilians, they are barred from participating in military combat.

Unlike the exclusion of the military from civilian law enforcement, this aspect of the bifurcation has received little, if any, attention from Congress, the courts, or scholars. The neglect is presumably attributable to the fact that there has historically been no real interest in bringing civilian law enforcement into military endeavors.

That may be changing. In the post-9/11 United States, the exclusion of civilian law enforcement from the general process of responding to external threats has been eroding. No official effort has been exercised to recruit civilian law enforcement personnel to participate in military combat in Iraq or anywhere else, and such an eventuality seems unlikely. There has, however, been an *ad hoc* effort to involve civilian law enforcement at least tangentially in the general process of responding to external threats.

In the twenty-first century, external threats are far more nuanced than they were even a century ago. In Chapter 4, we saw that military conflict has come to encompass both "war" and "military operations other than war" (MOOTW). There has been no effort to involve civilian law enforcement personnel in either category of formal military conflict, but these categories do not comprise all of the external threats modern nation-states confront.

---

737 *See* Protocol Additional to the Geneva Conventions of 12 August 1949, *supra* at Article 43 (defining combatants as "[m]embers of the armed forces of a Party to a conflict"). *See also* Geneva Convention Relative to the Treatment of Prisoners of War, art. 4, August 12, 1949, 75 U.N.T.S. 135 (1950), http://www.yale.edu/lawweb/avalon/lawofwar/geneva03.htm.

738 *See* Protocol Additional to the Geneva Conventions of 12 August 1949, *supra* at Article 43(1).

739 *See* Protocol Additional to the Geneva Conventions of 12 August 1949, *supra* at Article 43.

740 Under the Protocol, a civilian "is any person who does not belong" to one of the categories of persons defined elsewhere in the Protocol and in the Third Geneva Convention, i.e., anyone not a member of the armed forces of a nation-state engaged in conflict with another nation-state. *See* Protocol Additional to the Geneva Conventions of 12 August 1949, *supra* at Article 50(1). The Protocol also states that "[i]n case of doubt whether a person is a civilian, that person shall be considered to be a civilian." *Id.*

The umbrella term that best captures the contemporary universe of external threats is "national security."[741]

Though the general concept of national security is as old as the nation-state, the phrase has taken on a narrower, far more specialized meaning in the last six decades. As one author notes, the term "national security" was "not in common use before World War II."[742] The modern conception of national security was established in the United States when President Truman signed the National Security Act of 1947 on July 26, 1947.[743] This conception was the product of a "new view of the world," one that emphasized the inherent instability and precariousness of the United States' position vis-à-vis other countries.[744]

Unlike the old generic notion of national security, which focused on countering imminent, catastrophic threats from a specific nation-state, this concept of national security is fluid and forward-thinking. It seeks to frustrate external threats before they can attain imminence and is therefore predicated on a much broader definition of "external threat."[745] The modern approach to national security expands the concept of external threat by postulating the interrelatedness of political, military, and economic developments occurring at any moment and in any place on the planet.[746] Because the goal is to neutralize burgeoning threats before they can achieve any level of imminence, it follows that the national security effort must include the collection of relevant information and the

---

741 The term denotes "the safety of the U.S. government, territory, and people from external threats." Gregory E. Maggs, *The Rehnquist Court's Noninterference with the Guardians of National Security*, 74 George Washington University Law Review 1122, 1122 (2006)

742 Kim Lane Scheppele, *Law in a Time of Emergency: States of Exception and the Temptations of 9/11*, 6 University of Pennsylvania Journal of Constitutional Law 1001, 1016 (2004).

743 *See* "National Security," Wikipedia, http://en.wikipedia.org/wiki/National_security. *See also* National Security Act of 1947, Public Law Number 235, 80 Cong., 61 Stat. 496 (July 26, 1947).

744 Scheppele, *Law in a Time of Emergency, supra* at 1016–1017.

745 *Id. See also* Daniel Yergin, Shattered Peace: The Origins of the Cold War and the National Security State 194–198 (Houghton Mifflin 1977). The U.S. Code defines "national security" as "the national defense and foreign relations of the United States." *See* 22 Code of Federal Regulations § 171.20(i) (2005) ("National Security means the national defense or foreign relations of the United States"). *See also* 10 U.S. Code § 801(16); 18 U.S. Code Appendix 3 § 1. For a similar British provision, *see* Security Service Act-United Kingdom Statute 1989 ch. 5 § 1(2).

746 *See* Yergin, Shattered Peace, *supra* at 196.

implementation of preventative measures designed to mitigate or abrogate potential threats.[747]

In the United States, as elsewhere, both aspects of the national security effort are and have for decades been carried out by civilian and military personnel, even though it in effect involves responding to external threats, a task that historically belonged solely to the military.[748] The National Security Act of 1947 institutionalized civilian participation in this response process by creating a division of labor in which "overt warfare" remains the exclusive responsibility of military personnel and "covert intelligence-gathering" is carried out by agencies supervised by the president.[749] The premise was that the military would continue to conduct traditional military operations while civilian-led intelligence agencies implement the new "covert war" effort.[750] This arrangement ensures civilian control over both aspects of the expanded process of responding to external threats: The military remains under the command of a civilian Commander-in-Chief; and the civilian-led intelligence

---

747 *See*, e.g., "Mission," Central Intelligence Agency, https://www.cia.gov/about-cia/cia-vision-mission-values/index.html:

> We are the nation's first line of defense. We accomplish what others cannot accomplish and go where others cannot go. We carry out our mission by:
>
> - Collecting information that reveals the plans, intentions and capabilities of our adversaries and provides the basis for decision and action.
> - Producing timely analysis that provides insight, warning and opportunity to the President and decisionmakers charged with protecting and advancing America's interests.
> - Conducting covert action at the direction of the President to preempt threats or achieve US policy objectives.

748 *See*, e.g., "Introduction to DIA," Defense Intelligence Agency, http://www.dia.mil/thisis-dia/intro/index.htm ("With over 11,000 military and civilian employees . . . DIA is a major producer and manager of foreign military intelligence").

749 Harold Hongju Koh, *The Spirit of the Laws*, 43 Harvard International Law Journal 23, 33 (2002).

750 Harold Hongju Koh, *Why the President (Almost) Always Wins in Foreign Affairs: Lessons of the Iran-Contra Affair*, 97 Yale Law Journal 1255, 1280–1281 (1988). *See also* Anne Joseph O'Connell, *The Architecture of Smart Intelligence: Structuring and Overseeing Agencies In the Post 9/11 World*, 94 California Law Review 1655, 1660–1661 (2006) (16 U.S. agencies engaged in intelligence-gathering: Defense Intelligence Agency, National Security Agency, National Geospatial-Intelligence Agency; National Reconnaissance Office, Army, Navy and Air Force intelligence units, Directorate of Information Analysis and Infrastructure Protection, Coast Guard Intelligence, Department of Energy Office of Intelligence, Federal Bureau of Investigation, Drug Enforcement Administration, Department of State's Bureau of Intelligence and Research, Department of Treasury Office of Terrorism and Financial Intelligence, and Central Intelligence Agency).

agencies operate under the supervision of a president who is advised by the National Security Council, another civilian entity.[751]

As we saw earlier, "covert warfare" has two components: intelligence-gathering and the implementation of preventative measures of various types. Both components are used to protect the United States from "the following actual or potential threats to its security by a foreign power or its agents: (i) An attack or other grave, hostile act; (ii) Sabotage, or international terrorism; or (iii) Clandestine intelligence activities, including commercial espionage."[752] Both are covert in that the United States disclaims, and makes every effort to conceal, its role in them.[753]

This bring us back to law enforcement: In Chapters 2 and 3, we saw that some of the activities we approach as crime—most notably, terrorism and espionage—can be ambiguous in nature even when they occur wholly in the real-world. When domestic actors such as Tim McVeigh carry out acts of terrorism, the acts clearly constitute threats to internal order.[754] When foreign actors carry out such acts, the nature of the threat changes: If the perpetrators were acting on behalf of a hostile nation-state, the acts of terrorism constitute a threat to external order because the hostile state is using them in an effort to undermine the sovereign integrity of the victim state.[755] Here, the real-world acts of terrorism represent a foray into the non-zero-sum conception of warfare we examined in Chapter 4.

The nature of the threat becomes more uncertain when the foreign actors act not on behalf of a hostile nation-state but on behalf of a rogue, stateless political group, such as Al-Qaeda. Here, the threat is external because it comes from noncitizens bent not on enriching themselves (crime) but on undermining the sovereign integrity of the victim state (war?).[756] The threat

---

751 *See* 50 U.S. Code § 402(a).

752 39 Code of Federal Regulations § 233.3 (c) (9) (2005).

753 *See* 50 U.S. Code § 413b(e) (covert action is activity "of the United States Government to influence political, economic, or military conditions abroad, where it is intended that the role of the United States Government will not be apparent or acknowledged" but does not include diplomatic or law enforcement activity). *See also* 50 U.S. Code § 413(f) (intelligence activities includes activities defined in § 413b(e), as well as "financial intelligence activities").

754 *See* Chapters 2 & 3. *See also* "Timothy McVeigh," Wikipedia, http://en.wikipedia.org/wiki/Timothy_McVeigh.

755 *See* Chapters 2 & 3. The same conclusion holds when citizens of the victim state carry out terrorist acts on behalf of another, hostile nation-state. The critical issue is not the nationality of the perpetrators, *per se*, but the motivations for, and sponsorship of, the terrorist acts.

756 If citizens of the victim state carried out the terrorist acts in this scenario, the threat would be internal, notwithstanding their allegiance to the stateless political group.

is not external in the traditional sense because it does not come from a competing state, but it is not really internal, either.[757]

As we saw in Chapter 4, we currently lump all three types of terrorism (domestic actor, foreign state-sponsored actor, and foreign non-state-sponsored actor) into a single category and treat it as a type of crime. As we also saw in Chapter 4, the analytical ambiguity that results when what is nominally "crime" is committed by foreign actors also arises when foreign actors—acting on behalf of a hostile state or on behalf of an external, stateless political group—engage in economic espionage. Here, again, we ignore the ambiguities and treat all types of economic espionage as mere crime.

My point is that as the threat definition quoted above demonstrates, the contemporary universe of external threats nation-states confront include crimes, the response to which is a civilian law enforcement function. That means law enforcement agencies "are playing an increasingly significant role in national security."[758] This has for the most part been an *ad hoc* process: Law enforcement officers carry out their obligation to respond to "crime" of whatever type, but in so doing, they may be at least partially responding to an actual or potential external threat.[759] Here, traditional law enforcement activity fortuitously advances the national security effort. There are also plans to institutionalize civilian law enforcement's contribution to the effort by formally integrating civilian law enforcement personnel into intelligence-gathering activities that are specifically designed to protect national security.[760]

The activities of the group known as Al-Qaeda in Iraq (Tanzim Qaidat al-Jihad fi Bilad al-Rafidayn) fall into this category. *See* "Al-Qaeda in Iraq," Wikipedia, http://en.wikipedia.org/wiki/Al-Qaeda_in_Iraq.

757 *See,* e.g., David Kretzmer, *Targeted Killing of Suspected Terrorists: Extra-Judicial Executions or Legitimate Means of Defence?*, 16 European Journal of International Law 171 (2005) (does this scenario constitute a war on terror "to be pursued according to the laws of armed conflict" or is it a "struggle against a particularly pernicious form of criminal activity that should be managed according to a law-enforcement model?").

758 Matthew Silverman, Comment, *National Security and the First Amendment: A Judicial Role in Maximizing Public Access to Information*, 78 Indiana Law Journal 1101, 1123 (2003).

759 *See* 39 Code of Federal Regulations § 233.3(c)(9) (2005). The law has for centuries allowed law enforcement officers to investigate not only completed crimes, but crimes the commission of which is still in the process of being prepared. *See,* e.g.,Wayne R. LaFave, Substantive Criminal Law §§ 11.1–11.5 & 12.1–12.4 (2d ed., Thomson West 2007).

760 U.S. Department of Defense, Strategy for Homeland Defense and Civil Support 4, 16 (U.S. Government Printing Office 2005). *See also* Eben Kaplan, Fusion Centers, Council on Foreign Relations, http://www.cfr.org/publication/12689/ (describing the creation of fusion centers that promote law enforcement intelligence-collection and sharing for the purposes of law enforcement and counterterrorism efforts).

Integrating law enforcement into the national security effort is probably inevitable, given the realities of the twenty-first century. As the Department of Defense's Strategy for Homeland Defense and Civil Support noted, "transnational terrorists have blurred the traditional distinction between national security and international law enforcement."[761] Others agree that "distinctions between 'foreign' and 'domestic' are archaic and counterproductive when addressing modern national security threats."[762]

The exclusion of civilian law enforcement from the process of responding to external threats is definitely eroding but not in any calculated, absolute sense. The erosion seems to be the product of an intuitive but generally unarticulated realization that things have changed, that the threat dynamics of the new century are different from those that have prevailed since the triumph of the nation-state. We return to this issue in the next chapter, where we analyze the possibility of conflating a variety of law enforcement and national security functions. My goal here is merely to point out that the barrier between law enforcement and military response functions has become permeable.

Civilian Exclusion from Attack Response

The sections below examine civilian exclusion from the law enforcement and military response processes. The first considers law enforcement; the second analyzes the military.

Civilians Excluded from Law Enforcement

Until the nineteenth century, civilians not only participated in law enforcement, they essentially *were* law enforcement.[763] As I have explained in more detail elsewhere, until Sir Robert Peel established the first professional police force in early nineteenth-century London, law enforcement was an *ad hoc* process that relied heavily on the efforts of citizens. Pre-nineteenth century

761 U.S. Department of Defense, Strategy for Homeland Defense and Civil Support, *supra* at 23.

762 Grant T. Harris, *The CIA Mandate and the War on Terror*, 23 Yale and Policy Review 529, 554 (2005).

763 *See* Susan W. Brenner, *Toward A Criminal Law for Cyberspace: Distributed Security*, 10 Boston University Journal of Science & Technology Law 1, 65–76 (2004). The following discussion of policing is taken from this source.

England and the American colonies had laws that required able-bodied men to participate in apprehending criminals. American civilians, anyway, were initially reluctant to surrender this function to armed professionals for fear of government overreaching. Their reluctance finally disappeared, and by the twentieth century, policing had become the sole province of law enforcement officers.

The process of professionalizing policing has been so successful that civilians no longer need to assume any responsibility for controlling or preventing crime. Those tasks are now monopolized by professional police forces organized in a hierarchical, quasi-military fashion. Civilians' only roles in this model of crime control and prevention are as sources of evidence, that is, as witnesses or victims.[764]

Civilian exclusion from law enforcement is so complete that civilian participation has been given a distinct, pejorative descriptor: vigilantism. Vigilantism is essentially a civilian's "taking the law into her own hands," that is, engaging in action that would be lawful if it were carried out by a law enforcement officer.[765] Because the vigilante is not a law enforcement officer, she will be prosecuted for her conduct if it violates an established criminal prohibition.[766]

"Pure" vigilantism almost always involves "volunteers": rogue actors who take it upon themselves to "assist" law enforcement by pursuing their own investigations and making their own "arrests."[767] Societies have long deemed "pure" vigilantism intolerable for several reasons. One is that that the activities of "pure" vigilantes create unacceptable risks of error in offender identification and apprehension.[768] Another reason for not tolerating "pure"

---

764 *See id.* The model of "community policing" that emerged at the end of the last century seeks to incorporate a level of civilian participation into the law enforcement process, but here, too, the civilians function almost exclusively as sources of information about actual or potential crimes. *See id.* Even when they take a rather more active role in crime control, civilian participants in community policing do not participate in the processes of investigating crime and apprehending perpetrators. *See id.*

765 One scholar characterizes vigilantism as "lawless law." Lawrence M. Friedman, Crime and Punishment in American History 172 (1993).

766 *See* Kelly D. Hine, *Vigilantism Revisited: An Economic Analysis of the Law of Extra-Judicial Self-Help or Why Can't Dick Shoot Henry for Stealing Jane's Truck?*, 47 American University Law Review 1221, 1227–28 (1998).

767 *See*, e.g., "Vigilante," Wikipedia, http://en.wikipedia.org/wiki/Vigilante.

768 *See*, e.g., Holt v. State, 9 Tex. App. 571, 1880 WL 9215 (Tex. App. 1880) (prisoner being held in jail was abducted and "foully assassinated" by a local vigilance committee). *See also* Luke Dittrich, *Tonight on Dateline This Man Will Die*, Esquire (September 5, 2007),

vigilantism is that it tends to undermine legal guarantees designed to safeguard civil liberties.[769] It also undermines respect for lawfully established authorities, including law enforcement and the judicial system. For these and other reasons, societies have rigorously, and successfully, discouraged "pure" vigilante efforts for the last century or so, in large part as a function of professionalizing law enforcement.

Our suppression of "pure" vigilantism will be successful only as long as law enforcement is perceived as effective in combating crime.[770] This is not a problem as far as real-world crime is concerned, at least not in most countries, but it is becoming a problem for cybercrime. As we saw earlier, cybercrime is a challenge for law enforcement because it differs in several critical respects from the real-world crime that shaped the current law enforcement model.[771] Law enforcement is losing its battle with sophisticated, transnational cybercrime,[772] and will continue to do so unless and until we can adapt our current law enforcement model to an increasingly online environment.[773]

While many citizens are unaware of this, others realize online law enforcement is failing. Some of those in the latter category have become "pure"

---

http://www.esquire.com/features/predator0907 (joint law enforcement-vigilante operation resulted in suicide of one suspect).

769 *See*, e.g., United States v. Jarrett, 338 F.3d 339, 341–343 (4th Cir. 2003) (online vigilante violated Fourth Amendment by illegally, remotely accessing contents of suspect's hard drive).

770 Although the *perception* that law enforcement is effectively combating crime necessarily encompasses the premise that law enforcement *actually* enjoys a level of success in this regard, it does not mean law enforcement must apprehend the perpetrator of every crime it is unable to prevent. Modern societies rely on a crime-control, not a crime-negation, strategy to maintain the baseline of internal order they require to survive and prosper. *See* Brenner, *Toward a Criminal Law for Cyberspace*, *supra* at 65–76. Crime-control strategies maintain that baseline of internal order by persuading citizens that the risks of apprehension are high enough that they dissuade all but a subset of the population from engaging in criminal activity. *See id.* This keeps crime at an acceptable level. *See id.* There can be a disconnect between the actual and perceived risks of perpetrator apprehension, but the disconnect will be irrelevant to the efficacy of the crime-control strategy as long as the perceived risk of apprehension is significant enough to act as a default crime-deterrent.

771 *See* Chapter 4.

772 *See*, e.g., McAfee, Virtual Criminology Report-Cybercrime: The Next Wave 31 (2007), http://www.mcafee.com/us/research/criminology_report/default.html (representative of the United Kingdom's Serious Organized Crime Agency quoted as saying, "I don't think cybercriminals have any real fear of law enforcement").

773 *See* Chapter 4.

online vigilantes: rogue actors whose goals are to frustrate online criminal activity and/or initiate the apprehension and prosecution of online perpetrators. The incidence of "pure" online vigilante activity will almost certainly increase unless we improve the efficacy of online law enforcement because "pure" vigilantism emerges when citizens perceive that crime control is ineffective.[774] The already-notable online vacuum encourages "pure" vigilantism, as do several other factors. One is the ease with which online vigilantes can affiliate with like-minded others; websites and e-mail let them share information and join in collaborative vigilante activity targeting online offenders. Another factor prompting online vigilantism is that it is a relatively low-risk activity: Because they have no reason to be in physical proximity with those they pursue, online vigilantes run little risk of physical violence from their prey; a vigilante can be in a different city, a different state, or a different country from those he targets. And because online vigilantes can conceal their identities and their locations, they are unlikely to be identified and prosecuted for any crimes they commit.

The eroding efficacy of our current model of law enforcement is compounding the difficulty of maintaining order in cyberspace: The model's increasing inefficacy in controlling online crime erodes societies' disparate abilities to discourage criminal activity in cyberspace; this not only undermines the perception that social order is being maintained "in" cyberspace, it also erodes the perception that societies are maintaining order in the real-world.

Criminal laws are designed to prevent citizens of a society from preying on each other.[775] The problem we confront is that while the enforcement of these laws in their real-world societal context continues to be efficacious enough to control traditional crime, the inefficacy with which criminal laws are being enforced in cyberspace bleeds into the real-world, where it undermines our faith in our government's ability to protect us. That, in turn, encourages "pure" vigilantism, which itself threatens societies' ability to maintain internal order; while vigilantes claim to be acting on behalf of the law, their conduct actually erodes the fabric and integrity of the law.

It seems, then, that we must continue to exclude civilians from law enforcement because doing otherwise would at least implicitly sanction vigilantism. And that is true as far as it goes: We cannot tolerate "pure" vigilantism in the real-world, in cyberspace, or in the intersection of the two. But "pure"

774 *See* Lawrence M. Friedman, Crime and Punishment in American History 158–168 NEW YORK: BASIC BOOKS, 1993).

775 *See id. See also* Brenner, *Toward a Criminal Law for Cyberspace*, *supra* at 65–76

vigilantism—vigilantes substituting for law enforcement officers—is not our only option. We could reinvent the past by instituting a limited revival of the long-discarded Anglo-American system in which civilians cooperated with (rather than replaced) law enforcement officers in investigating crimes and apprehending criminals.

One of the reasons law enforcement is struggling with cybercrime is a lack of resources and trained personnel. Agencies operating essentially on budgets that barely sufficed for real-world crime must now respond to real-world crime *plus* cybercrime. And cybercrime increases the complexity, as well as the quantity, of the crimes with which officers must deal; because cybercriminals exploit computer technology in more or less sophisticated ways, law enforcement investigators need special training and equipment, both of which must be continually updated.

The obvious solution to this problem is to increase law enforcement budgets to a level at which they can support the personnel, resources, and training necessary enhance the efficacy with which law enforcement responds to cybercrime. Unfortunately, this solution is inherently impracticable; the cost would be prohibitive in terms of what taxpayers in the United States and elsewhere would be willing and able to bear.

There are several reasons why taxpayers would not—and probably could not—fund the personnel and other resources needed to maintain an effective law enforcement response to cybercrime. One is the sheer magnitude of the problem: Law enforcement in the United States takes place primarily at the state and local level; there are consequently more than 17,500 state and local law enforcement agencies in the United States.[776] Bringing all these agencies up to speed in the battle against cybercrime and keeping them there would require (a) hiring and training an appropriate number of officers in each agency and (b) equipping each agency with some to-be-identified quantum of specialized computer hardware and software.

The initial costs would be staggering. New hires would be necessary because of the need to maintain current force levels to deal with real-world crime and because regular officers often have neither the interest nor the aptitudes needed to pursue cybercrime.[777] We can only speculate as to how

---

776 *See*, e.g., FBI Releases Its 2006 Crime Statistics, Federal Bureau of Investigation, http://www.fbi.gov/pressrel/pressrel07/cius092407.htm ("more than 17,500 city, county, college and university, state" and tribal law enforcement agencies in the United States).

777 *See* Marc D. Goodman, *Why the Police Don't Care about Computer Crime*, 10 Harvard Journal of Law & Technology 465, 479–483 (1997).

many officers would have to be hired but if it averaged, say, four officers per agency (a modest estimate, given the accelerating pervasiveness of cyberspace), seventy thousand officers would have to be hired for this purpose. The initial costs of bringing the agencies up to speed would therefore encompass salaries, benefits, and entry-level training for the new hires, as well as the purchase of the hardware and software they would need in their work.[778]

If we were dealing with real-world law enforcement, the initial costs would essentially be a one-time expense. While officers do continue to train in the use of weapons and other tactics, the weapons, other equipment, and police vehicles can all be used for years. Cybercrime, on the other hand, is an almost exponentially evolving arms race;[779] the varieties of computer hardware and software evolve at an amazing pace, a reality cybercriminals exploit to the detriment of law enforcement. Cybercriminals tend to have the latest technology, while law enforcement often lags behind. Optimally, cybercrime investigators should be equipped with and trained in the latest technology; their efficacy as investigators necessarily declines to the extent they do not have access to current technology and are not trained in its use. But providing them with this is not merely an expensive proposition, it is a recurring expensive proposition: Hardware, software, and skills all need to be renewed regularly or they become obsolete. It is conceivable, but exceedingly unlikely, that taxpayers could and would bear the expense involved in keeping law enforcement competitive with cybercriminals—this expense, again, being added to the continuing costs of funding law enforcement personnel to deal with real-world crime.

We could achieve the same objective—enhancing law enforcement efficacy against cybercrime—by different means if we used the approach noted above, that is, if we incorporated a level of civilian participation into law enforcement. Instead of increasing law enforcement resources as such, this approach uses civilian contributions to supplement law enforcement resources. The Anglo-American practice of incorporating such participation

---

778 For one agency's estimate of the costs entailed in hiring and equipping two computer forensics examiners, *see* Office of the Maryland State Prosecutor, Computer Forensics Laboratory: Long Term Strategic Plan (2007), http://www.dbm.maryland.gov/portal.

779 *See*, e.g., Christopher Rhoads, *Web Scammer Targets Senior U.S. Executives*, Wall Street Journal (November 9, 2007), http://online.wsj.com/public/article/SB119456922698387317-8NgwGNUFMUMRACalcmtkuwOGCXY_20071208.html?mod=tff_main_tff_top ("'We are in an electronic arms race,' says Shawn Henry, deputy assistant director of the FBI's cyber-crime division. 'Every time our technology catches up with the latest [malicious software], the bad guys come up with another way to get in'").

derived from the historically acknowledged need to supplement meager law enforcement resources. Of course, when the practice developed, officer needs were far more modest: manpower, weapons, and horses. The principle, though, remains the same: civilian participation can serve as an in-kind supplement to formal law enforcement resources.

We will assume for the purpose of analyzing the viability of this approach that corporate and/or individual civilians are able and willing to contribute to the law enforcement response to cybercrime. The difficulty, if any, of implementing this strategy will therefore lie in (a) identifying precisely *how* civilians would contribute to that endeavor and (b) resolving any legal obstacles to their doing so. We will defer the first issue for now, and address it in Chapter 7. Our concern at this point is with legal obstacles that might impede incorporating civilian participation into what has long been a purely sovereign function.

Logically, reestablishing civilian participation in law enforcement would require us to resolve two threshold legal issues. One is the vigilantism problem: How can we integrate civilian participation into law enforcement without sanctioning vigilantism and its attendant evils? The other issue is more straightforward: What, if any, statutory or other obstacles currently ban civilian participation in law enforcement? We examine both issues below.

Vigilantism

The vigilantism issue is concededly problematic. We will explore the specifics of this issue in more detail in Chapter 7, when we analyze the mechanics of integrating civilians into cyberconflict response processes. For the moment, it is sufficient to note that the strategy we are considering involves using civilians to *supplement*, rather than *replace*, law enforcement efforts. It therefore would not require us to legitimize "pure" vigilantism. The critical distinction between "pure" vigilantism and the strategy we explore in detail in Chapter 7 is that in the latter civilians operate under the supervision of authorized law enforcement officers.[780] Unlike "pure" vigilantes, these civilians would not initiate or control the course of investigations or the apprehension of suspects, which should eliminate the evils associated with

---

780 Federal law already authorizes this with regard to the execution of search warrants. *See* 18 U.S.C. § 3105 (2000) (private citizen may assist officer in executing a search warrant). *See, e.g.*, United States v. Schwimmer, 692 F. Supp. 119, 126–27 (E.D.N.Y. 1988) (execution of search warrant by "computer expert" acting under supervision of federal agent was proper).

"pure" vigilantism. And since the civilians remain subordinate to law enforcement officers, the public will perceive that crime control—efficacious crime control—is being implemented by law enforcement.[781]

Existing Law

Does existing law create any obstacles to the strategy posited above? There is a federal statute—the Anti-Pinkerton Act—which seems to prohibit such an effort, but probably does not. To understand why it probably is not an obstacle to the implementation of the strategy outlined above, we need to review a bit of history.

When the Civil War began, the federal government had no law enforcement officers of its own. Written long before professional policing was invented, the Constitution requires that Congress create and maintain "Armies" but not law enforcement agencies. As a result, when President Lincoln's life was threatened in 1861, federal officials had to turn to a private agency for help.[782] Allan Pinkerton, founder of what became Pinkerton's National Detective Agency, was hired to guard the president.[783] And after being asked to do so, Pinkerton also created a new "Secret Service" that handled intelligence for the Union Army.[784] For the rest of the Civil War, the North relied on Pinkerton detectives to provide security and collect intelligence because there was no other, official alternative.[785]

After the war, Pinkerton and his agents went back to providing security for businesses, which led to their involvement in "strike-breaking" for companies opposed to unionization.[786] The Pinkerton agency's antilabor

---

781 This negates the "perceived law enforcement vacuum" which, as we saw earlier, tends to encourage the rise of "pure" vigilantism.

782 *See* David Sklansky, *The Private Police*, 46 UCLA Law Review 1165, 1212 (1999). *See also* James Mackay, Allan Pinkerton: The First Private Eye 106–120 (Castle Books 1996).

783 *See* James Mackay, Allan Pinkerton, *supra* at 106–120.

784 *See id.* at 97–110.

785 *See id.* at 106–170. *See also* Gregory L. Bowman, *Transforming Installation Security: Where Do We Go from Here?*, 178 Military Law Review 50, 55 (2003).

786 *See* Sklansky, *The Private Police*, *supra* at 1213–1214. After the war, Congress created the United States Secret Service as a new federal investigative agency. *See* "History," United States Secret Service, http://www.secretservice.gov/history.shtml. This Secret Service was not the descendant of Pinkerton's Secret Service, either in function or in form. It was instead created to deal with a "large national counterfeit problem" that emerged after the Civil War. *See*, e.g., Stephen M. Rochford, *To Protect and Suppress, Why a Protective Function Privilege Is Bad for America*, 20 Whittier Law Review 987, 994 (1999). It was

activities, combined with an infamous riot in which both Pinkerton guards and strikers were killed, generated a great deal of concern among the public about the use of private security forces.[787] This concern, combined with lobbying by labor unions, resulted in Congress' adopting the Anti-Pinkerton Act of 1893.[788] The Anti-Pinkerton Act, the text of which has changed very little since it was adopted, states that "[a]n individual employed by the Pinkerton Detective Agency, or similar organization, may not be employed by the Government of the United States."[789]

While the Anti-Pinkerton Act seems to bar the federal government from hiring private citizens to participate in federal law enforcement activities, this may not be the case. The only court so far to interpret the substance of the Act held that an organization "is not 'similar' to the . . . Pinkerton Detective Agency unless it offers quasi-military armed forces for hire."[790] In *United States ex rel. Weinberger*, the then-Fifth Circuit held that the Anti-Pinkerton Act was meant to bar the federal government from hiring "armed guards" of the type who precipitated injury and death in nineteenth-century labor riots, not from retaining the services of companies (or individuals) who merely provide investigative services.[791]

The *Weinberger* court's holding is one reason why the Anti-Pinkerton Act is presumably not an impediment to implementing a civilian participation strat-egy of the type postulated above, at least not at the federal level. Because this strategy would encompass civilian participation in law enforcement endeavors, instead of in quasi-military activities, the civilians' contributions should fall within the "safe harbor" the *Weinberger* court carved out for investigative services.

The other reason why the Anti-Pinkerton Act does not seem to preclude implementation of a civilian participation strategy of the type postulated above derives from the language of the Act itself: The Anti-Pinkerton Act

---

not until later in the nineteenth century that the Secret Service took on the task of protecting the President, and then other federal officials. *See id.*

787 *See* Sklansky, *The Private Police*, *supra* at 1214–1216.

788 *See id.* Its adoption was also partially attributable to public concern about the fact that 25 Pinkerton agents served as guards during the 1889 presidential inauguration. Sklansky, *The Private Police*, *supra* at 1214 n. 297. Hostility toward Pinkerton and its strikebreaking activities was a factor in changing American attitudes toward the professionalization of policing. *See id.*

789 5 U.S.C. § 3108 (2000).

790 United States ex rel. Weinberger v. Equifax, Inc., 557 F.2d 456, 463 (5th Cir. 1977).

791 *See id.* at 462–463.

bars the federal government from "employing" those who are "employed" by the Pinkerton Detective Agency or similar organizations. This prohibition would not apply to the strategy postulated because it does not contemplate "employing" the civilians who would participate in it. "Employing" individuals denotes paying them for their efforts, and that would be impracticable here for the same reasons increasing law enforcement budgets is impracticable.[792] Because the strategy is predicated on *volunteer* civilian participation, the Anti-Pinkerton Act is, once again, inapposite.[793]

No statutory obstacles seem to exist at the state level because states do not appear to have analogs of the Anti-Pinkerton Act.[794] In the late nineteenth century, at least twenty-four states adopted laws that prohibited the use of private armed guards from another state.[795] The laws had little effect on the use of private security personnel for two reasons: They did not bar the use of plainclothes detectives, only uniformed guards; and a company that wanted to do business in one of these states could simply open a branch there, which meant its guards were domestic and therefore legal.[796] The laws had little effect and have apparently disappeared.[797]

The de facto exclusion of civilians from the law enforcement process seems to be more a product of custom or culture than law—a byproduct of the professionalization of policing that emerged in the nineteenth century and evolved in sophistication in the last century.

Civilians Excluded from Military Operations

Civilian participation in military endeavors falls into two categories. In the first category, civilians surrender their civilian status and become members of the armed forces; a civilian who joins the military is not only authorized,

---

792 Hiring civilians to supplement law enforcement efforts could be even more expensive than increasing law enforcement budgets because civilian consultants would probably cost more, per hour, than would law enforcement investigators.

793 The Anti-Pinkerton Act would also be inapposite to the extent that the strategy involved the participation of civilians who were *not* employed by the Pinkerton Detective Agency or similar organizations.

794 *See* Sklansky, *The Private Police*, *supra* at 1215 n. 296.

795 *See* Frank Morn, The Eye that Never Sleeps: A History of the Pinkerton National Detective Agency 107 (Indiana University Press 1982).

796 *See id.*

797 *See* Sklansky, *The Private Police*, *supra* at 1215 n. 296.

but required, to participate in combat and other military endeavors.[798] The second, more problematic category involves participation by civilians who remain civilians, that is, who have not officially joined the military.

The law of war prohibits civilians from participating in military combat when nation-states are at war.[799] The prohibition, which is meant to protect

---

798 *See* Armed Services § 61, Corpus Juris Secundum (one who joins the military assumes all the obligations and functions of a member of the military on active duty, including the duty to bear arms). *See, e.g.*, Petition of Green, 156 F.Supp. 174, 176 (S.D. Cal. 1974). *See also* 10 U.S.C. § 802(a) (2000); 32 C.F.R. §§ 1624.9 & 1627.1 (2002).

799 *See* Chapters 3 & 4. Until the mid-nineteenth century, there was a notable exception: state use of privateers. A privateer was a privately-owned ship–usually a merchant vessel–that was authorized by state-issued letters of marque to attack "enemy" ships on the high seas. *See, e.g.*, "Privateer," Wikipedia, http://en.wikipedia.org/wiki/Privateer. The former American colonies used privateers during the American Revolution, and the Constitution gives Congress the power to "grant letters of marquee," i.e., to license privateers. *See* U.S. Constitution article I § 8 clause 11. It denies that power to the states. *See* U.S. Constitution article I § 10 clause 1. In the War of 1812, the United States' offensive naval effort was conducted by more than five hundred privateers and twenty-two U.S. Navy ships. *See* Nicholas Parrillo, *The De-privatization of American Warfare: How the U.S. Government Used, Regulated, and Ultimately Abandoned Privateering in the Nineteenth Century*, 19 Yale J.L. & Human 1, 3-4 (2007).

Letters of marque were issued to two kinds of ships: heavily armed raiders, which sought out ships for capture; and merchant ships, which were also armed and which carried the letters in case they encountered an opportunistic victim. *See* Eugene Kontorovich, *The Piracy Analogy: Modern Universal Jurisdiction's Hollow Foundation*, 45 Harv. Int'l L.J. 183, 212 (2004). Letters of marque came with instructions as to how privateers should conduct themselves; the constant in these instructions was the limitation that only ships from a specific nation—usually an enemy of the nation issuing the letter of marque—could legitimately be seized. *See id.*

The first mention of letters of marque appears in a 1354 English statute. *See* J. Gregory Sidak, *The Quasi War Cases—and Their Relevance to Whether "Letters of Marque and Reprisal" Constrain Presidential War Powers*, 28 Harv. J. L. & Pub. Pol'y 465, 468 (2005). At that time, they were apparently not limited to the seizure of naval vessels. *See id.* A letter of marque originally *see*ms to have been something broader—a license a sovereign granted to one of its subjects that authorized him to exact "reprisal" from the subject of an enemy state for injury he suffered at the hands of the enemy state's army. *See id.* According to *Black's Law Dictionary*, the "law of marque" is a rule of reprisal that lets someone who has been wronged by another "but cannot obtain justice" take property belonging to the wrongdoer, as satisfaction of the injury done him. Black's Law Dictionary 643 (8th ed., Thomson West 2004). *See also* Alice Beardwood, Alien Merchants in England, 1350–1377— Their Legal and Economic Position 60 (Carnegie Corporation of New York 1931). In the fourteenth century, the law of marque was used to redress thefts and other crimes committed in England by merchants from other countries. *See id.*

The use of privateers was an accepted practice from the sixteenth century until it was outlawed by the Declaration of Paris in 1856. *See* "Privateer," Wikipedia, *supra*. *See also* Declaration of Paris, April 16, 1856, http://www.yale.edu/lawweb/avalon/lawofwar/decparis.htm. Privateers were used by both sides in the Civil War, because neither had executed the Declaration of Paris. *See, e.g.*, United States v. Steinmetz, 973 F.2d 212, 218–220 (3d Cir. 1992). The United States did not execute the Declaration, though it was in favor of abolishing privateering because of the constitutional provision giving

civilians from being targeted by the military forces of an adversary state, [800] is becoming increasingly difficult to implement due to the evolving complexity of modern warfare.[801] As one author explains, a "prohibition that may have been easy to apply with simple weapons systems operating at short range does not provide clear guidance about the legality of civilians providing essential services in support of a state's warfighting efforts."[802]

As warfare becomes more sophisticated, and remote warfare becomes more common, the distinction between civilian noncombatants and civilian combatants continues to erode.[803] "In order to save money and obtain sophisticated technical expertise, nation-states are increasingly using civilians to operate or maintain sophisticated equipment and otherwise support combat operations."[804] The integration of civilians into military efforts can create uncertainty as to whether someone is acting as a "civilian" (noncombatant) or as a military actor (combatant).[805]

As we have seen, under the law of war a "civilian" is someone who is not a member of a country's armed forces.[806] Ambiguity as to one's status is resolved by construing him or her as a civilian.[807] Civilians who are involved in military efforts fall into two classes: employees and contractors.[808] Civilian employees are "hired and supervised by the armed forces and have an employment relationship with them."[809] Civilian contractors "work independently or for a private company and have a contractual relationship with the armed forces."[810]

In the U.S. military, the role of civilian employees has been limited to providing combat support for real-world military operations; they therefore

---

Congress the power to issue letters of marque. *See, e.g.*, Note, Accession of the United States to the Declaration of Paris, 28 Am. L. Rev. 615 (1894). The fear was that executing the treaty would represent an ad hoc amendment of the Constitution.

800 *See* J. Ricou Heaton, Civilians at War: Reexamining the Status of Civilians Accompanying the Armed Forces, 57 Air Force Law Review 155, 157 (2005).

801 *See id.*

802 *Id.*

803 *See id.* at 157, 159–163.

804 *Id.* at 157. *See also id.* at 191–192.

805 *See id.* at 157, 159–163.

806 *Id.* at 173.

807 *See id.*

808 *See id.* at 184.

809 *Id.* at 174.

810 *Id.*

work in areas such as weapons system maintenance, logistics, and intelligence.[811] Civilian contractors, on the other hand, are "involved in almost every aspect of military activity."[812] They feed, equip, and house soldiers; maintain weapons; gather intelligence; and provide security.[813] Contractors employed by private military companies such as Blackwater[814] train military units and may accompany them into combat; they may also become "actively involved" in planning combat operations.[815]

As these examples illustrate, the roles contractors are assuming in real-world military operations can conflict with the law of war, which bars civilians from directly participating in military operations.[816] Although civilian employee participation in U.S. military endeavors so far generally comports with this requirement, contractor participation may not, depending on how one defines "direct" participation.[817]

The extent to which civilians of either type can participate in cyberwarfare is even more uncertain:

> The law of war provides limited guidance to help determine when computer network attack and exploitation [CNAE] actions are considered combat. No treaties specifically regulate CNAE . . . Those aspects of CNAE which cause physical damage can be treated like attacks with more conventional weapons, with the consequence that carrying out such attacks is limited to combatants. Other types of CNAE, particularly those involving attacks on networks to steal, destroy, or alter information within them, do not necessarily constitute direct participation in hostilities and are arguably open to lawful civilian participation.[818]

Given nation-states' accelerating commitments to online warfare,[819] the law of war will have to be modernized so it incorporates the new realities of cyberwarfare.

---

811 *Id.* at 184.

812 *Id.* at 186.

813 *Id.*

814 *See* Blackwater Worldwide, http://www.blackwaterusa.com/.

815 *See* J Heaton, Civilians at War, *supra* at 188–189.

816 *Id.* at 192–193. *See also* Chapters 3 & 4.

817 *See* Heaton, *Civilians at War, supra* at 190–193.

818 *Id.* at 194 (notes omitted).

819 *See* Chapter 3.

That process must include a reassessment of the role civilians can legitimately play in cyberwarfare. Their role must be reassessed because the rationale for excluding civilians from traditional combat operations either (a) does not apply at all to cyberwarfare or (b) applies in a less compelling fashion. The rationale for excluding civilians from combat is to protect them from retaliatory attacks by an opposing military force. As we have seen, cyberwarfare tends to erase the distinctions between civilian and military targets. Civilian infrastructure components will almost certainly become targets in cyberwarfare, and civilians are equally likely to become involved in responding to these attacks. We must recognize these emerging realities and adapt the law of war, along with other components of our law and our institutional structures, to them. We explore ways we can do this in the next chapter.

SEVEN

How Did We Get Here?

[T]here's a reflexive grasp for old responses.[820]

THE SYSTEMS WE RELY UPON to protect us from crime, terrorism, and war evolved in a world that nation-states dominated.[821] They have for centuries been the unchallenged autarchs of the physical world; they carve its land and peoples up into segments, each of which becomes the exclusive dominion of a particular nation-state.

If they are to survive, nation-states must prevent their dominions from falling into chaos; organized social life requires internal and external order.[822] Our law enforcement model evolved as a response to the need for internal order; it has given nation-states an effective way to keep their subjects' behavior within acceptable bounds by controlling the incidence of crime and/or terrorism.[823] Our war model evolved as a response to the corresponding need for external order; it has sought to discourage nation-state conflicts and to control the losses of life and property such conflicts inflict when they erupt.[824]

The accelerating migration of human activities into cyberspace erodes the efficacy of both models, a process that will certainly intensify and is almost certainly irreversible.[825] Though the models will presumably continue to be effective in controlling chaos in the real-world, they will become increasingly ineffective in controlling what I call "cyb3rchaos." Cyb3rchaos is an umbrella term encompassing the three categories of online activity that

---

820 John Carlin, *A Farewell to Arms*, Wired (May 1997), http://www.wired.com/wired/archive/5.05/netizen.html.

821 *See* Chapter 2.

822 *See* Chapter 3.

823 *See* Chapter 2.

824 *See* Chapter 3.

825 *See* Chapters 3-6.

threaten social order in the real and virtual worlds: cybercrime, cyberterrorism, and cyberwarfare. Its atypical spelling reflects the evolving culture of the online world; in the online argot known as Leet, numbers or characters replace letters, hence, "cyb3rchaos."[826]

Cyb3rchaos is an inevitable byproduct of the disconnect between physical space and social interaction. As territory becomes irrelevant in the conduct of human endeavors so, too, do the territorially based institutions that evolved to support and structure those endeavors. This is, as we have seen, increasingly true for the institutions currently charged with enforcing criminal law and waging warfare; it may eventually be true for other institutions, as well—perhaps even for the nation-state itself.

Our concern, though, is limited to the institutions responsible for maintaining order within and among nation-states. We may, at some point in the future, discard this distinction between internal and external order; it might become irrelevant if our default governing institution ceased to be the nation-state and became, say, a nonterritorially based corporate entity.[827] But because it is reasonable to assume nation-states will be the dominant governing institution for some time,[828] we will continue to assume that the task of maintaining order has two components: internal and external. We will also continue to assume, for the most part, that these two components

---

826 *See* "Leet," Wikipedia, http://en.wikipedia.org/wiki/Leet:

> Leet . . . is a written argot . . . which uses . . . alphanumerics to replace . . . letters. The term . . . is a degenerate form of the word 'elite', and the language it describes resembles a highly specialized form of electronic shorthand. . . . With the mass proliferation of Internet use . . . Leet has . . . become a part of Internet culture and slang.

827 *See* Chapter 4. The correlate needs for internal and external order may not disappear even if our default governing institution became non-territorially based corporate entities or some other, as yet unforeseen institutional configuration. Absent a fundamental change in human nature, governance systems will always need to control crime if they are to maintain internal order. *See* Susan W. Brenner, ***Toward A Criminal Law for Cyberspace: Distributed Security***, 10 Boston University Journal of Science & Technology Law 1 (2004). And as long as individuals "belong to" one of various governance systems, the potential for external conflict exists; the conflict may not take the form of armed combat, but the methodology involved is irrelevant. What matters is that the constituents of a competing governance entity engage in activity the purpose and effect of which is to disadvantage the constituents of the challenged entity in a manner significant for their survival and/or prosperity.

It consequently seems reasonable to assume that the distinction will persist unless and until we adopt a unified system of global government. Although crime would no doubt continue to be a problem in such a system, warfare should be irrelevant; because there would be no competing governance systems, there should be no external threats and therefore no need for specialized institutions charged with maintaining external order.

828 *See* Chapter 4.

will be jointly or severally assigned to two institutions: law enforcement and the military.

The challenges cyberspace creates for the operation of these institutions as currently configured are, as we have seen, complex, but they essentially fall into two categories:

- As we have seen, many of the challenges derive from the fact that activities in cyberspace defy the constraints of the physical world: Territory and identity become meaningless and often irrelevant in an online environment; because the institutions with which we are concerned evolved to deal with territorially-based activity, the rules and strategies they rely upon become increasingly inapposite.
- The more obscure challenges comprising the second category derive from the fact that activities in cyberspace tend to defy a constituent element of a world dominated by nation-states: In that world, states effectively monopolize power; they have a near-absolute monopoly on the more extreme varieties of real-world weaponry and a rather less-decisive monopoly on other artifacts of power, such as wealth, information and the ability to control the hearts, minds and behavior of their subjects. The efficacy of both order-sustaining models—law enforcement and war—is dependent on nation-states' ability to sustain their monopolization of the various artifacts of power and use them to maintain order within and among states. The migration of activities into cyberspace is eroding nation-states' ability to sustain that monopoly, a phenomenon the influence of which was implicit in our examination of the more evident challenges comprising the first category.

We have not specifically examined either the challenges composing the second category or the phenomenon responsible for them. We have not done so because we have been focusing on the efficacy (or inefficacy) of the extant approaches to maintaining internal and external order and those approaches, as I noted above, assume a nation-state monopoly on artifacts of power.

Because the monopoly is eroding, we must assess its ultimate viability: Is the erosion we see a limited event, which means the monopoly will continue, albeit in a weakened form? Or is the erosion, in effect, the beginning of the end? Is it the onset of a shift in the distribution of power, a movement away from state monopolization of the artifacts of power and toward their diffusion throughout a civilian constituency? If it is the latter, what happens to the models we rely on to maintain order—will new models evolve to replace them or will chaos rule?

The answers to these questions depend on two issues: One is why the nation-state monopolization of power arose and why it is eroding. If we understand why it is eroding, we should be able to estimate the eventual scope of the erosion. The other, related issue goes to the consequences of the erosion. We already see that a decline in state monopolization of power undermines the efficacy of the models nation-states use to maintain order within and among their domains. Because those models are embedded in our experience and that of our recent and more remote ancestors, we implicitly assume they are inevitable. They may not be. We must consider whether, if the erosion of state monopolization of power accelerates, it will become necessary to replace these models with new approaches—approaches that can accommodate the diffusion of power among various constituencies, state (perhaps) and non-state.

In the remainder of this chapter, we analyze why the nation-state monopolization of power arose and why it is eroding. In Chapter 8, we then undertake two related analyses: In the first, we assess whether—and how—we can adapt the order-sustaining models we rely upon so they function effectively in a world in which power diffuses into the commons. In the second, more difficult analysis, we consider new approaches that might replace the models we currently rely upon.

Power and the Nation-State

We reviewed the emergence of the nation-state in Chapter 2. There we saw that it primarily evolved as a way to maintain external order, that is, to fend off threats from those outside a particular human grouping.[829] What we did not examine in Chapter 2 were the forces that led to the creation of this relatively recent and rather unprecedented approach to maintaining order.

Rise of the Nation-State

The nation-state grew out of the chaos that prevailed in Western Europe after the fall of the Roman Empire and during the subsequent centuries

829 *See* Chapter 2. *See*, e.g., Martin van Creveld, The Rise and Decline of the State 2–12 (Cambridge University Press 1999).

known as the Middle Ages. As one author notes, during that era "war, rather than being waged on behalf of nonexistent nation-states," was "embedded in society."[830] It was then the idiosyncratic vocation of wealthy men who built their own fortifications and fielded their own armies; those who were neither wealthy nor affiliated with the wealthy were incidental, spectators to, or collateral damage of, battles waged by the wealthy.[831]

This began to change in the fourteenth century, with the introduction of gunpowder into Europe.[832] The use of cannon not only altered the way war was waged, it had a more important effect: The older fortifications built by the warrior class were vulnerable to cannon fire and so had to be replaced with larger, much more expensive fortresses.[833] Only the wealthiest nobles could build these elaborate new fortifications, and even they tended to find it necessary to raise funds from the burgeoning commercial class.[834] The warrior class began allowing their vassals to pay money instead of serving in their armies; that, in turn, resulted in the warrior class' increasing reliance on mercenaries to wage their wars.[835]

This state of affairs continued until after the Thirty Years' War, which ended in 1648.[836] It began as a religious conflict between Catholics and Protestants in the Holy Roman Empire but grew into a general war that engulfed much of Europe.[837] Many parts of Europe were devastated by plague,

---

830 *Id.* at 155.

831 *See id.* at 155–157.

Japan had a similar system in place until the nineteenth century, when Commodore Perry and the U.S. Navy forced Japan to end its self-imposed isolation and interact with the Western world. That rather quickly led to Japan's becoming a Western nation-state. *See*, e.g., Andrew Gordon, A Modern History of Japan: From Tokugawa Times to the Present 9–15, 48–53 (Oxford University Press 2003). *See also* "Japan," Wikipedia, http://en.wikipedia.org/wiki/Japan; "Tokugawa Shogunate," Wikipedia, http://en.wikipedia.org/wiki/Tokugawa_shogunate.

In the nineteenth century China, too, moved from being an empire to becoming a nation-state, again under the influence of the West. *See*, e.g., Franz Michael, China through the Ages: History of a Civilization 175–209 (Westview Press 1986).

832 van Creveld, The Rise and Decline of the State *supra* at 156–158.

833 *Id.* at 156–158. *See also* Frederic J. Baumgartner, From Spear to Flintlock: A History of War in Europe and the Middle East to the French Revolution 166–168, 219–223 (Praeger 1991).

834 *See* van Creveld, The Rise and Decline of the State, *supra* at 156–158.

835 *See id.* at 156–158.

836 *See*, e.g., "Thirty Years War," Wikipedia, http://en.wikipedia.org/wiki/Thirty_Years'_War.

837 *See*, e.g., *id.*

by famine, and/or by the depredations of mercenary armies.[838] The conflict finally ended with the Peace of Westphalia: the signing of two treaties in 1648.[839]

The Peace of Westphalia was meant to end the religious wars that had torn Europe apart, culturally as well as politically.[840] As one author noted, the wars devastated the area:

> When Philip II of Spain sent his great Armada against England in 1588, all Europe hoped that the outcome would be a decisive victory for Catholicism or Protestantism that would end the frenzy of wars.... By the end of the Thirty Years War the aching desire for security of life and property overpowered the concern with whether a ruler was deemed to have the right to speak for God. . . . In the Treaty of Westphalia . . . the statesmen of Europe accepted the formula *Cuius regio eius religio* as the basis for ending the war. This formula translat[es] into "[t]he prevailing religion is that of the ruler."...[841]

The architects of the Peace of Westphalia adopted the formula to eliminate the justification—the alleged mistreatment of minority Catholics or Protestants—countries had used for invading each other and starting new wars.[842]

The Peace of Westphalia established the principle that each state could choose its own religion without fearing outside intervention and, in so doing,

---

838 *See*, e.g., *id.*

839 *See*, e.g., *id.*

840 *See*, e.g., Kamari Maxine Clarke, *Internationalizing the Statecraft: Genocide, Religious Revivalism, and the Cultural Politics of International Criminal Law*, 28 Loyola of Los Angeles International and Comparative Law Review 279, 299 (2006). *See also* Mark C. Anderson, A *Tougher Row to Hoe: The European Union's Ascension as a Global Superpower Analyzed through the American Federal Experience*, 29 Syracuse Journal of International Law and Commerce 83, 99 (2001):

> [T]he Reformation brought ... a long series of conflicts in which religious passion played a leading role. Initially a war of faith . . . , the end of the of the Thirty Years War saw realpolitik considerations grow in importance.... The ... Peace of Westphalia ... was intended to resolve conflicts between Catholic and Protestant groups.

841 Gray L. Dorsey, *Nations and Civilizations from the Perspective of Jurisculture*, 1997 St. Louis-Warsaw Transatlantic Law Journal 121, 133.

842 *See*, e.g., Paul A. Clark, *Taking Self-Determination Seriously: When Can Cultural and Political Minorities Control Their Own Fate?*, 5 Chicago Journal of International Law 737, 737 (2005).

created the conceptual foundation of the modern nation-state.[843] It established fixed territorial boundaries for the nascent states involved in the Thirty Years' War and set the stage for the boundaries of the fully realized nation-states that would evolve later.[844]

It also introduced a new and radically different approach to governing: Until the Peace of Westphalia, Europe was philosophically united in the ***Respublica Christiana***, the Christian state, a religious-based system in which there was "no supreme authority within a territory" and "no sovereignty."[845] All Christians, regardless of where they lived, were part of the ***Respublica Christiana***.[846] Christianity was the thread that bound them together, and Christian law, natural law, was the universal law.[847] There were political offices, and those who held them exercised some authority, but there was no secular sovereignty. [848] The church was the overarching, all-encompassing authority, both spiritual and temporal.[849] Christians were part of a single moral organism in which each had his or her assigned station and purpose.[850] It was, as one author noted, an era of "dispersed authority."[851]

The Peace of Westphalia changed that by creating what is known as the Westphalian system.[852] This system, which derives from provisions in the treaties of Westphalia, is based on five explicit propositions:

1. Humankind is organized . . . into discrete territorial, political communities that are called nation-states (territoriality).
2. Within these blocks of territory, states . . . claim supreme and exclusive authority over, and allegiance from, their peoples (sovereignty).

---

843 *See*, e.g., Mark Weston Janis, *A Sampler of Religious Experiences in International Law*, 22 Mississippi College Law Review 233, 235 (2003).

844 *See*, e.g., "Thirty Years War," Wikipedia, *supra*.

845 Daniel Philpott, Revolutions in Sovereignty: How Ideas Shaped Modern International Relations 77 (Princeton University Press 2001).

846 *Id.* at 76–79.

847 *Id.* at 76–79.

848 *Id.*

849 *See*, e.g., Zack Kertcher and Ainat N. Margalit, *Challenges to Authority, Burdens of Legitimatization: The Printing Press and the Internet*, 8 Yale Journal of Law & Technology 1 (2005–2006).

850 *See* Philpott, Revolutions in Sovereignty, *supra* at 76–79. *See also* Walter Ullmann, The Growth of Papal Government in the Middle Ages 219 (Methuen 1962).

851 Philpott, Revolutions in Sovereignty, *supra* at 80.

852 *See*, e.g., David Held, A Globalizing World? Culture, Economics, Politics 133 (Routledge 2004).

3. Countries appear as autonomous containers of political, social and economic activity in that fixed borders separate the domestic sphere from the world outside (autonomy).
4. States dominate the global political landscape since they control access to territory and the economic, human, and natural resources therein (primacy).
5. States have to look after themselves—it's a self-help world (anarchy). One of the main functions of the state . . . is to ensure the security and well-being of its citizens, and to protect them from outside interference.[853]

There is also a sixth, implicit proposition: There is no authority higher than the state.[854]

The Peace of Westphalia dismantled the Christian state and replaced it with a congeries of discrete, autonomous, self-governing units, each of which vies with the others, at some level, for political and economic advantage. Because there is no higher authority, each unit—each state—is responsible for its own survival and for the survival of its citizens.

Monopolies

The state's monopoly on power is the culmination of a process that began millennia ago. To understand the origin and evolution of this trend, we need to consider the problem of order.

The need to maintain internal and external order is not unique to human society; all intelligent, social species confront this problem.[855] In the animal kingdom, as in the early human groupings, dominance hierarchies are used to maintain order within a collective and to organize the members of the collective to fend off external threats. In wolf packs, for example, the alpha male rules the pack, which has a clear hierarchical structure and an accompanying division of labor. The alpha male enforces internal etiquette

---

853 Held, A Globalizing World?, *supra* at 133. *See also* Montevideo Convention on the Rights and Duties of States articles 2–8, December 26, 1933, 49 Stat. 3097 (entered into force December 26, 1934).

854 *See*, e.g., *id.* at article 3 ("The political existence of the state is independent of recognition by the other states").

855 *See*, e.g., Susan W. Brenner, *Toward A Criminal Law for Cyberspace: Distributed Security*, 10 Boston University Journal of Science and Technology Law 1, 1–49 (2004). The discussion that follows is taken from this source.

among pack members and leads them on hunts; he also leads the pack when they defend themselves from attacks by other packs or launch their own attack on another wolf pack. The alpha wolf's dominance is based on power: on his physical strength and confidence. He can be challenged at any time and will eventually lose his dominance to a younger, stronger, and more assertive challenger. The pack structure, though, survives intact because it is based on a stable, clearly understood hierarchical structure.

Wolf packs, like early human groupings and like other animal collectives, are organized around two principles: power and rules. In wolf packs, power is based on the characteristics of a single individual who takes power by succeeding in physical combat; but naked power is not the only organizing principle, even among wolves. The alpha wolf achieves pack dominance by besting his predecessor in a physical confrontation and will sometimes use force or the threat of force to keep pack members in line; he does not, though, have to rely only on the use of force. As I noted earlier, wolf packs have a clear hierarchical structure that is created and sustained by rules; young wolves learn these rules from older pack members and abide by them, unless and until they decide to challenge the alpha male for pack leadership.

The earliest human groupings—bands and then tribes—were organized in an almost identical fashion. The leadership of the group might be based purely on physical dominance or it might be hereditary—the institutionalized product of an alpha human male's triumphing in an ancient physical contest. Like wolf packs, these human groups were hierarchically organized and relied upon a simple set of well-understood rules to sustain the hierarchy and its attendant division of labor among group members. The leader, supported by others who shared to some degree in his power, led his band or tribe in defending itself against attacks by other bands or tribes and in launching their own attacks on "outsiders." The leader, like the alpha wolf, would also use his power to sanction any members of the band or tribe who violated the rules that sustained its internal order and division of labor, and prohibited the group from "harming" each other in unacceptable ways.

This simple organizational structure sufficed as long as humans were hunter-gatherers, but it proved inadequate once they developed agriculture. As humans shifted to an agricultural lifestyle, populations increased enormously in size and density. That, in turn, had certain consequences for the way collectives were organized and governed: Agricultural populations were large, stable, and tended to congregate in a single area; this resulted in the rise of urban civilizations. Urban civilizations were far more complex than their predecessors; there was a much more intricate, much more extensive division of labor in these societies than in the hunter-gatherer systems. The increasing size and complexity of these urban collectives created new problems: The simple, informal rules and governance by personal dominance characteristic of the bands and tribes were inadequate. These collectives developed far more complex systems of rules—civil laws—that defined rights (e.g., ownership, kinship), obligations (e.g., citizenship obligations, spousal duties), and one's place in the collective, among other things. There were also rules—criminal laws—that prohibited members of the collective from engaging in certain, severely proscribed activities. These formal, institutionalized rules replaced the *ad hoc* understandings that had been the organizing principles in the bands and tribes.

As in the bands and tribes, rules were only part of the organizational equation. These by now much-evolved rules were designed to keep order within a collective and, as with the old bands and tribes, were successful to the extent the population accepted them as legitimate. Acceptance of the rules as legitimate was not, however, enough: Like all intelligent species humans can decide to be contumacious, that is, they can simply decide not to follow rules, for whatever reason. A human collective must therefore implement a way to enforce the rules it uses to maintain internal (and, to a lesser degree, external) order. The second-stage human collectives relied on an enforcement system that was only somewhat more evolved than the system used by bands and tribes. Like bands and tribes, these agriculturally based collectives were governed by an individual—a man—whose dominance derived from his ancestry (the institutionalized dominance that emerged in some bands and tribes) or from his seizing power from a predecessor. The rulers—the sovereigns—therefore achieved dominance by using force directly or by relying on what was, in effect, an institutionalized memory of a prior use of force. They also tended to rely on shows of physical force to enforce the civil and criminal rules they relied on to maintain order within their collectives; and they necessarily relied on the threat and use of physical force to fend off attacks from other collectives (and launch their own). Because their tenure as sovereign, as well as their physical survival and that of their subjects, depended on their ability to control and exert physical force, these rulers aggressively monopolized all of the artifacts of power—for example, wealth, weapons, personnel, territory—that were available to them.

This approach to human governance continued essentially unchanged until the Peace of Westphalia. The empires and principalities that succeeded the early agricultural collectives were merely somewhat more evolved dominance hierarchies. That is, they were organized around the principle that systemic power was centralized in, and under the absolute control of, a single, individual sovereign. This individual's claim to sovereignty was based on personal authority: his either having seized power or—what became more common as the centuries passed—having inherited it from his ancestors.[856] His authority derived from who he was, not from the office he held.[857] It also derived from the principle noted above, that is, his control of the available artifacts of power. This intrinsically ***ad hoc*** principle was the product of historical practice and basic reality: The sovereign held a monopoly on power within a social system because such a concentration of power was essential for the survival of the system and of those who comprised it.[858]

The "power" these and later sovereigns exercised was a more articulated version of the dominance predicates the leaders of the bands and tribes relied on to maintain order within and among their systems. Here, "power" is a person's—and, later, an entity's—ability to ensure that a group of individuals will obey commands, which will take the form of general rules and direct orders.[859] As we have seen, power primarily derives from two sources: One is the ability to use coercive physical force to enforce commands.[860] The other source is the ability to establish rules people accept as "legitimate" and therefore obey without being physically coerced to do so.[861]

The evolved dominance hierarchy systems outlined above tended to assume chaos was, if not the default state of mankind, a constant possibility (as, indeed, it often was).[862] Chaos is, as we have seen, inimical to the states of order humans require if they are to survive and prosper. It therefore followed that these personal sovereigns had to be able to suppress chaos—internal or external—whenever it might erupt.[863] The application of power was then—as

---

856 *See* Max Weber, The Theory of Social and Economic Organization 341–348 (A. M. Henderson, trans., Talcott Parsons, ed., Oxford University Press 1947).

857 *See id.*

858 *See*, e.g., Thomas Hobbes, Leviathan 122–124 (A. R. Waller, ed., Cambridge University Press 1904).

859 *See* Chapter 2. *See also* Weber, The Theory of Social and Economic Organization, *supra* at 152.

860 *See id.* at 152-156. *See also* Mariano-Florentino Cuellar, *The Mismatch between State Power and State Capacity in Transnational Law Enforcement*, 22 Berkeley International Law Journal 15, 16 (2004).

861 *See* Chapter 2. *See also* H.L.A. Hart, The Concept of Law 198–199 (Clarendon Press 1961); Weber, The Theory of Social and Economic Organization, *supra* at 152.

862 *See*, e.g., Hobbes, Leviathan, *supra* at 115–135.

863 *See id.*

now—the only way to suppress chaos, either prospectively (discourage chaos) or reactively (defeat chaos).[864] It therefore seemed to follow that the most efficient, most effective organizational model was one that centralized power in a single source: the sovereign.

It is difficult for us to parse out why that seemed—and seems—the most effective model because it is all we know. It is all we know because it is all there has been. And it is all there has been because it works: When Germany invaded Russia in 1941, the only way the Russians could defeat the chaos the invasion initiated and restore order to their social system was by using power—physical force—to repel the invaders and, along with the other Allied powers, destroy Germany's ability to threaten other countries in the future.

This scenario has played out over and over again in human history. It exemplifies the problem of maintaining external order: Members of one social system try to destroy another and use its resources to enrich themselves. Power—in the form of physical force—has been the tool used to initiate such assaults and to repel them (just as it is with wolf packs and other species). Challenges to external order have, until recently, been synonymous with war: a full-scale, all-out assault on the viability of another human social system. Since war has been ***the*** catastrophic challenge to the survival of a social system, it is not surprising that the war model has shaped our approach to social organization. The war model—whether among humans or wolves or any other warlike species—assumes a hierarchical command and control system that is used to concentrate force to attack another system or to repel such an attack.

Since warfare has always been a zero-sum, all-or-nothing undertaking, the correlate assumption has been that human systems should be organized in a fashion analogous to the command and control system we use to wage war. This gives them the ability to mobilize for warfare, if and when that becomes necessary. That assumption—along with the perceived utility of being able to use centralized power to discourage internal chaos—explains why human social systems from antiquity until today monopolize power in the central governing authority—the sovereign. As human social systems and technologies have evolved over the millennia, that organizational structure has been adapted so that it is also used to discharge other essential functions. To understand that process, we need to review the evolution of nation-states after the Peace of Westphalia.

---

864 *See* Chapter 2.

As nation-states evolved, they rather quickly began to monopolize what I am calling the artifacts of power. The first thing these early states did was to create their own military forces. Most began by recruiting former mercenaries whose employability declined after the Peace of Westphalia.[865] Their employability declined because war was becoming a national responsibility. Until the Peace of Westphalia, the sovereigns who ruled Europe hired mercenaries to fight their wars because war was a personal matter between two individual rulers and, as such, could be contracted out (much as modern litigants hire lawyers to resolve disputes). [866] The Westphalian treaties changed this by predicating sovereignty on territory, instead of on the personal power of an individual. War now became a conflict between two territorially defined states and, as such, was properly waged by citizens of those states; after the Peace of Westphalia, therefore, nation-states waged war with domestic forces, that is, with men recruited from their own territory.[867] Because nation-states were now responsible for the conduct of war, they created war ministries, which were responsible for recruiting, training, equipping, housing, and feeding a nation's troops.[868] This further promoted the institutionalization of a domestic, national military force as an aspect of the nation-state.

It is important to note that the early national military forces were not necessarily full-time military forces. Until well into the nineteenth century, European countries and the United States relied in varying degrees on civilian militias.[869] Two systems developed: In some nation-states, civilians were drafted to serve for a specific period of time, which could be months or years or, in a few instances, a lifetime; in others, including the United States, they were drafted only when needed (as in the War of 1812).[870] As military science and technology evolved, civilian militias became increasingly ineffective;[871]

865 *See*, e.g., van Creveld, The Rise and Decline of the State, *supra* at 156–162.

866 *See*, e.g., *id.* at 156–162.

867 *See*, e.g., *id.*

868 *See*, e.g., *id.*

869 *See*, e.g., Jerry Cooper, The Rise of the National Guard 1-22 (University of Nebraska Press 1997); Hew Strachan, European Armies and the Conduct of War 56–72 (Routledge, 1991).

870 *See*, e.g., Cooper, The Rise of the National Guard, *supra* at 1–22; Strachan, European Armies and the Conduct of War, *supra* at 56–60. *See also* Chapter 6.

871 *See*, e.g., Harold S. Herd, *A Re-examination of the Firearms Regulation Debate and Its Consequences*, 36 Washburn Law Journal 196, 213 (1997) (noting the "ineffective" performance of the U.S. militia in the War of 1812).

as a result, by the mid-nineteenth century, most countries, including the United States, were moving to a full-time, professional national military force.[872]

As national armies evolved in size and sophistication, the use of mercenaries declined and finally ended (until recently).[873] The decline in the use of mercenaries is attributable to two factors that exacerbated the effects of the shift to national armies: One is that private armies could not compete with the resources and manpower states could summon; the other factor is that states began to realize that private armies could be difficult to control and could therefore threaten their sovereignty.[874] By the end of the nineteenth century, nation-states had come to rely exclusively on national military forces; the martial arts had evolved into a profession and, as we saw in Chapter 3, gave rise to a complicated set of rules governing the conduct of warfare.

As we also saw in Chapter 3, nation-states somewhat later took over sole responsibility for maintaining internal order. By the nineteenth century, states in Europe and America were transferring response authority for maintaining internal order from an essentially ***ad hoc*** collaboration of civilian-professional personnel to a professional, quasi-military cadre of law enforcement officers. These professionals assumed exclusive responsibility for enforcing the criminal rules states use to maintain internal order. [875] This model spread around the world and by the mid-twentieth century had become default approach to maintaining internal order.

Because nation-states rely primarily on their military and law enforcement agencies to maintain external and internal order respectively, it is not surprising that they allow these two constituencies, and only these two constituencies, to exercise the state's monopoly on the use of physical force.[876] And, therefore, it is not surprising that both constituencies wear uniforms that set them apart from the rest of the domestic population—the civilians

---

872 *See* Jonathan Turley, *The Military Pocket Republic*, 97 Northwestern University Law Review 1, 32–36 (2003); Michael S. Neiburg, Warfare in World History 49–52 (Routledge 2001).

873 *See* Todd S. Milliard, *Overcoming Post-Colonial Myopia: A Call to Recognize and Regulate Private Military Companies*, 176 Military Law Review 1, 10 (2003). *See also* Chapter 4.

874 *See id.* at 10–11.

875 *See* Chapter 2.

876 *See*, e.g., Eric W. Orts, *War and the Business Corporation*, 35 Vanderbilt Journal of Transnational Law 549, 566 (2002); Adam Crawford, The Local Governance of Crime: Appeals to Community and Partnerships 16 (Oxford University Press 1999).

whom they are charged with protecting.[877] They become the visible, ubiquitous symbol of the state's power and authority.

Centralization

We know states replaced their traditional civilian-based military and law enforcement systems with professionalized military and police forces, but why? Why has the centralization and professionalization of response authority been seen as superior to the citizen-based model?

As we saw in Chapter 2, the professionalization of law enforcement was at least in part the product of urbanization, which at once increased the incidence of crime and reduced the likelihood that perpetrators would be identified and brought to justice. And as we saw above, the professionalization and nationalization of the military was a reaction to the ineffectiveness of the militias and the inconstancy and uncertain quality of private armies. Technology also played a role in the move away from civilian-based systems: The professionalization of policing and of warfare let nation-states take advantage of technological and other advances in the processes of controlling internal and external threats.[878] And professionalization allowed for specialization: Modern detectives are far more effective than their counterparts of several centuries ago; and military technology long ago passed the stage in which it was efficient, and effective, to rely on civilian conscripts to operate the increasingly sophisticated implements of war.

The impetus for professionalizing law enforcement and military personnel also derived from the advantages states realize from having a reliable, available cadre of professionals who can respond to whatever threats may arise, internally or externally. The effectiveness of this stable cadre of response professionals is, as noted above, enhanced by its including specialists who concentrate on particular problems.[879]

---

877 *See*, e.g., van Creveld, The Rise and Decline of the State, *supra* at 156–158.

878 *See* The Past as Prologue: The Importance of History to the Military Profession 1–22, 150–216 (Williamson Murray and Richard Hart Sinnreich, eds.) (Cambridge University Press 2006); The Role of Police in American Society 76–77, 120–121 (Bryan Vila & Cynthia Morris, eds.) (Greenwood Press 1999); Geoffrey Parker, The Military Revolution: Military Innovation and the Rise of the West, 1550–1800 14–28 (Cambridge University Press 1988).

879 *See*, e.g., Areas of Military Specialization, Royal Military College of Canada, http://www.rmc.ca/academic/grad/calendar/p10milspec_1_e.html; Neil Munro, *Cyber Warriors*, Government Executive (October 29, 2007), http://www.govexec.com/story_page.

The professionalization of law enforcement and the military is but one aspect of a trend that began with the creation of the nation-state and accelerated over the last century or so: the centralization of state power. We saw earlier that sovereigns of whatever type have historically monopolized the artifacts of power because this has been seen as the optimum way to ensure that a social system can use force to maintain order by suppressing threats, both external and internal. Monopolization, though, is not necessarily synonymous with centralization.

In pre-nation-state social systems, sovereigns monopolized the power within a system because the sovereign ***was*** the social system. Until the Peace of Westphalia, there was no separation between ruler and state. In this world "there was no government in our sense of the term, only people who served" the sovereign.[880] The available artifacts of power—land, subjects, "money, armies, ministers and princesses"—were personal assets that "belonged to rulers, and were freely passed from one to another" as they jockeyed for advantages of varying types.[881]

Conceptually, the sovereign monopolized all of the power in the system. As a practical matter, however, power was decentralized—distributed throughout the system.[882] One author, for example, described governance in fifteenth and sixteenth century France as follows:

> Not only were the customs, laws, and privileges of each . . . territory recognized, but the kings established . . . courts in. . . . Burgundy, Languedoc, Brittany . . . and elsewhere from which there was no appeal to any higher court at Paris. . . .
>
> [This] . . . was paralleled by the decentralization of administration. There was a governor in each of the . . . provinces . . . who . . . assumed . . . regalian powers. . . . [A]bout two-thirds of the provinces had representative assemblies that voted, . . . collected taxes and attended to other administrative matters. Beneath the provinces were the bailiwicks and seneschalries.

---

cfm?filepath=/dailyfed/1007/102907ol.htm; "Behavioral Analysis Unit," Wikipedia, http://en.wikipedia.org/wiki/Behavioral_Analysis_Unit.

880 van Creveld, The Rise and Decline of the State, *supra* at 173.

881 *Id.*

882 *See*, e.g., J. Russell Major, *The Limits of Absolutism in the "New Monarchies"* in The New Monarchies and Representative Assemblies 78–84 (Arthur J. Slavin, ed., D.C. Heath 1964) (decentralized configuration of power in Renaissance and feudal monarchies). *See also* Philpott, Revolutions in Sovereignty, *supra* at 80.

> These jurisdictions were ruled by the bailiff or seneschal and a host of lesser officials who, like the governors, were as apt to follow their own desires as to obey the directives of the king.
>
> Alongside the . . . royal official there existed the seigneury and the town. The seigneurs were still a power in the villages . . . [T]he peasant rarely came into contact with royal authority. . . . [T]owns were largely self-governing, with their own elected officials, independent systems of taxation, and militia. . . .
>
> Closely associated with the decentralization . . . was the confusion of boundaries . . . and jurisdictions . . . in every branch of the government. The sea provided the only clearly defined boundary of the . . . state. Elsewhere much land was in dispute. . . .
>
> More serious was the confusion about the boundaries of the administrative subdivisions of the kingdom. . . . [I]t was impossible for any magistrate to know exactly what territory he was to administer.[883]

The disconnect in these systems between the sovereign's monopolizing power and the decentralized implementation of power was in large part a function of the limited transportation and communication technologies then in existence. Courts, governors, bailiffs, seneschals, and other functionaries were charged with implementing aspects of the sovereign's unitary power in places that were, in effect, out of sight and out of mind. The sovereign—the French King, in this particular instance—could exercise a considerable amount of control over the functionaries who toiled in close proximity to him. He, or his immediate subordinates, could monitor their activities and, if necessary, either order improvements or replace an inept official.

That kind of supervision was simply not possible for the host of functionaries who toiled at a substantial distance from the palace or wherever where a sovereign spent most of his time. The only way to communicate with distant functionaries was by sending a messenger to deliver a written or oral communication; and the only way such a messenger could travel to an outlying province, town, or village was on horseback or, in some instances, by boat. Either method could take days, both for the outbound trip and the return trip.[884] Though that kind of communication was possible, and

---

883 *Id.* at 78–80.

884 *See, e.g.*, Jeffrey L. Singman, Daily Life in Medieval Europe 214–217 (Greenwood Press 1999); Dena Goodman, The Republic of Letters: A History of the French Enlightenment 140–145 (Cornell University Press 1994).

certainly occurred, it simply could not be conducted with the frequency necessary to centralize routine decision-making in the central authority: the individual sovereign.[885]

The decentralization and—to us—haphazardness of the organizational structures that implemented sovereign power in pre-Westphalian Europe were also the result of a very different world-view. As we saw earlier in this chapter, until the Peace of Westphalia sovereignty was tied to an individual who, coincidentally, controlled certain territory. Sovereignty and territory were not linked as they are today because sovereignty was then a purely personal commodity; a sovereign "owned" certain territory, along with the people and resources in it, in the same way a modern property holder "owns" his property and its attendant appurtenances. The property holder is in a literal sense "sovereign" over his property, but that sovereignty is subsumed in the greater, nation-state sovereignty that encompasses him and his property.

Pre-Westphalia sovereigns had an analogous position in the ***Respublica Christiana***—the great Christian state that subsumed all territory and all human agency.[886] Their sovereignty was not atomized in the way modern sovereignty is, which is the conceptual explanation for why pre-Westphalian sovereigns and their administrators did not invent modern bureaucracy: They saw no need to. Sovereigns monopolized system power, and so could use it to repel assaults from other sovereigns and, if necessary, resolve internal disorder. Aside from maintaining a baseline of order, the implementation of power was inconsequential; the details involved in implementing power could have no conceivable effect on the sovereignty of the ruler because that came from God, part of the divine order.[887]

---

885 This assumes, of course, that the organizational structures charged with implementing power in these systems aspired to the rational, professionalized administrative culture to which we are accustomed. That is almost certainly not true.
A study of the seventeenth-century English civil service found that it exhibited few, if any, of the characteristics we associate with modern bureaucracies: Officeholders were chosen for their political connections, not their qualifications, and tended, as a result to be effectively immune from discipline or dismissal. Many officeholders took positions because they were perceived to offer opportunities for self-aggrandizement; because there was no distinction between personal and public interests, office holders used their positions to pursue their own interests. There was no clear, hierarchical organization and no clear lines of authority, though everyone was generally assumed to be loyal to the king. Overall, the system was disorganized and inefficient, in modern terms. *See* G. E. Aylmer, The King's Servants: The Civil Service of Charles I, 1625–1642 (Columbia University Press 1961).

886 Philpott, Revolutions in Sovereignty, *supra* at 76–79.

887 *See id.*

The Peace of Westphalia changed this by altering our conception of sovereignty and thereby triggered a movement toward the centralization of state power. As we saw earlier in this chapter, the Peace of Westphalia eliminated the ***Respublica Christiana*** and replaced it with an operationally random assortment of disconnected, potentially antagonistic nation-states. In this system, sovereignty is no longer personal; it becomes an abstraction, the joint product of territory and power. Nation-states are inherently garrison-states; each lives or dies on its own, according to how well it can create and exercise power.

It is consequently not surprising that nation-states do not content themselves, as did the pre-Westphalian sovereigns, with monopolizing power. In the three and a half centuries since the Peace of Westphalia came into existence, evolved nation-states have come to understand that they must not only hold power, they must also be able to exercise it effectively. This is one of the reasons why nation-states have over the last century or so made an increasing effort to centralize their power. Another reason is that societies have become much more complex, so the old approach of relying on essentially ***ad hoc*** systems to carry out essential functions such as education and transportation are not longer satisfactory. Yet another, related reason is the almost-exponentially evolving pace of technology, which, as we have seen, complicates the task of maintaining order. I also suspect state centralization of power is a phenomenon that tends to create its own momentum; that is, I suspect centralization breeds further centralization because it is be perceived as enhancing nation-state power (and the power of those charged with implementing centralization).

In the last section we saw how nation-states centralized their control of physical force by professionalizing law enforcement and the military—the agencies that, respectively, maintain internal order by suppressing crime and maintain external order by securing the state's territory from the depredations of other states. Physical force is not, of course, the only artifact of power states rely on to maintain order. As we saw earlier, they also rely on their power to make and enforce rules that are meant to channel their population's behavior in certain, desirable ways.[888]

These traditional artifacts of power—this reservation of rule-making authority and the concentration of power in professional response agencies—have over the last century or so become the source of new

888 *See* Chapter 2.

artifacts of power. As we saw, nation-states began the process of centralizing their power by concentrating the power to respond to internal and external threats in professional law enforcement and military agencies, respectively. That concentration of power having proved (and/or being perceived as) effective, nation-states moved to concentrate their power in other areas.

This effort did not take the form of increasing the overt concentrations of a nation-state's power to coerce civilian conduct; it was much more subtle. During the twentieth century, states increased their control over internal matters by using their rule-making power to bring various activities under the supervision—the regulation—of the nation-state. In this process, a state establishes rules civilian individuals and entities must abide by if they are to engage in certain activities (e.g., operating a nuclear power plant, flying a plane). This tactic in effect imports the law enforcement dynamic into the civil sector because it establishes a clear line of demarcation between what is "allowed" and what is not. Then, in a direct analogue of the approach it used to centralize response power, the state creates a series of agencies, each of which is charged with enforcing the rules governing the various activities it has chosen to regulate. The agencies may also have the power to create new rules and/or modify existing rules, modifications often being made to increase the scope of an agency's regulatory authority.

The result is the "vast . . . concentration of power" that is "the hallmark of the modern regulatory state."[889] The effects of this concentration of power are pervasive and profound:[890] The state controls the currency, taxation, education, transportation, economic affairs, health care, consumer affairs, utility and other services, property ownership and use, media, and, in some countries, religion.[891] This rule-driven, bureaucratically administered regulatory

---

889 Richard A. Epstein, *The Proper Scope of the Commerce Power*, 73 Virginia Law Review 1387, 1443 (1987). *See also* Harry N. Scheiber, *Redesigning the Architecture of Federalism–An American Tradition*, 14 Yale Law and Policy Review 227, 265 (1996) (noting that a "march toward centralization of both policy responsibilities and administrative power" occurred between 1945 and 1983).

890 *See*, e.g., Norman C. Bay, *Executive Power and the War on Terror*, 83 Denver University Law Review 335, 345–346 (2005):

> Modern American history has seen the rise of the regulatory state and . . . agencies with . . . lawmaking powers. In the last century, federal laws have created . . . a myriad of agencies. Congress has delegated rulemaking authority to those agencies. . . . [A]dministrative agencies reach into virtually every aspect of modern American life.

891 *See id. See also* van Creveld, The Rise and Decline of the State, *supra* at 156–170, 210–222, 224, 260.

system becomes another artifact of power or, perhaps more accurately, a constellation of artifacts of power.[892]

The process of maintaining external order, on the other hand, has not changed all that much since the Peace of Westphalia: In the last century or so, nation-states have formulated state-on-state rules in an effort to discourage warfare, but as we saw in Chapter 4, rules have so far played a minor role in maintaining order between nation-states. The use or threatened use of physical force has been much more important in this context. The primary artifacts of power used to maintain external order have consequently continued to be zero-sum, tangible commodities: control of territory, armaments, and warriors. Strategic alliances among states also serve as an artifact of power, and though they may not seem to be a zero-sum, tangible commodity, they are, for two reasons. One is that any strategic alliance among nation-states is ultimately predicated upon the states' respective control of the tangible artifacts of power cited above. The other reason is that while strategic alliances as such are not tangible artifacts, they are still zero-sum phenomena; an alliance either exists between two nation-states or it does not.

The concentration of nation-state power that has taken place over the last century and a half is analogous to the refinement of rudimentary dominance hierarchies that took place when humans shifted from being hunter-gathers to urban farmers. As we saw in the previous section, when humans adopted a settled, agricultural lifestyle, populations exploded and the process of governing necessarily became more complex. The increased complexity of the social system and the attendant complexity of governing such a system resulted in the development of more sophisticated dominance hierarchies that relied on intricate, formalized rules and the overt use of physical force to maintain order, both internal and external. The accelerating concentration of nation-state power over the last century and a half is arguably, at least, a response to a similar phenomenon: the increased size

---

892 Some cite evolved technology a related artifact of power:

> [T]he effective sovereignty of modern states [was] transformed by a series of technical changes which profoundly altered the state's capacity for surveillance and control. New forms of administration, new techniques for record-keeping, new technologies for the transmission and processing of . . . information gave the modern state powers to govern which were simply unavailable to more traditional states. It was one thing for the pope to assert his authority as the head of all Christendom, but something else for officers of the state to have more or less instantaneous access to the personal details, criminal records and credit status of each of its citizens.

Christopher Pierson, The Modern State 13 (Routledge 2004).

and complexity of the internal systems that comprise modern nation-states. If that is the impetus for progressively concentrating power in nation-states, then we should expect to see the trend continue because human social systems will, absent some countervailing tendency, almost certainly continue to increase in complexity (if not in size).

Eroding Monopolies

We have answered the first question posed earlier: Nation-states monopolize the artifacts of power to maintain internal and external order and thereby ensure that they survive and prosper. Now we must consider the second question: Why is the monopoly eroding?

The answer lies in the changing nature of the artifacts of power. As we saw above, they have historically been tangible, zero-sum commodities: Many of the artifacts nation-states use to maintain internal order—such as currency, armed state agents, and rules governing matters such as property ownership and use, utility and other services, economic affairs, transportation, health care, consumer affairs, and aspects of the media—have existed primarily, or exclusively, in tangible, physical form. And as we also saw above, the same is true of the artifacts of power states use to maintain external order.

The tangible, zero-sum nature of these artifacts has made it relatively easy for states to control access to and use of them. Most states, for example, make a sustained and generally successful effort to prevent the corruption of their currency and control ingress into and egress from their territory. Most states also make a concerted effort to control access to and use of the armaments and other instruments of physical force they rely on to maintain internal and external order. They have, as we saw above, maintained control over these and other artifacts of power by promulgating rules designed to implement the necessary level of control and by establishing agencies to enforce these rules.

States have enjoyed similar success in controlling the less tangible artifacts of power—such as education, religion, and broadcasting—by employing the same approach they use for tangible artifacts. States adopt and enforce rules that are designed to ensure that the conduct of these activities furthers the maintenance of order or, at the very least, does not undermine it. These rules control access to the activities by establishing requirements for participating in them and prescribing how they are to be conducted; the agencies charged with enforcing those rules then ensure that they operate to control access to and the conduct of these activities.

There are two reasons why nation-states have historically been successful in controlling access to and use of the various artifacts of power. One is that the process of gaining access to and using *any* artifact of power was necessarily visible or, in the current vernacular, transparent. Until recently, all our activities were carried out in the real, physical world; this meant, as we saw in Chapter 2, that our actions ***could*** be observed by others and ***would*** leave physical evidence of our presence in a specific physical place and even, perhaps, of what we had done while there. The unavoidable transparency of our conduct in the real, physical world consequently promoted nation-state control of the artifacts of power in two ways: It created disincentives for engaging in efforts to compromise nation-state control of a particular artifact; because conduct in the real-world is visible and therefore "public," those engaging such efforts risked social condemnation and/or sanctions.[893] The transparency of conduct in the physical world also created a trail of evidence those charged with protecting the artifacts could use to find, and sanction, those who sought to access them without being authorized to do so.[894]

The other reason nation-states have historically been successful in controlling access to and control of the artifacts of power is that there was in effect only one game in town. That is, there was only one arena of human activity; as noted above, until recently our activities were necessarily carried out in the physical world because there was no alternative. This meant that if a state could effectively control (a) movement into and out of its territory and (b) conduct within its territory, it could control access to and use of the artifacts of power. Nation-states used the approaches outlined earlier in this chapter to control activity within their territory: the rule-based systems we reviewed earlier operated to prevent the loss of control over individual and entity activity, and the law enforcement response authority we also reviewed

---

893 The psychology of these disincentives has both a positive and a negative aspect. *See* Chapter 2. As to the positive aspect, a state adopts rules that encourage its citizens to honor its control of the artifacts of power; most accept the rules as legitimate, and consequently abide by them. *See id.* The negative aspect targets those who are not sufficiently well-socialized or biddable to accept and obey the rules that encourage citizens to honor the state's control of its artifacts of power. *See id.* This aspect is therefore predicated on the notion of deterrence, which requires the use of criminal rules and criminal sanctions. *See id.*

894 This, as we saw in Chapters 2 and 3, simply means that it would not be particularly difficult for agents of the nation-state to identify and apprehend those who violated the rules designed to ensure the state's unchallenged control of the artifacts of power.

dealt with instances in which a loss of control occurred. The state's exercise of control over its artifacts of power has never been perfect, but a nation-state monopolizing power within the territory under its dominion has been able to exert enough control to maintain its monopoly over the artifacts it relies on for its existence.

Until very recently, every human being—and every artificial entity composed of human beings—was inevitably, inextricably situated in the territory of a particular nation-state. They either conformed to the nation-state's dictates or suffered the penalty for refusing to do so. It was a bounded, closed system and, as such, worked very well for the nation-states who were masters of their respective domains.

As we saw in earlier chapters, cyberspace changes this by essentially putting a virtual overlay over our physical reality. The overlay becomes an alternate, intangible forum for human endeavors and, in so doing, makes the heretofore fixed borders of territorial states permeable in a way they have never been. It also gives those who populate nation-states the ability to do something humans could never do before: Manufacture competing artifacts of power—purely virtual artifacts that can be used to erode or even overcome a state's efforts to control activity within its territory.

The ability to manufacture competing artifacts undermines the influence of another factor that facilitates nation-state control of the available artifacts of power: scarcity. Many artifacts of power—especially core artifacts such as wealth, territory, property, armaments, and warriors—are tangible, and therefore zero-sum, commodities. A given quantum of each exists in a state; if the state controls the available quantum of an artifact, it monopolizes that artifact and prevents others from using the artifact to challenge its dominance, its sovereignty. This is the principle of scarcity: Many of the tangible artifacts of power (such as territory) can be controlled but cannot be created; and although some can be created as well as controlled, the process of creating a tangible artifact of power (such as armaments or wealth) entails effort in the physical world that can itself be tracked and controlled. Finally, while nation-states also rely on intangible artifacts, most notably rules that prescribe certain behaviors and proscribe others, they exercise absolute control over these artifacts of power because the nation-state, alone, is the rule-maker and the rule-enforcer.[895]

895 *See* Chapter 2.

As we essentially saw in earlier chapters, cyberspace effectively eliminates the principle of scarcity: Unlike tangible artifacts of power, virtual artifacts are neither actually nor functionally zero-sum in nature. They can be created and shared around the world with comparatively little effort. Virtual weapons can be created and shared with impunity. Information control becomes challenging, if not impossible. Wealth can be generated by legal and illegal means and moved invisibly through cyber- and real-space. Communication occurs with a speed and anonymity unknown in the physical world. The availability of wealth, unfiltered information, and essentially unrestricted communication encourages the formation of often transient, anonymous alliances among state and/or non-state actors alike; it also facilitates the sharing of data that can be used to attack likely targets, both state and non-state, and thwart response efforts by the military or law enforcement.

So far, anyway, cyberspace does not give us the ability to fabricate the civil and criminal rules that serve as intangible artifacts of power. As we saw above and in Chapter 2, states, and only states, make and enforce the rules that govern the conduct of human beings in the physical world. But as we saw in Chapter 6, cyberspace does give us the circumscribed ability to evade state-proscribed, territorially applicable rules that are designed to maintain order in the physical world. I may have to abide by rules promulgated by the state in whose territory I reside when I conduct activity in the real-world but if I am adept, I can ignore them when I go online. Though we cannot counterfeit civil and criminal rules, we can elude them in the virtual world.

Nation-state monopolies on the various artifacts of power are eroding, and there is no reason to believe that erosion will do anything but accelerate; absent some catastrophic event, technology will continue to evolve in sophistication and pervasiveness. As technology evolves in sophistication and pervasiveness, the impact of the virtual overlay cyberspace has given us will only become more profound.[896] And that creates the problem we now need to address: How can our social systems maintain order in a world in which the limitations of physical reality and the influence of territorial boundaries become increasingly irrelevant? As we have seen, our conceptual and operational approaches to the problem of maintaining order of whatever type are inextricably bound up with the strictures of physical reality and territorial sovereignty.

---

896 *See* Susan W. Brenner, Law in an Era of Smart Technology (Oxford University Press 2007).

It is apparent that we need to develop new strategies to maintain order in the new world we confront—a world in which the physical and virtual realities interact. Logically, there are two ways we can go about doing this:

- We can modify the approaches we now use to maintain order so they accommodate an intertwined physical-virtual reality. We would maintain the current governance structure, in which nation-states monopolize power, but improve the processes of responding to internal and external threats so they become effective in neutralizing both.
- The other option is to implement a twenty-first Peace of Westphalia, that is, to devise a new governing configuration, one that is not predicated on, and limited by, territory.

As I noted earlier, we will examine both options in the next chapter.

EIGHT

Where Do We Go From Here?

No more was I part of a world . . . in which the civilian and military establishments each had its distinct role.[897]

AS WE HAVE SEEN, NATION-STATES use a bipartite response system to maintain order:[898] Professional law enforcement officers overtly maintain internal order by responding to violations of the criminal rules that control chaos within a state; and professional military personnel overtly maintain external order by responding to assaults from other nation-states. The activities of both implicitly maintain order by discouraging internal rule violations and external attacks.

As we have also seen, this system is not adept at dealing with threats vectored through cyberspace. We therefore must improve our ability to respond to these threats.

As we saw in Chapter 7, there are two ways we can go about doing this: One is to retain the default nation-state governance structure but modify our current threat response processes so they are effective in both the real and virtual worlds. The other option is to replace the threat response processes we presently rely upon with an entirely new approach to maintaining order; this might, or might not, involve replacing the nation-state with a governing configuration that is not predicated on, and therefore limited by, territory. We explore both options below.

Modifying the Present System

This is the simplest, and consequently most likely, option, at least in the short term. We may at some point move beyond territorially based nation-state

897 General Rupert Smith, The Utility of Force xiii (Alfred A. Knopf 2007).

898 In this chapter, I generally use phrases such as "response authority" or "response process" or "response effort" in a generic sense, i.e., to encompass both attribution and response.

governance into systems that resolve the problem of maintaining order in new and very different ways. Later in this chapter, we speculate a bit about what post-nation-state governance might look like and how it might approach the presumably eternal task of maintaining order.

At this point, our concern is with how we can improve the contemporary law enforcement and military institutions' ability to deal with technologically mediated threats. In considering this issue, we do not limit our analysis to the institutions as they are currently configured; that is, we do not assume that the only correct approaches are those that rely exclusively on professional responders. We also consider the possibility of restoring some level of civilian participation to the threat response processes.

The analysis we will pursue in the remainder of this chapter is therefore predicated on three assumptions: The first is that nation-states will continue to rely primarily on professional response authorities—law enforcement and military personnel—to maintain order. The second is that a segmented system in which only law enforcement officers respond to internal threats and only military personnel respond to external threats is not a viable approach in a world in which technology blurs, if it does not erase, the distinction between "internal" and "external." The third assumption is that states can no longer rely exclusively on professionalized response personnel but must also rely, at least to some extent, on the efforts of civilians.

These assumptions structure the analyses we undertake below: In the first section, we consider the possibility of replacing the current, segmented system with one that integrates the response efforts of law enforcement officers and military personnel. In the next section, we consider the permissibility, and utility, of adding civilian participation to this integrated effort.

Military-Law Enforcement Integration

Before we begin, I should reiterate that our goal is to ***integrate*** the efforts of these two distinct institutions, not fuse them into a single entity. There are very good reasons to maintain their institutional separation.[899] We are therefore only concerned with how to achieve a specific, limited level of operational integration.

---

899 *See* Chapter 6.

Cyberwarfare

The initial task here—as with crime and terrorism—is what-attribution: determining the nature of an attack.[900] Integrating the efforts of military and law enforcement personnel into this process does not present significant conceptual or practical difficulties: All we are concerned with at this point is (a) determining if there has been an attack (or if an attack is in progress) and (b) identifying the nature of the attack.[901] As we will see below, what-attribution does not implicate the military-law enforcement division of response authority because it necessarily precedes the formulation and implementation of any "response" to an attack.

What-attribution

What-attribution is, in effect, a prefatory stage that precedes the response process; it encompasses the initial, threshold task of deciding which institution—law enforcement or the military—should respond to an attack. We implicitly assume it is part of the response process because it so rarely becomes problematic in the physical world. When the Japanese bombed Pearl Harbor, there was no doubt it was warfare; and when Bonnie and Clyde robbed banks, there was no doubt it was crime. As we saw earlier, what-attribution becomes a problematic, and therefore conscious, endeavor in the cyber-context because of the inherent ambiguity of cyber-attacks.[902] Our awareness of the task does not, however, alter its nature: What-attribution is merely the winnowing process by which we direct attacks to the proper response authority—military or law enforcement. Because what-attribution is distinct from—is a precondition for—the response process, we can achieve an effective level of military-law enforcement integration at this stage and still maintain the institutional integrity of both entities.

How, precisely, might we go about doing this? The obvious way is to let military and law enforcement personnel share information about actual, or suspected, attacks. I suspect they have always shared information on an *ad hoc* basis when it became necessary.[903] In the United States, statutes specifically

---

900 *See* Chapter 4.

901 *See id.*

902 *See id.*

903 *See*, e.g., Tobias Barrington Wolff, *Political Representation and Accountability under Don't Ask, Don't Tell*, 89 Iowa Law Review 1633, 1656 note 89 (2004) ("law enforcement authorities often share information with military officials").

authorize military-law enforcement information-sharing in certain instan-ces,[904] and efforts are underway to formalize a more general sharing of information among law enforcement agencies and between civilian law enforcement agencies and the military.[905]

If those efforts do not sufficiently authorize timely, cross-institutional information sharing for the purpose of what-attribution, we could use specific statutes and, if necessary, regulations to establish this type of activity and define the parameters within which it is permissible. Because the purpose is to allow law enforcement and/or the military to answer the questions noted above—whether there has been an attack and, if so, what kind of attack it is—the information that can legitimately be shared at this stage should be limited to data that can be used to answer these questions. This would encompass data about the structure and target of the apparent attack; it could also encompass the apparent author(s) of the attack because, as we saw earlier, that can be dispositive in determining whether an attack is crime, terrorism, or warfare.[906]

Should this information sharing be reciprocal or should it be limited to law enforcement's providing information of the type described above to the military? At this point, anyway, I would argue that it should be limited to law enforcement's providing the military with information it has lawfully collected that is, or may be, pertinent to determining whether an attack is cyberwarfare. The reason for limiting ii in this fashion is that what we are concerned with to this point in our analysis is cyberwarfare, the identification of and response to which do not fall within civilian law enforcement's responsibilities. Given that, the only reason to allow information sharing between law enforcement and military personnel is to improve the latter's ability to identify cyberwarfare attacks. As long as we are concerned solely with what-attribution for cyberwarfare, there is no reciprocal justification for having the military share information with law enforcement.

---

904 *See*, e.g., 10 U.S. Code § 371(a); 18 U.S. Code § 2517(8).

905 *See*, e.g., U.S. Department of Justice, Fusion Center Guidelines (2006); U.S. Department of Justice, The National Criminal Intelligence Sharing Plan 9, 35 (2003). *See also* Richard A. Best, Jr., *Sharing Law Enforcement and Intelligence Information: The Congressional Role*, Congressional Research Service (Library of Congress 2007).

906 *See* Chapter 4. *See*, e.g., 18 U.S. Code § 2517(8) (A "law enforcement officer . . . who . . . has obtained . . . the contents of any . . . communication . . . may disclose such contents . . . to any appropriate Federal . . . or foreign government official to the extent that such contents . . . reveal[] a threat of . . . attack or other grave hostile acts of a foreign power . . . within the United States or elsewhere, for the purpose of preventing or responding to such a threat").

How should this unilateral information sharing work? The first requirement is a military that is receptive to the information provided by law enforcement; unless the military is prepared to accept, and act upon, the information law enforcement provides, the entire exercise will be futile. The best way to guarantee this receptivity is to ensure that military personnel, of whatever type and at whatever level, are alert for potential cyberwarfare attacks. This particular vigilance should be expected of all military personnel who interact with cyberspace as part of their official duties, not just those assigned to special "cyber commands."[907] Those in both categories should be trained to recognize the indicia of cyberwarfare attacks and to report any evidence of such attacks to their superiors or the appropriate, designated agency.

The next task is conceptualizing how civilian law enforcement shares information with the military. We begin with the proposition that law enforcement personnel will merely transmit is information they have collected in the routine course of their official duties. Law enforcement personnel will not be tasked with gathering information specifically for the purpose of assisting the military with the process of identifying cyberwarfare. To do that would erode the institutional and operational distinction between the two institutions and, as I noted at the beginning of this discussion, my goal is to integrate their efforts, not to fuse them into a single entity.

Law enforcement personnel will therefore provide only information they have routinely and legitimately collected. That raises another issue: Should law enforcement personnel filter information before they give it to the military? Logically, they have two options: They can filter information they have collected in an effort to narrow its focus to likely indicia of cyberwarfare; or they can transmit all of that information to the military. The argument for filtering is that selective reporting reduces the risk of overwhelming the military with extraneous data. The argument against filtering is that computers can analyze large amounts of data, and their capacity in this regard will only increase; this dramatically reduces the likelihood that transmitting raw, unfiltered data would overwhelm military analysts.

---

907 This obligation could also extend to off-duty military personnel's encounters with cyberspace, just as off-duty law enforcement officers who observe criminal activity are expected to alert on-duty officers as to what is occurring. If we decided to extend the obligation in this fashion, we could then encourage military personnel to report their observations by immunizing from suit for reasonable actions they take in an effort to ascertain if a cyber-event constitutes cyberwarfare. *See generally* Alaska Statutes § 09.65.330(a)(1) (2006) (off-duty police officer is immune from a suit for injury caused while engaging in "official duties"); Wisconsin Statutes § 175.40(6m)(a) (2006) (off-duty law enforcement officers may arrest a person in certain circumstances).

Probably the best approach is to require some of both. If the circumstances of an attack warranted, law enforcement officers could initially vet the attack, using a set of criteria supplied in advance by military personnel. If they decided there was a fair probability that the attack was cyberwarfare, the officers would transmit the filtered information supporting their conclusion to the military expeditiously and flag it as priority data. If, on the other hand, officers saw nothing to indicate that an event implicated cyberwarfare, they would transmit information about that event routinely, as part of a mass of data to be incorporated into a more general analysis. Law enforcement agencies would transmit this routine attack data with a pre-determined frequency, perhaps daily.

Law enforcement's sharing information in the second category might well give rise to concerns about eroding the partition between civilian and military authority. The information in the first category (likely cyberwarfare) poses little risk of violating the partition because here law enforcement is merely giving the military something to which it is entitled: information relevant to the cyberwarfare what-attribution process. Since this information presumptively concerns warfare, it has operational relevance to and value exclusively for the military. Sharing this information with the military should therefore pose little, if any, threat to the segregation of civilian and military authority.

Logically, the same conclusion holds for information in the second category because it is being provided not as domestic operational data—that is, not as information to be used against civilians—but as external operational data. Its value lies primarily, if not exclusively, in the fact that it is information the military can use in an effort to identify cyberwarfare. But logic should not be dispositive in this context, given the potential for this aspect of our information-sharing endeavor to be perceived as having sinister purposes. The civilian populace might come to believe law enforcement was involved in a cabal with the military, the purpose being to spy on domestic activities for sinister, but no doubt nebulous, purposes. One way to address this concern would be to adopt legislation designed to ensure that the military's use of the second category data is limited to the purpose for which is it provided, i.e., for cyberwarfare attribution.[908] The legislation might also create a monitoring

---

908 The military's use of the information would not necessarily be limited to identifying cyber-attacks. Because cyberwarfare response is the exclusive province of the military, and because the data law enforcement shares with the military cannot be used for domestic purposes, there seems to be no reason why the military cannot also use information lawfully shared by the process outlined above to respond to cyberwarfare, as well as to identify it.

body that would review the military's use of this information to ensure that it was, in fact, conforming to the strictures imposed by the legislation.

Fourth Amendment and Related Protections

Doctrinally, information-sharing for the purposes of what-attribution should not implicate the heightened legal protections many countries, including the United States, impose upon law enforcement or other government-initiated evidence collection. The bedrock protection under United States law is the Fourth Amendment to the Constitution, which creates a right to be free from "unreasonable" searches and seizures. Under the Fourth Amendment, law enforcement officers must either obtain a search warrant or invoke an exception to the warrant requirement before they can "search" or "seize" property owned by a citizen.[909] In some areas, the basic protections of the Fourth Amendment have been expanded by statutes, such as Title III, the federal statutory scheme that imposes additional requirements upon investigators who intend to intercept telephone or electronic communications.[910]

Information sharing for the purpose of what-attribution should not, though, implicate the provisions of the Fourth Amendment or Title III or the related statutory schemes that are in place in the United States. These provisions all evolved to protect U.S. citizens from a specific evil: overreaching by law enforcement officers collecting evidence for use in a criminal prosecution.911 They consequently apply only when agents of the government are proactively investigating or otherwise gathering information about a citizen; in other words, they only apply when the state targets someone for criminal prosecution.912 The what-attribution process does not implicate this *de facto* law enforcement dynamic because unless and until the first

---

909 *See* Wayne R. LaFave, Search and Seizure § 4.1 (4th ed., Thomson West 2007).

910 *See*, e.g., Daniel J. Solove, *A Taxonomy of Privacy*, 154 University of Pennsylvania Law Review 477, 492–493 (2006).

911 *See* United States v. Verdugo-Urquidez, 494 U.S. 259, 266 (1990) ("the purpose of the Fourth Amendment was to protect the people of the United States against arbitrary action by their own Government").

912 This is certainly true of the Fourth Amendment. *See*, e.g., Bureau v. McDowell, 256 U.S. 465, 475 (1921). It is true to a lesser extent of Title III and other, related provisions. Title III, for example, might allow someone whose communications were unlawfully intercepted to file a civil suit for redress, though it does not provide for the suppression of improperly obtained electronic evidence. *See*, e.g., United States v. Forest, 355 F.3d 942 (6th Cir. 2004).

what-attribution question (Was there an attack?) is answered affirmatively, there is no reason to target anyone.

What-attribution involves a very different dynamic—a binary, reactive dynamic. Instead of aggressively gathering information about a targeted person, those involved in what-attribution collect information in order to make two binary determinations:

1. Has there been an attack? Yes/no.
2. What kind of an attack was it? Crime/warfare.

Because the process of answering these questions does not target a person (or entity) for criminal prosecution or some other government response, it is outside the paradigm that gave rise to, and structures, the procedural protections of the Fourth Amendment and analogous statutes.

It is also outside the scope of those protections as long as it is limited to collecting data about the circumstances of an event—a potential attack—that is directed at a domestic entity or person.[913] Nothing in the Fourth Amendment or analogous statutory provisions prevents law enforcement or the military from investigating activity occurring in "public."[914] If a police officer sees what seems to be a bank robbery in process, he can monitor the scene and take other appropriate measures to complete the what-attribution process, that is, to determine if a robbery is, in fact, in progress. The officer's collecting information about the possible attack-event does not implicate interests protected by the Fourth Amendment or the statutes noted above because the collection occurs in a public place, not in the possible robber's home or some other private place. This same logic applies to cyber-attacks: One who attacks a computer system cannot expect the circumstances of that attack to be private and consequently protected by the Fourth Amendment, Title III or other, similar provisions.

There is still another reason why law enforcement-military information sharing for the purpose of what-attribution is outside the scope of the Fourth Amendment and of comparable statutory provisions, at least for certain types of cyberwarfare. These protections only apply (a) to individuals who are U.S. citizens and/or are in U.S. territory and (b) to artificial entities

---

913 The Fourth Amendment does not apply to searches and seizures that are conducted outside the territory of the United States and are directed at non-U.S. citizens. *See* United States v. Verdugo-Urquidez, 494 U.S. 259 (1990).

914 *See* Wayne R. LaFave, Search and Seizure, *supra* at § 2.1.

in U.S. territory.[915] They most assuredly do not apply to nation-states that are launching attacks on U.S. territory or property.916 This suggests that *none* of these provisions would be implicated by law-enforcement-military information-sharing for the purpose of what-attribution in the context of cyberwarfare, and that may or may not be true.

It depends on how we define "cyberwarfare." If we define it in the traditional sense—as an attack by one nation-state on another—then none of these provisions would be implicated by the program of information-sharing we are postulating because they were, and are, not intended to encompass the activities of foreign governments. That conclusion changes if we broaden our conception of cyberwarfare so that it also includes the non-nation-state type of cyberwarfare we analyzed in Chapter 4. As we saw in that chapter, non-nation-state actors—individuals—could launch cyberattacks on a nation-state that are factually indistinguishable from warfare; the only differentiating factor is that the attacks are not launched by another state.

Because this scenario is almost certain to manifest itself in the future (if, indeed, it has not already done so), we shall have to decide how to respond to it. We will consider that issue at this point—even though we are still concerned with the issue of attribution—because in this context, anyway, attribution and response are inextricably intermingled. We cannot ascertain whether the Fourth Amendment and the related protections provided by U.S. law apply to the non-nation-state architects of cyberwarfare unless we know whether those responsible for such attacks will be treated as individuals or, in effect, as hostile nation-states.

Response: Non-nation-state Cyberwarfare

If non-nation-state cyberwarfare attacks purely as warfare, then the U.S. military would presumably be in charge of identifying and responding to

---

915 *See* G.M. Leasing Corp. v. United States, 429 U.S. 338 (1977); Silverthorne Lumber Co. v. United States, 251 U.S. 385 (1920); Santa Clara County v. Southern Pacific Rail Road, 118 U.S. 394 (1886).

916 This proposition is derivable from the language of the Fourth Amendment, which states that "the right of the people to be secure.... against unreasonable searches and seizures" is not to be violated. *See* U.S. Constitution, amendment iv. The Supreme Court has long held that the amendment's reference to "the people" encompasses only individual and corporate citizens of the United States and alien individuals who are present within U.S. territory. *See* United States v. Verdugo-Urquidez, 494 U.S. 259, 266 (1990).

these attacks; that approach, though, presents certain problems. In warfare, nation-states prevail by destroying or conquering their enemy, that is, by capturing their territory, disassembling the governance structure responsible for the war and either establishing a new, native governance structure of simply incorporating the conquered territory and populace into their own system. How would the U.S. military go about defeating non-nation-state actors who were engaging in cyberwarfare with the United States?

They would certainly defend the U.S. targets from the attacks; they would also, no doubt, launch offensive attacks designed to prevent the attacks from pursuing their war campaign. What form would these offensive attacks take? Would they only be directed at crippling the hardware and software the attackers were using? If so, that is likely to be a transient solution, since it is not likely to be difficult to replace or restore either, even in poorer countries. Given that, would the U.S. defenders seek to personally neutralize the attackers themselves? If so, how would they do this? Would the U.S. defenders kill them (which could presumably create tensions with the country from which they were operating) or capture them? If they capture these civilian warriors, what would the U.S. military do with them? If it took our hypothesized non-nation-state cyberwarriors captive and treated them as captured enemy soldiers, then the Geneva Conventions and other provisions of the law of war would protect them.[917] If we decide to respond to non-nation-state cyberwarfare in this fashion, then what I said earlier still applies: Neither the Fourth Amendment nor the related statutory protections for individuals would apply to these civilian cyberwarriors; they would be in the peculiar position of being treated as if they were the agents of a foreign nation-state, when they were not. That, however, is an issue I will leave for the future or for others to resolve. We have resolved the issue we needed to resolve: If we treat non-nation-state cyberwarfare as warfare, then information-sharing conducted as part of the what-attribution is outside the scope of the Fourth Amendment or the statutory protections cited earlier because they only apply to internal, law enforcement investigations.

The alternative would be to treat the architects of non-nation-state cyberwarfare not as enemy soldiers, but as criminals. It would be predicated on the premise that both the identity of the perpetrators (individuals instead of nation-state *per se*) and the nature of the "harm" inflicted (financial and other property loss, possible injury and even death to individuals) are functionally

---

917 *See* Chapter 5.

more characteristics of crime/terrorism than of warfare. And this alternative would preserve the historic, categorical distinction between crime/terrorism and warfare, that is, individuals perpetrate one, nation-states, the other. This approach might also be easier to implement. As I noted earlier, having one state's military forces respond to acts of war that were launched from the territory of, but not at the behest of, another nation-state gives rise to certain difficulties. If we do not define non-nation-state warfare as a type of crime, then the extradition treaties and other approaches nation-states use to obtain custody of perpetrators located in the territory of another state are inapplicable. That, as I noted above, presumably leaves the offended state with no option but to invade the territory of the state hosting the perpetrators and kidnap them, both of which could be construed as direct offenses to the sovereignty of the host state.[918] So, this type of response to a non-nation-state cyberwarfare attack might result in the offended host state's launching retaliatory warfare (cyber or traditional) against the state whose military forces invaded its territory and kidnapped individuals who were lawfully present there.

We could, I suppose, devise alternative, cyberwarfare-specific treaties that encompass this situation. These instruments would presumably require one country to surrender individuals present within its territory to a requesting state if the latter established, by the requisite level of evidence, that the individuals were responsible for launching cyberwarfare attacks against it and its citizens. That, however, seems an unnecessarily cumbersome process, since it is, in effect, reinventing extradition. It would, therefore, probably be more logical and more efficient to define non-nation-state cyberwarfare as a type of crime (analogous to cyberterrorism), the response to which will be handled by the victim state's law enforcement and criminal justice system.

If we adopt this approach to non-nation-state cyberwarfare, the protections of the Fourth Amendment and the U.S. statutes noted earlier *would* apply to the response process. Does that mean they would also apply to the what-attribution process? I do not think they would because, as I explained earlier, I think that process necessarily precedes, and is distinct from, the process of responding to any type of threat.

Overall, therefore, neither the Fourth Amendment nor statutory provisions that further restrict the evidence-gathering activities of law enforcement officers should limit information sharing between law enforcement

---

918 The same could be true of the attacks the defending state—the United States in the scenario we used earlier—launched on the computer and other systems being used by the architects of the non-nation-state cyberwarfare directed at the U.S.

and the military for the purpose of cyberwarfare what-attribution. Later in this chapter, we consider whether that conclusion also holds when the focus is on identifying cybercrime or cyberterrorism. First, though, we need to consider who-attribution in the context of cyberwarfare.

Who-attribution

Once an attack is identified as cyberwarfare, the focus shifts to who-attribution—that is, identifying those responsible for the attack—and to the need to respond.[919] Who-attribution is not a unitary phenomenon: It is either an independent inquiry or a subsidiary component of what-attribution. When it is apparent an attack is cyberwarfare, who-attribution becomes an independent inquiry because it is not bound up with the process of what-attribution.[920] When it is not apparent an attack is cyberwarfare, who-attribution *may* become a subsidiary component of what-attribution; identifying the attackers may be an essential part of determining the nature of the attack.[921] When it is not apparent an attack is cyberwarfare but the nature of the attack can be ascertained without considering the identity of those responsible, who-attribution does not become part of the what-attribution process. It again stands alone as an independent inquiry.

To the extent who-attribution becomes part of the what-attribution process, it should be encompassed by the analysis outlined above; that is, information about the identity of those responsible for an attack should be included in the information-sharing processes described in previous sections. When who-attribution is an independent inquiry, that is, when it is clear an attack is cyberwarfare, it becomes part of the response process and should be treated as such.

Response Nation-state Cyberwarfare

In an earlier section we examined two possible ways of responding to a special case: non-nation-state cyberwarfare. As I noted there, I think it is

---

919 *See* Chapter 5.

920 *See* Chapter 4.

921 *See* Chapter 5.

probably more appropriate to treat non-nation-state cyberwarfare as a type of crime, just as we treat cyberterrorism as a type of crime.[922] In a world where threats are categorically divided according to whether they come from nation-state or individual actors, it is both conceptually and pragmatically reasonable to assign non-nation-state cyberwarfare to the criminal category. This at once blurs and maintains the related distinction between internal and external threats: It blurs the distinction because the non-nation-state authors of cyberwarfare are very likely to be operating from outside the territory of the victim-state, which means they are literally an external threat. Treating non-nation-state cyberwarfare as a type of crime maintains the distinction between internal and external threats because it equates external threats with attacks from nation-states, an element that is absent in non-nation-state cyberwarfare.

For this purposes of this analysis, we will assume that non-nation-state cyberwarfare is defined as a type of criminal activity, the response to which will be handled exclusively by law enforcement officers from the victim-state(s). Can military personnel assist law enforcement officers who are responding to non-nation-state cyberwarfare, or must the officers act on their own? Because we are treating non-nation-state cyberwarfare as crime, it follows that any response to non-nation-state cyberwarfare attacks must come solely from law enforcement officers. To allow military personnel to assist in this endeavor would undermine, if not erase, the partition between civilian law enforcement and the military.[923]

What about the very significant residual category—nation-state cyberwarfare? Since this is merely the online manifestation of traditional warfare, the actual response process, as with real-world warfare, would be handled exclusively by the military.[924] That does not, though, preclude law enforcement from playing some role in the process: Law enforcement agencies should share information they acquire that may be relevant to prior or ongoing nation-state cyberwarfare incidents; this information-sharing, unlike the information sharing we examined earlier, would be part of the response process. That distinction is significant for two reasons.

One has to do with the information being shared: The information to be shared here is not limited to filtered or unfiltered data that may be relevant

---

922 *See* Chapter 3.

923 *See* Chapter 6.

924 *See* Chapters 3 & 6.

to identifying the nature of an attack, as was the case for what-attribution information sharing. The nature of the attack has already been ascertained. This information sharing instead focuses on data that may help the military respond to confirmed nation-state cyberwarfare being conducted by what may or may not be an identified nation-state. The information law enforcement can legitimately share with the military consequently falls into two categories: data that may help the military respond to completed or ongoing cyberattacks; and data that may help the military identify the nation-state author(s) of attacks. It is at this point that who-attribution becomes part of the response process

The other reason goes to the very different postures law enforcement personnel assume (a) for information sharing in support of the binary what-attribution process analyzed above and (b) for information sharing in support of the nation-state cyberwarfare response process. As we saw earlier, law enforcement personnel play a purely passive role in information sharing for the purpose of what-attribution; they merely transmit data they routinely collected while performing their normal duties. Their role here must be passive if we are to maintain the military-civilian law enforcement bifurcation. The bifurcation, as we saw earlier, is based on a division of responsibility: law enforcement deals with internal threats, the military with external threats. Each institution has its own dynamic: law enforcement responds to crime by apprehending domestic perpetrators who are sanctioned via the civilian criminal justice system; the military responds to warfare by using physical force to engage external forces outside the territorial confines of the nation-states they serve.

If law enforcement officers were to assume an active role in information-sharing for what-attribution, their dynamic would change; instead of focusing exclusively on responding to internal threats, the law enforcement dynamic would implicitly expand to encompass external threats, as well. This might seem a trivial concern because all we would in effect be doing is letting law enforcement officers investigate cyber-bogies—unidentified cyber-events that might be cyberwarfare or cybercrime/cyberterrorism. The problem is that this apparently trivial result eradicates the distinction between the law enforcement and military dynamics by allowing law enforcement officers to investigate when they have no legitimate reason to do so, that is, when they have no reason to believe a crime has been committed.

That may seem inconsequential because law enforcement officers often investigate to determine if a crime has been committed; they frequently investigate, for example, to determine if a death was homicide or the result of natural causes. In these instances, though, the officers are investigating purely to determine if a crime has been committed; if law enforcement were to take an active role in gathering information for what-attribution, officers could be investigating an incident to determine if a crime had been committed (only), but they could also (or alternatively) be investigating to determine if an attack was cyberwarfare. The factual distinction between the scenarios (crime-only versus crime-and/or-cyberwarfare) is likely to be slight, but the conceptual distinction is not; letting law enforcement officers investigate the possibility that anything other than a crime has been committed would incrementally expand their responsibility to encompass external threats and thereby begin the process of eroding the partition between the military and law enforcement. It would begin that process by authorizing them to do something that does not fall within their institutional responsibilities; and it would violate U.S. statutes to the extent that their non-crime investigative activities impacted on U.S. citizens or on others lawfully within the territory of the United States.[925]

These considerations do not militate against allowing law enforcement personnel to take a more active role in collecting and sharing information with the military concerning what have been ascertained to be nation-state cyberwarfare attacks. That does not mean we would bring law enforcement into the military response process; doing so would erode, if not destroy, the partition we are trying to maintain. By taking a more active role in the response process, I mean that law enforcement officers could be alert for information that might be relevant to completed or ongoing cyberwarfare attacks once the military had ascertained that a particular nation-state had embarked upon a cyberwarfare campaign. They could function in a capacity analogous to how I imagine U.S. law enforcement officers functioned during World War II: If a New York City police officer observed what seemed to be a German U-boat in Long Island sound, I am sure he reported what he had seen to the military.[926] Our hypothetical New York officer was not acting as a law enforcement officer when he collected the information about the submarine; military submarines do not fall within the responsibilities of civilian law enforcement. He gathered the information because he knew the

---

925 As we saw earlier, the Fourth Amendment and comparable statutory provisions only apply to U.S. citizens and to persons within the territory of the United States.

926 *See*, e.g., Life on the Home Front—Oregon Responds to World War II, http://arcweb.sos.state.or.us/exhibits/ww2/threat/bombs.htm (police officer reported Japanese balloon carrying a bomb to military authorities).

United States was in a state of war and had, no doubt, been asked to "keep his eyes open" for evidence of enemy military activity. His efforts were clearly intended to assist the military in responding to an external threat; as such, they fall outside his institutional responsibilities. But unlike the active information gathering for what-attribution we examined above, they do not erode the military-law enforcement partition because they are directed at an ascertained external threat to which the military is responding. The ascertained external threat focuses law enforcement's information acquisition efforts and thereby reduces, if it does not eliminate, the possibility that they will improperly gather evidence in a fashion and for a purpose outside the legitimate scope of their institutional responsibilities. And, finally, letting law enforcement share information relevant to the military response to cyberwarfare would not violate any U.S. laws; as we saw in Chapter 6, U.S. law does not restrict law enforcement's ability to share information about nation-state warfare—cyber- or otherwise—with the military.

Should information sharing be law enforcement's only contribution to the process of responding to nation-state cyberwarfare? To answer that question we need to consider what will be involved in responding to this type of cyberwarfare.

Nation-state cyberwarfare conforms to the traditional model of warfare in one respect—it is conducted by a nation-state—but deviates from that model in another notable respect. As we saw in Chapter 3, our model of real-world warfare involves conflict between the military forces of opposing nation-states, exclusively. Civilians play no role in real-world warfare; indeed, the law of war requires that they be sheltered as much as possible from its effects.[927]

Nation-state cyberwarfare will conform to our expectations of warfare in that it will be conducted at the behest and the direction of a nation-state and will presumably be carried out primarily by that state's military forces. It will deviate from our expectations by eroding, if not erasing, the noncombatant-combatant distinction that is a fundamental premise of the evolved, twentieth-century conception of warfare. As we saw in Chapter 3, civilian computer systems are a likely target of nation-state cyberwarfare, and an attacking state may use some of its civilians in launching cyberattacks on a victim state. These and other deviations from the traditional model alter the dynamic to which we are accustomed: Cyberwarfare battles will almost

927 *See* Chapter 3.

certainly not consist of simultaneous, usually large-scale clashes between opposing military forces (only); they are far more likely to take the form of prolonged, intermittent skirmishes carried out by a fluctuating mix of military and civilian forces from the opposing nation-states.

The changes in the dynamic are inevitable given the altered nature of the arena in which battles will be waged and the weaponry that will be used. Though our concept of war has moved far beyond the medieval and post-medieval collision of opposing massed forces, it still involves a simultaneous clash between the military forces of opposing states that is predicated on the use of physical force and continues until one side prevails. This is, and has been, our model of warfare because war has always been a physical conflict, and physical conflicts are zero-sum phenomena: They have a beginning, a middle and an end—an outcome, a victory for one of the participants. Like any physical conflict, real-world warfare requires mutual, simultaneous engagement in some real-space, and that requirement structured the dynamic to which we are accustomed.

Unlike real-world warfare, cyberwarfare does not occur in physical space; and it while it does involve the use of a type of force, it does not involve the use of tangible, physical force.[928] There is therefore no reason to believe it will conform to the essentially zero-sum expectations we have of warfare. Cyberwarfare can, and probably will, be sequential and intermittent; this is a more logical dynamic because as we saw in Chapter 4, cyberwar may approach hostilities in a very different fashion. The goal in cyberwarfare may be to whittle away at the infrastructures of the victim state and debilitate it to the point at which it ceases to be a significant competitor and can consequently be ignored or subjected to physical conquest.[929] This attenuated approach to waging war has certain advantages for an attacking state: It is much less expensive, because it does not require the mobilization of massive, heavily armed military forces; the attacking state runs little risk of losing personnel, territory, or equipment to an enemy state; the attacking state has plausible deniability; and a prevailing state does not become the caretaker of conquered territory and populations.[930]

Because there is no operational need for cyberwarfare to conform to the zero-sum conflict model of traditional warfare, and since the attenuated

---

928 *See* Chapters 3 & 4.

929 *See* Chapter 4.

930 *See id.*

approach to warfare outlined above offers an attacking state several advantages, we can expect cyberwarfare to manifest itself in new and unexpected ways. One of those ways, as I noted earlier, is civilian involvement. As we saw in Chapters 3 and 4, there are two, related reasons why we exclude civilians from traditional, real-world warfare: One is that the task of waging warfare became a profession, in large part due to advances in military technology; the other, related reason is that civilians became increasingly vulnerable to the activities of heavily armed, professional military forces. The world resolved the tension between these two circumstances by dividing warfare into "players" and "non-players." Because they are not equipped to participate on an equal basis and are therefore sitting ducks for a hostile military force, civilians became "non-players;" the law of war removes them from the contest because it would not be sporting to let professional militaries ravage civilians.

This exclusion of civilians is not likely to be an aspect of the law of cyberwarfare, if and when it evolves. As we have seen, civilian-operated systems are likely to be among the primary targets of cyberwar; and because cyber-infrastructures are to a great extent owned or operated by civilians, many civilians will be as adept at using the tools of cyberwar as are cyberwarriors. This not only erodes the operational distinction between civilians (noncombatants) and military personnel (combatants), it will encourage nation-states to use civilians in launching and in responding to cyberattacks. Civilian participation as such is not, however, our concern at this point in our analysis; we take up the issue of integrating civilians into the response processes later in this chapter.

Our concern here is with whether we should let law enforcement play a more substantial role in the nation-state cyberwarfare process. The arguments for and against letting them do so derive to some extent from the combatant-noncombatant dichotomy outlined above. If we apply that dichotomy, we arrive at the conclusion that law enforcement personnel are civilians and, as such, are excluded from participating in combat under the current law of war. That conclusion, though, may not be valid in this context. As we saw above, civilians of all types are excluded from participating in real-world military combat because they are not trained and equipped for such an activity; they are so operationally disadvantaged it is both rational and merciful to bar them from participating.

That is not true for cyberwarfare; civilians, including law enforcement personnel, may be as adept at cybercombat as are trained cyberwarriors. This is especially likely to be true of law enforcement personnel who are charged with responding to cybercrime, cyberterrorism and the non-nation-state cyberwarfare we analyzed earlier. They should have substantially the same tools and substantially the same skill-sets as military personnel who are charged with waging cyberwar. Another factor also militates in favor of allowing civilian law enforcement personnel to participate actively in the military response to cyberwar: Civilian law enforcement is charged with protecting civilian' lives and property; to the extent cyberwarfare attacks jeopardize either of those, they implicate concerns that are directly encompassed by law enforcement's institutional response obligations.

Although these, and perhaps other, factors support allowing civilian law enforcement personnel to become actively involved in the response to nation-state cyberwarfare, I do not think that is advisable, for several reasons. One is the point I have made several times before: If we were to allow law enforcement to respond to cyberwarfare in any capacity and at any level, we would be taking the first step toward dismantling the law enforcement-military partition. As long as we are committed to maintaining a rigid division of response authority between these institutions, we cannot mix metaphors; we cannot sanction a law enforcement-military response to cyberwarfare and maintain, at the same time, that the institutions operate in a mutual state of perfect insularity.

Another reason I think we should not try to incorporate law enforcement personnel into the actual process of countering cyberwarfare attacks is more pragmatic: I suspect that any attempt to incorporate civilian law enforcement personnel into online combat would create at least some risk of impeding the effectiveness of the military responders. As long as the two are institutionally discrete, their command and controls structures would be separate, as would their facilities and, perhaps, other operational components. That institutional segregation means the military personnel who bear primary responsibility for responding to nation-state cyberwarfare attacks would either have to (a) devote resources to coordinating their efforts with the civilian law enforcement personnel who were authorized to participate in that endeavor or (b) allow the law enforcement personnel to operate with a great deal of autonomy. The first option would almost certainly reduce the effectiveness of the military personnel by dividing their focus; they would be focusing both on responding to the attacking nation-state and on monitoring the activities of the law enforcement personnel chosen to participate in that response process. The second option does not directly create the risk of reducing the effectiveness of the military response personnel, but it creates other, equally undesirable risks. If the civilian law enforcement personnel who are participating in the response process do not have real-time access to the attack data available to the military responders, they might err in their response efforts: At its worst, this could result in their directing defensive measures at an innocent nation-state; at the least, it would result in wasting their time and expertise (and could give rise to the perception that the defending state, the United States in this instance, was at least somewhat inept in responding to cyberwarfare).

I, therefore, do not think civilian law enforcement personnel should play any direct role in the offensive or defensive processes of responding to nation-state cyberwarfare attacks. I do think they can play a useful role in working with "regular" civilians to help prevent and withstand such attacks. We will take up this issue later in this chapter, after we consider the possibility of integrating the military and law enforcement in the processes of responding to cybercrime and/or cyberterrorism.

Cybercrime and Cyberterrorism

Because the processes of identifying and responding to cybercrime and terrorism are the institutional responsibility of law enforcement personnel, the issue we will need to analyze is the propriety of integrating the military into either or both processes. Here, as with cyberwarfare, we will begin with what-attribution.

It is reasonable to assume that while military personnel perform their designated function of identifying cyberwarfare attacks and attackers they will encounter attacks that are clearly not cyberwarfare. Unless and until we parse cyberassaults into new categories, these attacks will by default constitute cybercrime or cyberterrorism. Since it is also reasonable to assume the military's ability to scan cyberspace for attacks is superior to that of civilian law enforcement, it is logical to conclude that the military will acquire information about cybercrime and cyberterrorism events that may not be available to civilian law enforcement. It is both logical and prudent to let the military share this information with law enforcement: It is not information the military can act upon; and letting the military share the information with law enforcement personnel is likely to enhance the latter's ability to identify and respond to cybercrimes and acts of cyberterrorism.

If we decide this is an appropriate strategy, we will need to define the parameters of the military's authority to transmit attack information to law enforcement. Here there is no reason to filter the information being shared according to the type of attack involved; that is, there is no reason why the military cannot periodically send law enforcement all the unclassified information it has collected concerning cyberattacks on the United States.[931] Such an unfiltered transmission would be over-inclusive insofar as it would provide information about nation-state cyberwarfare, for which law enforcement has no response authority, but there seems to be no downside to omnibus information sharing as long as the information being shared with law enforcement is not classified.

Any information the military shares that concerns nation-state cyberwarfare will be operationally irrelevant because law enforcement has neither the authority, the resources nor the inclination to respond to that type of cyberwarfare.[932] This information might, however, be useful in what-attribution; at the very least, it would help law enforcement personnel exclude identifiable cyberwarfare attacks from the universe of cybercrime and cyberterrorism events for which they have response authority.[933] There is also a reason to allow unfiltered information sharing: The more data law enforcement personnel have about cyberwarfare attack signatures, the more effective they can be in identifying potential acts of cyberwarfare and sharing that information with the military.

Should information sharing be the way the military supports the law enforcement effort? As we saw in Chapter 3, one of the major challenges U.S. law enforcement faces in identifying and responding to cybercrime and

---

931 The argument for excepting classified information about cyberwarfare and non-cyberwarfare attacks is that even information in the latter category could implicate national security concerns by, say, revealing the technology military personnel used to identify an attack and/or collect data about the nature of that attack. Unless and until we give security clearances to law enforcement personnel, which seems unlikely, we cannot let the military routinely share classified information with them.

932 As we saw in the previous section, we are assigning law enforcement personnel the response authority for non-nation-state cyberwarfare, which we are treating as a type of criminal activity. Letting the military share all the information it collects about cyberattacks with law enforcement should help law enforcement personnel identify and respond to non-nation-state cyberwarfare.

933 The cyberwarfare-related information might also play a role in who-attribution for cybercrime, cyberterrorism, and non-nation-state cyberwarfare. There could, for example, be some overlap between cybercriminals and nation-state cyberwarfare because nation-states may hire domestic or foreign cybercriminals to help them develop a cyberwarfare capacity or to help them implement their initial efforts in that regard. Law enforcement personnel might be able to use data about distinctive attack signatures in the "excess" cyberwarfare information they had received from the military to identify a specific attacker and link him or her to cybercrime, cyberterrorism and/or non-nation-state cyberwarfare. The likelihood of this seems particularly pronounced in the case of non-nation-state cyberwarfare because those who have expertise in cyberwarfare may hire themselves out to state and non-state actors.

cyberterrorism (and now, non-nation-state cyberwarfare) is the lack of nonpersonnel resources, such as hardware, software, and training for the officers assigned to cybercrime/cyberterrorism units.[934] While the military cannot provide personnel,[935] it could alleviate this challenge by providing technical training to law enforcement officers and by donating superfluous or out of date equipment to law enforcement. As we saw in Chapter 6, in the 1980s, Congress authorized precisely this type of assistance to improve law enforcement's ability to combat the illegal drug trade; there seems, therefore, to be no doctrinal reason why the military could not also provide similar assistance to improve law enforcement's ability to combat cybercrime, cyberterrorism and non-nation-state cyberwarfare. The argument for allowing this becomes even more compelling when we note that cybercrime, cyberterrorism and non-nation-state cyberwarfare are directly analogous to the drug trade in that all four tend to encompass transborder criminal activity. The same policy considerations that justified allowing the military to provide non-personnel resources to enhance law enforcement's effectiveness in combating the drug trade consequently seem to militate in favor of allowing similar assistance in the cybercrime/cyberterrorism/non-nation-state cyberwarfare context.

Arguably, anyway, this type of assistance poses little, if any, risk of eroding the military-law enforcement partition. Congress has already, and uneventfully, authorized such assistance in law enforcement's battle against a different transborder threat to internal order. This does not mean that the issue would not be raised; it probably would, and that probability requires that we analyze whether, and how, the contribution of nonpersonnel resources by the military could undermine the authority partition.

The first argument one could make is that an erosion of the partition would result from law enforcement's effectively becoming indebted to the military. The erosion would result not from *quid pro quo* indebtedness, but from a subtle shift in allegiance, as law enforcement began to look to the military instead of to civilian authority for support. The premise is that support creates bonds between individuals, and bonds can transmute into

---

934 *See*, e.g., National Institute of Justice, Electronic Crime Needs Assessment for State and Local Law Enforcement 16–19 (2001), http://www.ojp.usdoj.gov/nij/pubs-sum/186276.htm.

935 Letting military provide its own personnel to supplement law enforcement's resources would almost certainly violate the Posse Comitatus Act. *See* Chapter 6. It would also raise serious, legitimate concerns about eroding the partition between civilian and military authority.

allegiance. Though that is no doubt true, it does not necessarily mean the partition would be in jeopardy; based upon our experience to date with this rather minimal level of support, it seems unlikely it could transform law enforcement personnel into military vassals. I am not, however, sure we can dismiss this concern out of hand; the possibility that further institutionalizing this type of nonpersonnel assistance could have unforeseen consequences is something we should certainly consider.

Another argument against the nonpersonnel resource assistance effort outlined above is that the risks associated with providing assistance to combat the drug trade were much less than the risks that could ensue from providing assistance to combat the triad of cyberthreats: cybercrime, cyberterrorism, and/or non-nation-state cyberwarfare. U.S. law enforcement efforts to combat the drug trade focused to a great extent on offshore activities and non-citizens; its efforts to combat the triad of cyberthreats are likely to focus to a greater extent on activity that occurs in the territorial United States and is conducted by U.S. citizens. The premise of this argument, then, is that what was acceptable when law enforcement was concentrating primarily on "them" is not acceptable when it is concentrating primarily on "us." I can see the empirical logic behind this argument, but I am afraid I do not see how it has any conceptual support.936 It seems to me that we want to avoid is *any* erosion of the military-law enforcement partition, not merely an erosion that focuses on non-U.S. citizens (or non-citizens of whatever other

---

936 In distinguishing between what is appropriate for U.S. citizens and non-U.S. citizens, this argument is reminiscent of the Supreme Court's approach to the Fourth Amendment. As we saw in the previous section, the Supreme Court has held that the Fourth Amendment applies to law enforcement activity that targets U.S. citizens wherever they may be and persons or places within the territorial United States but does not apply to extraterritorial law enforcement activity directed at non-citizens. The proponents of the argument noted in the text might be relying on these cases as doctrinal support for their argument, because they in effect hold that U.S. law enforcement personnel have more operational latitude when they conduct investigations targeting non-U.S. citizens outside the territory of the United States.

If that is the doctrinal predicate for this argument, then it must also be based on the empirical assumption that allowing the type of assistance postulated above would undermine the protections U.S. citizens enjoy under the Fourth Amendment. I see two problems with that argument: One is that I do not think it is valid. The effectiveness of the Fourth Amendment and similar measures in protecting citizens from overreaching by law enforcement officers would in no way be diminished by law enforcement's relying on alternate sources of material support. The other problem I have with this argument is that I do not see how it has anything to do with eroding the military-law enforcement partition; it seems to me this is, instead, an argument about maintaining the integrity of the protections the Fourth Amendment provides for U.S. citizens and for others within the territory of the United States.

country might undertake a similar effort). I consequently do not see this as a viable argument against instituting the type of non-personnel resource assistance I outlined above.

Actually, I think that the concern about military assistance's eroding the civilian-military authority partition is more compelling with regard to the military's sharing information with law enforcement than they are for the resource assistance effort outlined above. The military's providing information to law enforcement in the strategy outlined at the beginning of this section could create the perception—if not the reality—that the military was spying on U.S. citizens to assist law enforcement.[937] The reality is not a concern, at least not under current U.S. law: If the military were to move from merely passing on information it collected in the course of carrying out its obligation to respond to nation-state cyberwarfare to actually compiling information about cybercrime, cyberterrorism, and non-nation-state cyberwarfare for the express purpose of facilitating law enforcement investigations, its actions would violate the Posse Comitatus Act.[938]

The perception, on the other hand, could jeopardize the information sharing strategy we examined at the beginning of this section unless there was a reliable way to ensure that the military's information collecting was conducted solely for legitimate military purposes. In the previous section, I said that the version of this issue that arises for law enforcement's sharing information with the military could be addressed by adopting statutes which limit the recipient's—the military's—use of data provided by law enforcement. A similar approach could work here, but here it should target the provider (the military), instead of the recipient (law enforcement); the goal here, after all, is to preclude the military from having any incentive to collect data for the express and illegitimate purpose of assisting law enforcement. Statutes barring the military from sharing any data with law enforcement except data military personnel collected as an essentially inadvertent byproduct of carrying out their legitimate military functions should accomplish that by acting, in essence, as an exclusionary rule. It would probably also be advisable to have a court or some other government agency conduct

---

937 *See* Bruce Schneier, Giving the U.S. Military the Power to Conduct Domestic Surveillance, http://www.schneier.com/blog/archives/2005/11/giving_the_us_m.html ("The police and the military have . . . different missions. The police protect citizens. The military attacks the enemy. When you start giving police powers to the military, citizens start looking like the enemy.")

938 *See* Chapter 6.

periodic reviews of the information the military was providing, to ensure it was staying within constitutionally appropriate bounds.[939]

There is one final issue we need to consider before we move to our next topic: bringing civilians into an integrated military-law enforcement effort. That final issue is the possibility of allowing the military to contribute more directly to the law enforcement response to cybercrime, cyberterrorism, and non-nation-state cyberwarfare. If military personnel and military technology could provide operational support for the law enforcement response to those threats, it would certainly enhance the efficacy of that response in at least two ways. The military generally has more advanced technology than law enforcement, so the military's participation could enhance the tools law enforcement could use in responding to the triad of non-nation-state cyberthreats. And the participation of military personnel would augment the often limited staff available to law enforcement agencies. Such collaboration might have other advantages for law enforcement, as well.

It is not, however, a possibility, at least not in the United States. As long as the Posse Comitatus Act is in effect, military personnel cannot actively participate in the law enforcement process.[940] Absent a change in the law, military-law enforcement integration would necessarily be limited to the relatively passive efforts outlined above.

Civilian-military-law Enforcement Integration

Logically, the next step in using an integration of constituencies to improve the way we respond to cyberthreats is to bring civilians into the process.

---

939 Measures designed to limit law enforcement's use of data obtained from the military would not be as effective because they would only prohibit on-record use of the data in the investigation and prosecution of cybercrime, cyberterrorism and non-nation-state cyberwarfare. Such an approach would be underinclusive, as law enforcement could still use the information for strategic purposes, such as for developing initiatives or attack profiles. The Supreme Court long ago recognized that the exclusionary rule is ineffective in controlling police behavior "where the police either have no interest in prosecuting or are willing to forego successful prosecution in the interest of pursuing some other goal." Terry v. Ohio, 392 U.S. 1, 14 (1968).

940 *See* Chapter 6. As I noted in Chapter 6, the Posse Comitatus Act does not apply to members of the National Guard when they have not been called into federal service. They therefore can, and do, participate in state law enforcement activities. *See*, e.g., Doggett v. State, 791 So.2d 1043, 1052 (Ala. Crim. App. 2000); Wallace v. State, 933 P.2d 1157, 1159 (Alaska App. 1997). So members of a state's National Guard could work with state law enforcement in dealing with cyberattacks.

In this section, we analyze the possibility—and the mechanics—of incorporating a level of civilian participation into the military-law enforcement integration analyzed above. Before we begin, though, I need to define a term we will use in the analysis and note a premise that implicitly structures the analysis.

The term is "pure" civilian. By "pure" civilian, I mean a citizen of the United States (or of any other country that decides to implement an institutionally integrated strategy for responding to cyberthreats) who is neither (a) directly employed by the military, by a government agency that is integrally involved in the military function or by a law enforcement agency nor (b) works as a consultant or contract employee for the military or either type of agency. This definition includes corporate and other artificial entities that are recognized as U.S. citizens. "Pure" civilians are completely outside the military and law enforcement institutional structures; they have no official role in, and no responsibility for, maintaining internal or external order.

The premise is simply that we are exploring the potential for integrating "pure" civilian participation into an integrated military-law enforcement effort of the type hypothesized above. Our analysis to this point has been based on the premise that an appropriately circumscribed integration of these two constituencies can enhance the efficacy of national efforts to address external (military) and internal (law enforcement) cyberthreats. In the sections below, we pursue an analysis based on the secondary premise that the selective incorporation of "pure" civilian participation can further enhance the efficacy of these efforts.

Why, one might ask, is there any need to incorporate "pure" civilian participation into this already-integrated effort? Why not incorporate "pure" civilian participation into the efforts of law enforcement (only)? Additively, or alternatively, why not incorporate "pure" civilian participation into the efforts of the military (only)? The answers to these questions lie in the different roles, and different cultures, of the two institutions.

Integrating the efforts of "pure" civilians into the law enforcement function essentially entails orchestrating collaboration between civilian constituencies. While law enforcement officers play an institutional role that differentiates them from "pure" civilians in their professional capacity, their status remains, at base, that of civilians.[941] Law enforcement officers work in the civilian world with civilian personnel. Their institutional responsibility is

---

941 *See*, e.g., Judith Berkan, *Mano Dura—Official Police Department Bias Takes A Hit*, 69 Revista Juridica Universidad de Puerto Rico 1267, 1274 (2000) ("police officers are civilians and the military is not"). *See also* Robert M. Perito, Where Is the Lone Ranger

to maintain order within civilian society; and when law enforcement officers are not performing their professional duties, they resume "pure" civilian status.[942] There is, for these and other reasons, less of an institutional and cultural gulf between law enforcement officers and "pure" civilians than there is between "pure" civilians and military personnel.[943]

Military personnel are to a great extent governed by laws different from those that apply to "pure" civilians.[944] They work and often live in environments that are spatially and culturally distinct from the general civilian culture that is the shared experience of "pure" civilians and law enforcement officers.[945] Their institutional posture is another differentiating factor: The military's professional role is to confront and overcome external threats to the nation-state to which its members have sworn allegiance; to accomplish this, military personnel are authorized to use methods and machineries that are not allowed in civilian society.[946] The activities they engage in are therefore alien to and segregated from civilian society; and information concerning some of these activities is strictly denied to civilians of all types.

Logic and pragmatism therefore suggest we should not concentrate on integrating "pure" civilian efforts discretely into law enforcement and into the military. The institutional and cultural divide between "pure" civilians and the military would make it difficult to design and implement a standalone integration of their respective efforts. It seems the best approach is to use law enforcement as the gateway for incorporating a level of "pure" civilian participation into the law enforcement-military integration outlined above. This is the approach we analyze below.

When We Need Him?: America's Search for a Postconflict Stability Force 85–86 (United States Institute of Peace Press 2004).

942 In some states, off-duty law enforcement officers can make arrests for offenses committed in their presence. *See*, e.g., State v. Brown, 672 P.2d 1268, 1269 (Wash. App. 1983). Of course, in some states civilians can make arrests under certain circumstances. *See*, e.g., "Arrest" § 56, American Jurisprudence 2d (2006).

943 *See*, e.g., Perito, Where Is the Lone Ranger When We Need Him?, *supra* at 85–86.

944 *See id.* at 85–86. *See*, e.g., 10 U.S. Code § 654(a)(9) ("The standards of conduct for members of the armed forces regulate a member's life for 24 hours each day beginning at the moment the member enters military status and not ending until that person is discharged"). *See also* 10 U.S. Code § 654(a)(10) ("Those standards of conduct, including the Uniform Code of Military Justice, apply to a member of the armed forces at all times . . . , whether the member is on base or off base, and whether the member is on duty or off duty").

945 *See*, e.g., Perito, Where Is the Lone Ranger When We Need Him?, *supra* at 85–86.

946 *See*, e.g., *id.*

The pivotal issue in this analysis is the conceptual and doctrinal gap that separates the military and law enforcement from "pure" civilians. In the United States, this gap is the product of two established dichotomies: One is the constitutionally mandated partitioning of civilian and military authority; the other is the *de facto* and *de jure* distinction between "pure" civilians and civilian law enforcement officers. The cumulative effect of these dichotomies is to segregate "pure" civilians from military personnel and law enforcement officers. Given that, how can we incorporate "pure" civilian efforts into the integrated law enforcement-military strategy outlined above without undermining the integrity of either or both of these dichotomies? That is, how can we do this without eroding the institutionally essential distinctions between "pure" civilians and military personnel and/or law enforcement officers?947

There are two ways to approach this task: formal and informal. The formal option would entail institutionalizing the "pure" civilian effort; we would create a new societal institution to act as the conduit for incorporating "pure" civilian participation into the integrated law enforcement-military strategy outlined above.948 The informal option involves relying on voluntary, essentially *ad hoc* participation by "pure" civilians. We will analyze both below.

Formal Integration

This alternative should probably be captioned "more formal" because it does not actually encompass the creation of a "real" institution analogous to, say, law enforcement, the military or public education. A defining characteristic of "real" institutions is that each has an independent presence in society (facilities, personnel) and is the occupational focus of a cadre of individuals who belong to that institution.[949]

There is one compelling reason why we cannot use a "real" institution as the conduit for integrating "pure" civilian efforts into campaigns against cyberattacks. Formally institutionalizing civilians' efforts would effectively

---

947 At this point, we are assuming that we will retain these distinctions and the dichotomies that give rise to them. Later in this chapter we will consider the possibility of modifying or even abandoning this partitioned approach to maintaining internal and external order.

948 We would need to create a new institution under this approach because we cannot formally integrate "pure" civilians into the military and/or law enforcement institutions without eroding the professionalism and effectiveness of either or both institutions.

949 *See*, e.g., Michael Hechter, Karl-Dieter Opp & Reinhard Wippler, Social Institutions: Their Emergence, Maintenance and Effects 13–16 (1990).

eliminate their status as "pure" civilians; they would become more or less professionalized constituents of that new institution. Such a result would defy both logic and pragmatism.

It would defy logic because the constituency of this hypothesized institution would be impossibly over-inclusive. If we were to create the Civilian Cybercorps and make every "pure" civilian in the United States a (possibly recalcitrant) constituent of that institution, we would, as I noted above, essentially eliminate the concept of the "pure" civilian. This outcome is fatally over-inclusive for several reasons. One is that the gravamen of a societal institution is specialization; institutions such as the military, government, and education each exist to perform a specialized task that is essential for the survival of a nation-state.[950] Each fulfills its task by using specially trained personnel and a carefully structured division of labor.[951] Institutionalized divisions of labor and task specialization have become standard features of modern societies for good reason; they are effective ways to organize the implementation of essential tasks. [952] An ecumenical institution of the type postulated above could not be effective; its size and the lack of selective recruitment based on certain qualifications preclude specialization and an efficient division of labor. If anyone were in favor of this approach, they could argue that we could remedy these deficiencies by instituting training programs for the erstwhile "pure" civilians who became the constituents of this institution. That is certainly a logical possibility but, as we will see below, it founders on pragmatic considerations.

However we were to go about it, mandating participation in the Civilian Cybercorps would require establishing governance and enforcement structures to ensure that the erstwhile "pure" civilians were doing their part in the effort against cyb3rchaos. That brings me to the second problem with this approach: the pragmatic problems. Unless we wanted the effort to disintegrate into anarchy, we would need recruitment (At what age does one become obliged to participate? At what, if any, age is one excused from participating?), training, governance, and enforcement structures. We would need facilities to house those responsible for implementing these functions, equipment for them to use in carrying our their responsibilities and funds with which to pay them for their time. All this would require resources that

---

950 *See*, e.g., Peter L. Berger & Thomas Luckmann, The Social Construction of Reality 47–92 (Anchor Books 1966).

951 *See id.*

952 *See id.*

are not available. As I noted earlier, perhaps the most significant challenge law enforcement confronts in its battle against cybercrime and cyberterrorism is a lack of resources. Because we do not have the resources to meet the needs of an existing, far more modest institution, it is exceedingly unlikely that we could find the massive additional resources needed to create and sustain a new institution of the type postulated above.

And there are other pragmatic objections to this approach. One is its effect on a nation-state's ability to carry out the other essential tasks, such as maintaining a viable economy and educational system, supplying its populace with food, clothing, shelter and the other necessities for sustaining life, not to mention the requisites for enhancing the quality of life. The creation of an institution of the type postulated above would directly impede—if not significantly erode—a populace's ability to attend to these matters because a notable portion of their time would be dedicated to the anti-cyber3chaos effort. Another objection is that the approach ignores the issue of aptitude: Some percentage of any modern population will have expertise and skills that could be very useful in a coordinated effort to resist cyb3rchaos; most of the population will not. It is therefore both illogical and eminently impractical to dragoon all of the population into such an effort. It would be more reasonable, and more effective, to recruit only those whose talents will allow them to contribute markedly to such an effort.

We will therefore not explore this, rather literal version of the formal option. We will, instead, analyze something far more modest in scope and ambition: a formal institution that would recruit, train and coordinate the activities of "pure" civilians willing to volunteer to support the integrated law enforcement-military effort outlined above. The use of such an organization has certain advantages, including the following:

- Institutional leaders could implement a vetting process for applicants in an effort to ensure that only committed, serious individuals are allowed to participate.
- Because participation would be voluntary, this institution would not have to monitor the civilian participants to be sure they were "doing their part." The vetting process should further ensure that only willing, committed civilians participate.
- Representatives of this institution could work with the military and law enforcement to create taxonomies and other operational criteria that would structure the efforts of participants in consistent, optimally effective ways.

- Representatives of the institution could also develop and implement training programs to ensure that (a) new participants had the skills they needed to be able to participate effectively and (b) all participants were regularly instructed in new tactics and new issues.
- Members of the institution could establish consistent, regularized standards for the civilian participants, so they would know what was expected of them (and what was forbidden to them).
- Use of such an institutional structure would facilitate the process of establishing routine, reliable channels of communication between institutional participants and law enforcement-military personnel.

I am essentially proposing a larger-scale, less focused analog of the training and support programs used in community policing.[953] Community policing programs involve "pure" civilians in the process of maintaining internal order. They do this in several ways: In some community policing programs, civilians play an active role in the crime detection and prevention processes by patrolling their neighborhood or otherwise providing direct operational support to professional law enforcement officers.[954] In other programs, the civilian contributions are limited to providing police with information; the neighborhood watch systems that have been implemented in many urban areas are an example of this approach to community policing.[955]

We could import either or both alternatives into the effort to control cyb3rchaos. In our earlier analysis of an integrated military-law enforcement effort, we essentially limited the cross-institutional contributions to providing information. We decided (a) that law enforcement's only contribution to the military's efforts against cyberwarfare would be providing information about incidents that might constitute cyberwar and (b) that the military's

---

953 *See*, e.g., Volunteer/Intern Application, Department of Police, Montgomery County, Maryland, http://www.montgomerycountymd.gov/poltmpl.asp?url=/content/POL/ask/communityservices/volunteer/Appl_cover.asp; San Antonio Police Department, Cellular on Patrol, http://www.sanantonio.gov/saPD/cop2.asp. *See also* Wesley G. Skogan and Susan M. Hartnett, Community Policing, Chicago Style 110–93 (Oxford University Press 1997).

954 *See* Nigel Fielding, Community Policing 25–26 (Clarendon Press 1995). *See*, e.g., Community Assistance Patrol, Frederick County, Maryland, http://www.co.frederick.md.us/index.asp?NID=733#Book1; Chilliwack Citizens on Patrol, Chilliwack, Canada, http://www.chilliwack.com/main/page.cfm?id=1073.

955 *See* Fielding, Community Policing, *supra* at 63, 76–77, 108–109, 205–213. *See*, e.g., Neighbor-hood Watch, City of Tigard, Oregon, http://www.ci.tigard.or.us/police/community_policing/neighborhood_watch.asp.

primary (and perhaps only) contribution to law enforcement would be providing information about possible cybercrime, cyberterrorism or non-nation-state cyberwarfare. We restricted cross-institutional contributions in that context because of the legal and pragmatic constraints that derive from the institutional separation of the military and law enforcement.

Here, we are essentially writing on a blank slate: creating an institution with an unfamiliar purpose and as-yet undefined legal status. We could incorporate a level of civilian participation that goes far beyond mere information sharing. We could design this hypothesized institution so that it let "pure" civilians take an active role in investigating cyberattacks; we could involve them in apprehending cyberattackers and even in retaliating against such attacks.956 Such an approach would restore the level of "pure" civilian

---

956 Retaliation involves "striking back" at a cyberattacker: An online victim launches retaliatory strikes—viruses, worms, or denial of service attacks—against someone who has attacked him or her, or it, in the case of a corporation. *See*, e.g., Tim Mullen, *Strikeback, Part Deux*, Security Focus (January 13, 2003), http://www.securityfocus.com/columnists/134.

There is a historical precedent for such an approach. As I noted in an earlier chapter, privateers—civilians operating under a license from a nation-state—played a major role in naval warfare from the sixteenth century until it was outlawed in 1856. *See* Chapter 6, note 183, *supra*. Naval privateers received letters of marque from, say, Nation-State A that authorized them to conduct certain activities against vessels operating the flag of Nation-State B, with which Nation-State was at war. *See id.* Letters of marque also came with instructions—guidelines for how a privateer was to operate. *See* Eugene Kontorovich, *The Piracy Analogy: Modern Universal Jurisdiction's Hollow Foundation*, 45 Harv. Int'l L.J. 183, 212–213 (2004). The instructions specified the country or countries whose ships were fair game; all others were could not be seized under the authority of the letter of marquee. *See id.* The instructions also required that captives be treated humanely and established procedures for selling a ship captured by a privateer. *See id.* When a seized ship was sold, the state issuing the letter of marquee to the privateer took a share—usually 10%—and the rest would be divided among the owner of the privateer ship, its officers and crew. *See id.* Profit, not patriotism, was the incentive for privateering.

One could argue for importing an analogue of this system into the cyb3rchaos context, on the premise that cyberspace is analogous to the high seas. Absent a state of active cyberwarfare, the analogy does not seem particularly apt. Would we issue letters of cyber-marque to the civilian users of cyberspace and authorize them to . . . do what and to whom? The privateer analogy is likely to have a superficial appeal for those who favor self-help, i.e., letting civilian targets strike back against the cyberattackers who victimize them, but the analogy is flawed in many respects: First, it would require a declared state of cyberwarfare between two states; if such a state of warfare existed, letters of cyber-marque could only authorize cyber-privateers to do "something" to attackers who were citizens of or otherwise affiliated with the enemy state. That, of course, simply transposes the issues of attribution and response to the civilian context, without resolving any of the issues we examined in earlier chapters. And if we could resolve the attribution issues, what type of response would we authorize a cyber-privateer to exact? It is not possible for an online privateer to seize the physical system—the cyber-vessel—being

participation in maintaining internal and external order that was common until the nineteenth century and, in so doing, *might* significantly enhance the efforts of professional military and law enforcement personnel. We know this approach would be consistent with historical practice, but we cannot be certain as to the impact it would have on the efforts of the professionals whom we would retain as our first line of defense against the various manifestations of cyb3rchaos. For the purposes of this analysis, however, I assume that an approach of this type would enhance the efficacy of our military and law enforcement personnel. I make that assumption because I believe I can demonstrate that we should not institute such an assumption, even if it would bring certain benefits.

---

used in the attack. And what about the profit motive? Would we authorize cyber-privateers to raid the bank accounts or other assets of those whom they believed had attacked them under the aegis of cyberwarfare? I could note other objections, but I think this should illustrate my point: I do not see how the privateer notion can be transposed into the cyb3rchaos context.

There might be an alternative: Letters of marque were an evolved, specialized version of a practice that was apparently common in fourteenth century England. *See* Alice Beardwood, Alien Merchants in England, 1350–1377—Their Legal and Economic Position 60 (Carnegie Corporation of New York 1931). The practice was an application of the law of marquee, which allowed someone who had been "harmed" by the actions of another but who could not obtain justice by other means to take property that was owned by the person who inflicted the "harm" and located in the victim's "precinct." *See* Black's Law Dictionary 643 (8th ed., 2004). *See also* Alice Beardwood, Alien Merchants in England, *supra.* The principle was applied to foreign merchants who caused "harm" to an English subject while in England. *See* Alice Beardwood, Alien Merchants in England, *supra.* The English victim would file a complaint with the authorities, which would investigate and request for reparation to the government of the country where the foreign merchant lived. *See id.* If reparations were not forthcoming, English authorities would seize the offending foreign merchant's property that was in England or in English reach; they would seize enough property to recompense the victim for the "harm" done, and had that property over to the injured English citizen. *See id.*

Could we utilize a variation of that principle to incorporate a measure of civilian participation in the process of dealing with cyberthreats? This possibility also has a superficial appeal, but I do not see how it could be implemented in practice. As we have seen, ""authorities" in the United States (and in other countries) are overwhelmed dealing with cyberattacks; requiring them to play their role in implementing an analogue of the law of marquee system would simply further burden an already overburdened system. We also have the problem I noted above: How would someone go about identifying "property" that could legitimately be seized under a cyber-law of marquee and used to make the victim of a cyberattack whole?

I could add other, specific objections to such a practice, but implementation details are not the real reason to avoid this option. The ultimate problem with attempting to utilize an evolved cyber-law of marque system to reinforce our defenses against cyberattacks is that, like the privateer system, it lets civilians help themselves to the property of others. Such a system is inevitably going to be abused, especially in the privacy and secrecy of cyberspace. We could create a solution that is worse than the problems we currently confront.

Ninety-one years ago, concerns about German spies and saboteurs led to the creation of the American Protective League ("the League").[957] The League was a "volunteer organization to aid the Bureau of Investigation of the Department of Justice" in identifying, apprehending, and generally frustrating the efforts of foreign agents who were, or were believed to be, operating within the United States.[958] It was created because neither federal nor state law enforcement had the personnel or other resources to mount effective investigative and enforcement campaigns against what was perceived to be a serious threat of espionage and sabotage.[959]

A well-meaning group of civilians—mostly businessmen—organized what became a national effort to supplement the official resources available to combat spies and saboteurs and carry out related security efforts.[960] By June of 1917, the League had nearly 100,000 members and "branches in almost six hundred cities and towns"; by 1918, its membership had grown to nearly 250,000.[961] According to a reliable estimate, until the League was disbanded in February of 1919, its members investigated 3,000,000 cases for the War Department and "perhaps another" 3,000,000 cases for the Department of Justice.[962] One author summarized the climate that gave rise to the League and the methods they employed:

> "Hysteria" is hardly too strong a word for the sentiment that swept the country after . . . Germany became an official enemy of the United States. . . . Attorney General . . . Gregory called upon Americans to serve as the government's eyes. . . . To aid that effort, the Justice Department organized . . . the American Protective League. . . . League members wiretapped telephones, impersonated federal officers, opened mail, broke into and searched offices, ransacked the homes of "suspects," and made illegal arrests. . . . In September 1918, New York City members . . . mounted a series of raids designed to capture draft dodgers, dragging an astonishing 50,187 suspects out of offices and street cars.[963]

---

957 Joan M. Jensen, The Price of Vigilance 22 (Rand McNally 1968).

958 *See id.*

959 *See id.* at 17–32.

960 *See id.*

961 *See id. See also* Homer Cummings & Carl McFarland, Federal Justice: Chapters in the History of Justice and the Federal Executive 421 (Macmillan, 1937).

962 *See* Jensen, The Price of Vigilance, *supra* at 155.

963 Philippa Strum, *Brandeis: The Public Activist and Freedom of Speech*, 45 Brandeis Law Journal 659, 664–665 (2007). *See also* Jensen, The Price of Vigilance, *supra* at 189–213.

As this excerpt suggests, League members had a great deal of latitude both in terms of the activities they chose to investigate and the methods they employed.[964] So although League investigations initially focused on the German spy-saboteur threat, they soon drifted into other areas, such as disrupting union activities and enforcing blue laws.[965] The drift in the League's activities was attributable to the fact that the nature and scope of its responsibilities were not well defined when it was created, and the understanding of what was and was not authorized only deteriorated as time passed.[966] Some of the drift is also attributable to a lack of supervision; the Department of Justice did not, and perhaps could not, assign enough of its professionals to ensure adequate supervision of local League activities. As a result, League members not only exceeded the scope of their authority as civilians, they went far beyond what professional law enforcement officers were authorized to do.[967] As the excerpt above indicates, League members freely violated the Fourth Amendment by conducting illegal searches and arrests.[968]

This brings us back to the institution we are in the process of designing. As I noted earlier, our goal is to supplement the resources available to law enforcement and to the military by incorporating a level of "pure" civilian participation into the integrated professional effort outlined earlier in this chapter. That is, of course, the same goal that was responsible for the creation of the American Protective League.

---

964 It was routine for League members to make "arrests" even though, as civilians, they were not authorized to arrest anyone. *See* Jensen, The Price of Vigilance, *supra* at 93–94, 213–219.

965 *See*, e.g., State v. Rogers, 145 Minn. 303, 177 N.W. 358, 358 (Minn. 1920):

> The American Protective League, an organization formed to aid the United States Department of Justice i . . made the investigations which resulted in this prosecution. Two members of the league testified . . . that on the evening of December 19, 1918, they registered at the hotel and were assigned rooms; that they informed the clerk at the desk that they wanted two women; and that the women came to their rooms a few minutes thereafter.

See also Jensen, The Price of Vigilance, *supra* at 180–184.

966 *See id.* at 17–31. This happened even though the League had a detailed, complex organizational chart and an operations manual. *See id.* at 130–134. *See also* American Protective League, University of North Carolina at Asheville, http://toto.lib.unca.edu/findingaids/mss/biltmore_industries/american_protection_league/default_league.htm (League operations manual and organizational chart).

967 *See* Jensen, The Price of Vigilance, *supra* at 105–130.

968 *See id.* at 93–94, 213–219.

As I also noted earlier, we can allow "pure" civilians to contribute to this effort in either or both of two ways: They can, in effect, act as an online neighborhood watch by passing along information the observe may be relevant to actual or potential cyberattacks; they can also play a more active role in the processes of responding to cyberattacks by assisting with investigations, helping to apprehend perpetrators and contributing to law enforcement and/or military efforts to retaliate directly against online attackers. Members of the American Protective League did both: As we saw above, they collected information about "enemy" activities and gave it to state and federal law enforcement officers; they also conducted investigations, arrested suspects and, in some tragic instances, administered their own form of "justice."[969]

The League, though, was far from being a success. It serves as an object lesson in the need to take great care in formally incorporating any level of civilian participation into what have been government-monopolized activities. The presumptive advantage of creating an institution like the League is that it solves the vigilante problem;[970] because those who participate in such an institution operate on behalf of and with the approval of a government agency, they occupy a position midway between that of professional military and law enforcement personnel and "pure" civilians. The official involvement implies that the civilian participants in such an institution are trained and supervised sufficiently to prevent their engaging in the excesses associated with "pure" vigilantes.[971] That, sadly, was not true of the League; in many instances, the activities of League members were indistinguishable from those of "pure" vigilantes.[972]

Creating an institution like the League also has distinct disadvantages: As the League's short history demonstrates, one disadvantage of an institution such as this is that it can shield members from liability for violating law in their ostensible efforts to assist with its enforcement.[973] A related disadvantage is that the quasi-official status membership in such an institution

---

969 *See id.*

970 *See* Chapter 6.

971 *See id.*

972 *See* Jensen, The Price of Vigilance, *supra* at 147 ("Vigilantism may not have been an official tenet, but as long as members engaged in it, as long as the League did not oppose it but instead on occasion encouraged it, the APL appeared to . . . sanction to practices inimical to the very principles the United States was supposed to be defending).

973 *See id.* at 17–32.

confers can, as the history of the League again demonstrates, encourage excess and lawlessness.[974]

Our goal is to identify how "pure" civilian participation can be used to increase the effectiveness of the integrated law enforcement-military effort outlined above. An essential aspect of this endeavor is incorporating civilian participation in such a way that it does not undermine the integrity and professionalism of our attribution and response processes for cyberthreats. We seek to improve, not degrade, the methods we use to protect ourselves.

An approach modeled on the American Protective League would create a voluntary civilian organization that *actively* works to support law enforcement (or, for our purposes, an integrated military-law enforcement effort). I believe it is inherently inadvisable to allow active civilian participation in law enforcement or joint military-law enforcement efforts. The organizers of the League created a detailed set of rules and operating standards for their members, along with a complex national organization to enforce these rules and standards,975 but things still went tragically awry.

They went awry because League members actively engaged in law enforcement without being trained or supervised by law enforcement professionals.[976] The federal government did not allocate resources for training League members and halfheartedly assigned a few officials to supervise their activities. [977] The size and geographical diversity of the League's membership, coupled with the then rather primitive state of transportation and communication technologies, consequently made effective supervision impossible.[978]

Transportation and communications technologies have vastly evolved in sophistication, as has our understanding of the need to train and supervise those whom we trust to defend us from the varied manifestations of chaos. I, though, still believe we would see things go awry if we were to institutionalize active civilian participation in the integrated military-law enforcement effort outlined earlier in this chapter. I assume they would not go as seriously awry as they did with the League; I doubt, for example, that the members of

---

974 *See id.*

975 *See id.* at 130–150. *See also* American Protective League, University of North Carolina at Asheville, http://toto.lib.unca.edu/findingaids/mss/biltmore_industries/american_protection_league/default_league.htm (League operations manual and organizational chart).

976 *See* Jensen, The Price of Vigilance, *supra* at 130–50.

977 *See id.*

978 *See id.*

our Civilian Cybercorps would drag suspected cyberattackers off of trains and airplanes. But I think things would inevitably go awry because the factors I cited above—the ones that contributed to the League's disintegration into vigilantism—would again come into play.

We are assuming civilian participation in the integrated military-law enforcement effort outlined earlier in this chapter would be purely voluntary. Even if participation were voluntary, as it was with the League, such an effort would attract hundreds of thousands, even millions, of participants in the United States. As we saw earlier, the League attracted a quarter of a million members in the first year of its existence. That is an astonishing number given the substantially smaller size of the U.S. population in that era.[979] It is also astonishing given that would-be League members had to discover the organization, find the local office, show up at the local office, sign up, and then attend meetings and go out into the community to conduct their own investigations. Being a League member consumed a fair amount of time, effort, and probably some of one's own money, as well. It was also potentially fraught with the possibility of physical danger.

Compare that with participation in the again voluntary but online effort we are exploring: The government would disseminate information about the effort online, via emails and websites, which means it would be readily available; instead of having to seek out information about the effort, most people would either receive it directly or encounter it as they conducted their online activities. To participate, they probably would not have to go anywhere; they could enlist in the effort online, and most, if not all, of their training might be conducted online. They would carry out their volunteer efforts online because that is, after all, the point: Volunteers in the Civilian Cybercorps would carry out their neighborhood watch function by collecting information online and passing it along to their law enforcement-military contacts; they would also, for the purpose of this particular analysis, engage in more active efforts such as participating in investigations and supporting official retaliatory efforts against cyberattackers. They could, and would, do all of this from home and/or from their place of employment (if their employer approved). They could participate on their lunch hour or while relaxing at

---

979 There were approximately 106,000,000 people in the United States in 1920, so the 1918 figure would have been somewhat less. *See* U.S. Census Bureau, Selected Historical Decennial Census Population and Housing Counts, Urban and Rural Populations 1790-1990, http://www.census.gov/population/www/censusdata/hiscendata.html. The 2007 estimate for the U.S. population is 301,621,157. U.S. Census Bureau, Population Finder, http://factfinder.census.gov/servlet/SAFFPopulation?_submenuId=population_0&_sse=on.

home or on a boring Sunday afternoon; the essentially opportunistic nature of participation would no doubt make it appealing.

I suspect participation could become very popular—become "the thing to do." I can see participation in this effort going viral for several reasons: One is that the American public is infatuated with the exploits of the military and law enforcement; their fictional efforts have long been an integral feature of the popular media. This effort would let citizens participate, however minimally, in real-life versions of those exploits and then talk about what they had done (unless confidentiality was imposed). So instead of watching television shows such as *CSI* or *24*, citizens could play a role in combating evildoers who threaten the stability of their communities and their country. I can also see peer pressure coming into play; those who were not initially inclined to participate might feel obliged to join in—to do their part.

For these and probably other reasons, the effort we are currently postulating should, therefore, draw millions of Americans. We could restrict the number of participants by being selective in terms of who was allowed to join the effort, but that would be counterproductive to the extent our goal is to enhance the efforts of professionals by bringing in a substantial number of civilians. If part of our goal is to recruit an effective online neighborhood watch, this approach would clearly be counterproductive because it would limit the pool of observers. And selective recruiting would, as we saw earlier, require the expenditure of a great deal of effort; the government would have to implement a complex vetting process for people who applied to participate in the endeavor. The process would involve verifying identity (which should be an element of any version of such an effort) and stringently assessing the applicant's pertinent expertise and skills; it might also involve a background check to identify problematic personality characteristics, habits, associations, and/or experiences (such as criminal convictions). Selective recruiting would, in other words, require us to create a formal, free-standing institution with staff, facilities, and its own budget.

And that would be self-defeating: We are exploring how we can use civilians to support the integrated military-law enforcement effort outlined earlier in this chapter because we do not, and probably cannot, allocate the resources necessary to make these institutions effective in the battle against cyb3rchoas. Our goal is to devise an alternate strategy: a non-resource intensive informal effort that can enhance the joint and several efficacy of these institutions. We seek an alternative strategy because our resources are limited and because cyberthreats present new and distinct challenges for those charged with maintaining order; we will return to this last issue later in this chapter. My point here is that both from a resource perspective and from a tactical standpoint, our solution is not to create another traditional, hierarchical institution, not even as an adjunct to a nontraditional effort. Selective recruiting does not, then, seem feasible.

Selective recruiting, however, would be one way to try to prevent mission drift among the members of our volunteer Civilian Cybercorps. Another, at least equally important way to do this would be to assign military and law enforcement personnel to supervise the efforts of the volunteers; that, however, would incrementally reduce the number of professionals who could respond directly to cyberthreats. The alternative is even less acceptable: Not assigning professionals to supervise the volunteers could invite a degradation of effort; like the members of the American Protective League, participants in our Civilian Cybercorps might descend into spying, harassment, public humiliation and misplaced retaliation against those they believe are responsible for cyberattacks of whatever type.

Without supervision, non-professional volunteers can drift toward vigilantism out of an excess of zeal. Most, if not all, of the League members who descended into vigilantism did so because they truly believed they were protecting their communities and their countries from a serious menace; they did the wrong things for the right reasons. That possibility is a serious concern in orchestrating an effort of the type we are analyzing because the threat of vigilante drift may actually be even more pronounced in this context: In the League era, whatever the League members did occurred in real-space and was consequently "public" to a great or lesser degree; carrying out activities in public tends to inhibit us somewhat because we realize others can observe what we do and judge the propriety of our actions. Our inhibitions decrease as our actions become less observable or, to be more precise, less directly attributable to us, which is a primary cause of the phenomenon known as "road rage." When we operate an automobile, we feel more powerful and less visible; we know the vehicle is visible, but we implicitly assume that we are less identifiable.[980] So people who would never cut in front of another in a grocery line routinely cut other drivers off in traffic.

This phenomenon is even more pronounced online: My activities online may be visible—observable—to others, but I can conceal my identity by remaining anonymous or by assuming another identity. Those who study

---

980 *See*, e.g., Deborah Lupton, *Road Rage: Drivers' Understandings and Experiences*, 38 Journal of Sociology 275, 280 (2002); Kent Walker, *The Costs of Privacy*, 25 Harvard Journal of Law & Public Policy 87, 101 (2001).

cyberspace note that it can have an effect analogous to, but even more pronounced than, operating a motor vehicle. Many people become aggressive when they are online because of the presumed lack of identifiability and accountability.[981] I believe that phenomenon significantly increases the possibility that vigilantism will manifest itself in an online effort of the type we are postulating. The invisibility of a volunteer's actions coupled, as I noted earlier, with the authority assumed to derive from participating in the Civilian Cybercorps could easily lead to excess among a substantial number of the volunteers, particularly if they were not well-supervised. We would, in a sense, be revising the failed League experiment.

I believe, then, that sanctioning *active* civilian participation in an effort of the type we are exploring creates a potential for abuse and over-reaching which is, simply, unacceptable. I also believe, as I noted earlier, that creating a formal institution to implement such an effort in whole or in part is operationally counterproductive, not to mention fiscally impracticable. That leaves us with the residual possibility we take up in the next section: creating an essentially unstructured, voluntary organization the contributions of which are limited to providing information military and law enforcement personnel can use in responding to cyberattacks.

Informal Integration

I believe the residual possibility is the better path: Instead of using a formal institution as the conduit for civilian contributions, we rely on a loosely structured, voluntary organization in which the civilian participants' role is limited to reporting potential cyberattacks. Civilians could send their information to law enforcement, which passes it along to the military when this seems appropriate, or send it directly to law enforcement and the military. I suspect the best approach would be to have law enforcement act as the intermediary between the civilians and the military; the military would then decide whether law enforcement should vet the civilian-provided data in an effort to identify cyberwarfare-related information before passing it on to the military or send all he raw, unfiltered data to the military. This system might include one or

---

981 *See*, e.g., M. E. Kabay, *Anonymity and Pseudonymity in Cyberspace: Deindividuation, Incivility and Lawlessness Versus Freedom and Privacy*, Address Before Annual Conference of the European Institute for Computer Anti-virus Research 8 (March 1998), http://www2.norwich.edu/mkabay/overviews/anonpseudo.htm. *See also* Kent Walker, *The Costs of Privacy*, 25 Harvard Journal of Law & Public Policy 87, 101 (2001).

more exceptions that let civilians bypass the law enforcement intermediary when information involved what seemed to be exigent circumstances, such as a cyberwarfare attack. In those instances, the civilians could either send their information directly to the military or they could send it to designated personnel in an organization such as the one I outline below; the designated personnel in this organization could then pass the information along to their contacts in the military, which should ensure that it was not overlooked.

The civilian organization that implements this information-sharing process should, as I noted above, be as loosely structured as possible. Our goal is, as I explain later in this chapter, to avoid the resource-intensive, hierarchical organization and attendant evils that are inevitable in any formal institution: We want, among other things, to replace insular vertical hierarchies with a lateral, distributed network of interested parties and to substitute speed and flexibility for the strictures and dilatoriness of bureaucratic "channels."

To that end, the organization should be as virtual as possible; instead of having a fixed location and physical facilities, it should consist of a web of civilians who are networked by e-mail and secure websites. It should use its websites and e-mail to train volunteers and make their contributions as effective as possible. Volunteers should be trained when they join the effort and thereafter on a continuing, probably as-needed basis. The initial training has two purposes: One is educating volunteers about the type of information that is likely to be useful to law enforcement and/or the military; the other is educating them about the procedures they will use to report a suspected cyberattack. The organization should also supply them with a dynamic set of operating standards and cyberevent identification criteria. The identification criteria, which should be updated as often as possible, alert the volunteers to the kinds of attacks and attack signatures they can expect to encounter online. By focusing the volunteers' efforts in this way, the criteria should at once make them more effective in identifying real attacks and reduce the extent to which they report non-attacks.

Operationally, the role these volunteers would play in the cyberattack attribution and response effort is analogous to the role civilians played in providing the military with information about enemy aircraft during World War II. In the United States, the Civil Air Patrol "enrolled civilian spotters in reconnaissance. . . . [They] were trained to recognize enemy aircraft, so as to report if any were seen."[982] Throughout the war, civilians all over the

---

982 "United States home front during World War II," Wikipedia, http://en.wikipedia.org/wiki/Homefront-United_States-World_War_II#Civilian_support_for_war_effort.

United States watched the skies, but most never saw an enemy aircraft; because World War II military aircraft had a relatively short flying range, few made it onto U.S. territory.[983] The Observer Corps played a much more important role in wartime Britain, which was not as well insulated by geography. As one author noted, the civilian members of the Observer Corps "were an integral part of the defense system of Great Britain. One official history described the Observer Corps as 'the primary source of intelligence for the whole defence system about the movement over Britain of hostile aircraft.'"[984]

Like these and other, similar efforts, the organization I am postulating would recruit civilians to provide authorities with information about potential threats that are virtual, rather than physical. As with the analogous efforts we examined earlier in this chapter, the primary virtue of this approach is that it gives law enforcement and the military access to information they have not yet received or that they might not otherwise receive. The incrementally improved access to information should enhance their respective abilities to identify and respond to cyberattacks.

The additional information should, for example, benefit law enforcement by alleviating the underreporting of cybercrime that makes it difficult, if not impossible, for law enforcement personnel to identify patterns and trends in cybercrime and cyberterrorism.[985] If the information provided by civilian volunteers can improve law enforcement's ability to identify patterns in the incidence of these cyberthreats, it will also improve their ability to respond to those threats; as we saw earlier, identifying patterns in threats to internal order lets law enforcement allocate its resources more effectively.[986] Law enforcement personnel can concentrate resources in areas where threats are likely to manifest themselves. For real-world crime, this means focusing law enforcement personnel and other resources in certain geographical areas; for cyberthreats,

---

983 This effort proved successful "almost to a fault," as in the "Plains states where many dedicated aircraft spotters took up their posts night after night . . . in an area of the country that no enemy aircraft of that time could possibly . . . reach." *Id.* Japan was able to launch a few air strikes in the U.S.: In September 1942, Japanese aircraft twice dropped/tried to drop bombs in Oregon; and in an effort to start forest fires or cause other damage, Japan launched 9,000 fire balloons toward North America between November 1944 and April 1945. See "Attacks on North American during World War II," Wikipedia, http://en.wikipedia.org/wiki/Attacks_on_North_America_during_World_War_II.

984 W. Hays Parks, *Air War and the Law of War*, 32 Air Force Law Review 1, 123 note 379 (1990) (quoting Terence O'Brien, Civil Defence 4 (HMSO and Longmans 1955)).

985 *See* Chapters 2 & 3.

986 *See* Chapter 2.

this would probably mean focusing personnel and resources on likely targets, such as financial institutions, known vulnerabilities, and so forth. The goal is to give law enforcement the ability to be at least somewhat proactive, instead of simply responding after crimes, or cybercrimes, have been committed.

The additional information generated by this approach should provide similar benefits for the military. As we saw earlier, cyberwarfare, unlike its real-world counterpart, is likely to be directed at civilian targets.[987] As we also saw, cyberwarfare is not likely to begin with a dramatic, Pearl Harbor-style attack; it is more likely to begin with a series of probes, smaller attacks that test security on particular, disparate systems.[988] The owners or operators of the systems may not realize anything untoward is occurring; since attacks on computer systems are not uncommon, those responsible for these targeted systems may assume the probes are routine forays by run of the mill cybercriminals. They will therefore either ignore the probes or report them to a local law enforcement agency; that agency will probably assume the probes are "mere" cybercrime. The likelihood of that occurring is enhanced by the extent to which the probes target systems that are geographically and/or functionally disparate; if our hypothesized probes target systems that are in different parts of the country and are used in different endeavors (e.g., banks, online retailers, and air traffic control systems), local law enforcement officers may well not realize that what appear to be discrete cybercrimes are really aspects of a cyberwarfare assault.

My point here is the one I made earlier: a country could be undergoing cyberwarfare attacks but not realize it. The integrated military-law enforcement effort we postulated earlier in this chapter should help a victim nation-state to realize when cybercrime is really cyberwarfare in disguise because it ensures that information which comes to law enforcement is shared with the military (and vice versa). I do not, though, believe we can rely solely on that effort because it depends on information that comes through formal channels; law enforcement and the military obtain information only if a victim reports a cyberattack to law enforcement (or, perhaps, to the military). If a victim does not report an attack, which is often the case, neither constituency is likely to learn of it, at least not in a timely manner. To remedy this operational deficiency, we must adopt an approach—such as the voluntary

---

987 *See* Chapter 3.

988 *See id.*

effort we are currently analyzing—that brings as much additional, current information as possible to the attention of both constituencies.

We could try to do this by encouraging civilians to report every attack they experience, but this is not likely to be effective for several reasons. One is that, as we saw earlier, there are disincentives for corporate and other commercial victims to report their having been attacked; if information about the attacks becomes public, it can erode confidence in the victim's ability to secure its systems and protect those with whom it does business.[989] Under the current system, this information may well become public: When victims report attacks, the information they provide is likely to become public as an incident of law enforcement's investigating the attack or as a result of the perpetrator's being identified and charged with the attack. Another reason that victims—both corporate and individual—do not report cyberattacks is they do not believe law enforcement can, or will be able to, do anything about them;[990] this perception should change if we can implement procedures that visibly improve law enforcement's effectiveness against cybercriminals. Other victims, especially individual victims, may not realize that what happened to their system was a crime that could, and should, be reported to the authorities. They may dismiss the attack as an aggravation and assume it is not anything a law enforcement officer (or a military officer) would take seriously. And, finally, many victims do not report their having been victimized because they do not realize it has occurred.[991]

The organization we are postulating could institute measures designed to overcome these disincentives and encourage civilians to report information they have about a possible cyberattack. It could, for instance, provide assurances that any information someone shares with the organization would be kept as confidential as possible for as long as possible. Some law enforcement agencies have implemented similar measures: A few years ago, anyway, the U.S. Secret Service's New York Electronic Crime Task Force would not release information about its investigations "without the express consent of the . . . corporate victim because" doing otherwise would "discourage the victims of electronic crimes from reporting."[992] This seems to be a continuing

989 *See id.*

990 *See id.*

991 *See id.*

992 Robert N. Weaver, Assistant Special Agent in Charge, New York Electronic Crimes Task Force, Testimony Before the House Science Committee (October 10, 2001), http://www.nymissa.org/documents/testimony.htm.

aspect of the Electronic Crime Task Forces that are being established around the United States.[993] In December of 2002, the United Kingdom's National High Tech Crime Unit (which would later be disbanded) implemented a Confidentiality Charter that was designed to "increase the reporting of electronic crime" by overcoming businesses' fears that the information they provided would become public.[994] In the Charter, the NHTCU pledged to keep information provided by victims confidential and to sanitize information before sharing it with those outside the NHTCU.[995] And the Federal Bureau of Investigation's InfraGard program, which encourages information-sharing between and among law enforcement and members of the private sector, allows victims of cyberattacks to sanitize information they give the FBI before it is distributed to InfraGard members.[996]

We could also try to encourage victims to report cyberattacks by offering them some assurance that the information they provided would not be used to initiate criminal proceedings against a perpetrator unless they agreed to that step. Although the NHTCU did not include a pledge to this effect in its Confidentiality Charter, it seems to have been willing not to pursue charges in certain instances, presumably to encourage a victim to share information

---

993 *See*, e.g., Laura Taylor, *Secret Service Cyber Forensics Team Available to Assist Corporate America*, Relevant Technologies (June 7, 2007), http://www.relevanttechnologies.com/secserv_070607.asp (Secret Service agents who "work on information security intrusion understand that confidentiality is important, and they don't disclose any information to the press"). *See also* United States Secret Service, Electronic Crimes Task Forces and Working Groups, http://www.secretservice.gov/ectf.shtml.

994 *See*, e.g., Iain Thomson, *First E-Crime Congress Meets*, vnunet.com (December 12, 2002), http://www.vnunet.com/News/1137500. *See also* National High Tech Crime Unit, The Confidentiality Charter: NHTCU Working with Business (November 2002).

995 *See id.*

996 *See* Testimony of Ronald L. Dick, Director, National Infrastructure Protection Center Before the Senate Judiciary Committee—Subcommittee on Technology, Terrorism and Government Information (July 25, 2001), http://www.fbi.gov/congress/congress01/rondick072501.htm:

> A key element . . . is the confidentiality of reporting by members. The reporting entities edit out the identifying information about themselves on the notices that are sent to other members of the InfraGard network. This . . . protects the information provided by the victim of a cyber attack. . . . This measure helps to build a trusted relationship with the private sector and . . . encourages other private sector companies to report cyber attack to law enforcement.

InfraGard also operates a secure website that lets InfraGard members share information about attacks. *See* About InfraGard, InfraGard, http://www.infragard.net/about.php?mn=1&sm=1-0.

that could be used in countering other, similar attacks.[997] The NHTCU's interest in obtaining information for purposes other than prosecution is evident in its Confidentiality Charter, which differentiates between providing information for "intelligence purposes" and for "investigative purposes."[998] The Charter also notes that the purpose for which information is being provided—intelligence or investigation—should be established in the NHTCU's initial contact with the victim.[999]

We could incorporate a similar component into the reporting structure of the organization we are hypothesizing. As I see it, the organization would collect information from two sources: civilians who are affiliated with, and have been trained by, the organization; and civilians who are not affiliated with the organization but have cyberattack information they wish to share, so it can be passed along to law enforcement (and/or the military). I am assuming corporate entities are as likely to participate in this organization as are individuals; and while either could become the victims of cyberattacks, I suspect corporate victims are more likely to be concerned about confidentiality than individual victims. The confidentiality measures I outlined above would, of course, apply to corporate and individual victims, and to "outsiders," as well as to members of the organization. A commitment not to pursue charges in exchange for providing information about an attack could also apply to those who are, and are not, affiliated with our hypothetical organization,

---

997 *See* Thomson, *First E-Crime Congress Meets, supra* ("In some cases the police may decide not to prosecute offenders so long as no major crime or violation of the [European Union's] human rights legislation has taken place").

998 *See* National High Tech Crime Unit, The Confidentiality Charter: NHTCU Working with Business (November 2002). The distinction between "investigation" and "intelligence" is not always clear, but in this context it is essentially based on the purposes for which information is being gathered: "Investigative" information is information law enforcement personnel collect for the purpose of initiating criminal proceedings against the perpetrator of a crime, or cybercrime. "Intelligence" information is information military and/or law enforcement personnel collect for several reasons, such as monitoring what particular individuals or groups are doing, determining whether they may have been involved in particular attacks and attempting to determine what they might intend to do in the future. As one author noted, "intelligence fathering is. . . . less targeted and more programmatic than law enforcement collection." William C. Banks, *And the Wall Came Tumbling Down: Secret Surveillance after the Terror*, 57 University of Miami Law Review 1147, 1152 (2003). Or, as another source put it, law enforcement investigations focus "on determining causative factors explaining *past* events", while intelligence gathering focuses on "the extrapolation of current events to provide plausible representations of the *future*." Strategic intelligence versus investigations, Strategic Insights (November 27, 2007), http://strategic-insightsblog.com/strategic-intelligence-versus-investigations/.

999 *See* National High Tech Crime Unit, The Confidentiality Charter, *supra*.

but I suspect it would be more likely to appeal to the "outsiders," to those who have fortuitously become the victims of cyberattacks and want to share information about what happened to them.

I suspect it would be more likely to appeal to "outsiders" because their nonaffiliation with our organization is, in many instances, likely to be a function of a general disinclination to report attacks for fear of negative publicity. The confidentiality assurances I proposed above might not be enough to persuade those in this category to report their victimization. They might be willing to provide information about attacks they sustained if they were confident it would be used for general, "intelligence" purposes rather than for a criminal investigation and prosecution. If we allowed victims to condition their providing information about an attack on a guarantee that it would not be used for "investigative" purposes, we could probably encourage some to come forward who would not do so otherwise. The question is, is it acceptable to enter into such an agreement?

My initial inclination is to say it is not, but that is because I am the product of a culture that implicitly, and almost instinctively, tends to equate "attack" with "crime" and to value the apprehension and conviction of those who perpetrate crimes over many things, including the privacy and dignity of crime victims.[1000] That calculus may be an historical artifact, the product of the millennia we have spent struggling to maintain internal order; we implicitly assume internal order justifies our demanding certain sacrifices from those who have been victimized, and that is true, as far as it goes. The calculus is limited in that it only balances a victim's sacrifice against a system's sacrificing its ability to sanction one whose actions undermine internal order. Here, the greater good clearly demands the victim's sacrifice; to do otherwise would allow perpetrators to evade prosecution by ensuring their victims would refuse to cooperate with law enforcement.

Because it is the product of a nonwired world, this calculus does not incorporate the possibility that another, even greater, good could weigh in favor of allowing a victim to veto prosecution. The calculus does not incorporate that possibility because it assumes a closed system: an absolute segregation between threats to internal order and threats to external order. As we saw earlier, such a segregation was the norm prior to the rise and proliferation of modern technology, particularly computer technology.[1001]

1000 *See*, e.g. Rachel A. Van Cleave, ***Rape and the Querela in Italy***, 13 Michigan Journal of Gender and Law 273 (2007); Cheryl Hanna, ***No Right to Choose: Mandated Victim Participation in Domestic Violence Prosecutions***, 109 Harvard Law Review 1849 (1996).

1001 *See* Chapters 2 & 3.

As we also saw earlier, our use of cyberspace erodes this segregation and, in so doing, introduces a new factor into the traditional calculus: Instead of merely balancing a victim's sacrifice solely against a system's sacrificing its ability to sanction one whose actions threaten internal order, we must balance the victim's sacrifice against the system's ability to sanction one whose actions threaten internal order and/or external order.[1002] More precisely, in the expanded calculus we must balance a victim's sacrifice against a system's ability to respond to one whose actions either (a) threaten internal order (only) or (b) threaten internal order and external order.[1003] When an attacker's actions only threaten internal order, that is, are directed solely at one or more individual victims, the traditional calculus applies. When an attacker's actions threaten both internal and external order, we must apply the expanded calculus, which inferentially creates the possibility of allowing a victim to provide information about an attack in exchange for the state's promising not to pursue criminal charges against the attacker, if and when she is apprehended.

It seems to me that we should incorporate this option into the information-sharing system we are exploring in this chapter. The option could, and should, be used selectively; whether the state will make such a promise in a given instance should depend on a number of factors, the most important of which is the nature and extent of the threat the attacks pose to the state's ability to maintain external order and the value the victim's information has for those who are attempting to nullify this threat. As to who would make that decision, it would most certainly not be committed to the members of the voluntary, civilian organization we are currently exploring. It seems only logical to have the decision made by representatives of the integrated military-law enforcement effort we examined earlier in this chapter; representatives from the military would be in the best position to assess the value of the victim's information, while representatives from law enforcement would be able to assess the relative "cost" of entering into a promise not to use the information to prosecute. And the promise, if given, should only be a promise not to use the information provided by the victim to prosecute; it should not be a promise not to prosecute at all. We should preserve the

---

1002 *See id.*

1003 There is of course a third category: instances in which the perpetrator's actions only threaten external order. That scenario is not implicated in the calculus outlined above because there is no discrete, individual victim; the victim, as such is the nation-state that is being attacked.

option to bring a prosecution based on information from extraneous sources. If we were to let the possessor of useful information veto any possibility of prosecuting those responsible for an attack, we would give the possessor of that information more than that to which he or she is entitled and we would open up the possibility that cyberattackers could use information about their attacks to, in effect, extort absolute immunity from prosecution.

Operationally, we could implement the option by designating particular representatives from law enforcement and from the military, who would then become the points of contact whenever someone approached the voluntary organization we are hypothesizing with an offer to provide information in exchange for an offer not to use it to prosecute.[1004] Doing this should also help to ensure that "bad guys" of whatever strike (criminals, terrorists and/or nation-state actors) cannot exploit the availability of this option to gain insight into the nature and quantity of information the organization has collected on specific attacks or to inject corrupt information into the system.

Conversely, we could encourage other victims, and witnesses, to report by letting them know their information has value for law enforcement and/or the military. The strategy here is to encourage citizens to report by letting them know they will be taken seriously and by putting them on notice that by providing information, they are participating in efforts to maintain order and security in their country. That aspect of this approach is directly analogous to the appeal that was used to recruit civilians into the World War II enemy aircraft spotting efforts; it is also used to encourage citizens to participate in Neighborhood Watch programs.[1005]

Finally, an effort such as this could increase the likelihood that attacks will be reported because it creates the possibility that reports can come from those in the best position to identify an attack: the computer staff. As I noted above, commercial entities, in particular often do not report cyberattacks for fear of negative publicity and its corrosive effect on their organization. We have already postulated two measures that are designed to reduce corporate hesitance in this regard: a promise to keep information confidential for as long as possible and a promise not to use the information provided to prosecute.

1004 The law enforcement representatives designated for this purpose could, and probably should, include a prosecutor, as well as law enforcement personnel. Prosecutors, after all, are the ones who decide whether or not charges should be brought in a particular instance.

1005 *See*, e.g., National Sheriffs' Association, Neighborhood Watch: A Manual for Citizens and for Law Enforcement 5–8, http://www.usaonwatch.org/AboutUs/Publications.php.

Both target the entity decision makers who are ultimately responsible for whether the entity itself will provide information to our hypothetical organization and the military-law enforcement effort it supports. These measures are likely to be effective in some instances, but will certainly prove ineffective in others. We could address the latter, residual category of non-reporting by incorporating an additional measure into the system we are postulating: We could allow citizens to provide information anonymously.

That option might be of relatively little significance for individual or other noncorporate victims, but it could increase the likelihood that relevant information about attacks on corporate entities, especially commercial corporate entities, would be reported to our organization: It would, in effect, give the computer staff of such organizations a backchannel they could use to provide information when it would otherwise not be forthcoming. The computer staff is likely to understand the peculiar need to report information about a given attack; they may realize that it significant for its unique attack signature, for its apparent point of origin, for its being repeated or for any of a variety of reasons. Computer staff may lobby the organization's decision makers for permission to report an attack, but be denied the ability to do so. In the system we have so far outlined, as in society today, if the corporate decision makers decide not to release information, it will not be released. The question here is whether we should allow personnel within such an organization—typically its computer staff—to bypass the organizational leadership and report information anonymously.

The advisability of implementing this option depends on how we resolve a more general issue: What are the advantages and disadvantages of allowing anonymous reporting?[1006] The primary, and perhaps only, advantage is the one we have already seen: Anonymous reporting will encourage those who would not otherwise provide information about cyberattacks to do so. Allowing anonymous reports would—to some unknown and probably unascertainable extent–increase the quantity of information our organization

---

1006 There are, I am sure, other policy arguments that can be made against instituting this option; it does, after all, encourage institutional disloyalty. I am not a corporate lawyer and therefore cannot speak to the commercial or other civil consequences of giving corporate staff the option of disregarding institutional policy and reporting information that may ultimately be detrimental to the interests of the organization. I therefore limit my analysis to the legal and policy issues the implementation of such an option would have for the organization we are hypothesizing. As to the general propriety of encouraging employee faithlessness, I would only note that there are a number of U.S. statutes that expressly protect corporate whistleblowers. *See* "Whistleblower," Wikipedia, http://en.wikipedia.org/wiki/Whistleblower#Legal_protection_for_whistleblowers.

could collect. The obvious disadvantage of allowing anonymous reporting is the possibility—the probability—of introducing error into the information that is collected.

Error could arise in either of two ways: A person acting in good faith who was afraid to identify himself (a corporate employee, say) could provide information anonymously, believing that it was accurate and that it would be helpful to the effort we are exploring in this chapter; the person could, though, be wrong. The information he provided could be inaccurate for any of a number of reasons, for example, it was incomplete, it was out of date, it is taken out of context, it was distorted in the collection and/or transmission process. This would be a good-faith error, but it is still an error. Because the person who provided the information chose to remain anonymous, we might have no way to determine its accuracy. When an identified victim or witness provides information about an identified attack, the law enforcement and/or military personnel to whom the information is ultimately transmitted should be able to verify its accuracy by contacting the person who submitted the information and analyzing the system that was allegedly attacked. They do not have that option here: If the person who provides information about an attack elects to remain anonymous—and endeavors to obscure the location of the attack—the law enforcement and/or military personnel to whom the information has been transmitted *might* be able to determine the accuracy of the information to a greater or lesser extent by analyzing it in the context of verified information they have received concerning similar and/or related attacks. Even if they could do this, some level of uncertainty—some potential unreliability—would still be associated with the information.

The other, more troubling type of error anonymous reporting could allow is the advertent, intentional reporting of inaccurate information about an attack. Cyberattackers could attempt to frustrate law enforcement and/or military efforts to identify them and respond to their attacks by providing inaccurate information to our hypothetical information-gathering organization. They could, in other words, use the anonymous reporting option in an effort to corrupt the information in the system and thereby frustrate efforts to respond to past attacks and prevent future attacks. Because the error here is intentional, it would probably be much more difficult for law enforcement and/or military personnel to determine the accuracy of the information.

I cannot say that the advantage of allowing anonymous reporting overcomes these distinct and serious disadvantages. Our choice, though, may not be zero-sum: Instead of deciding whether to allow or not allow anonymous reporting, we could structure the reporting process used by our organization so that anonymous reports were accepted, but they would be flagged and quarantined until law enforcement/military personnel could determine whether they were reliable enough to be incorporated into attribution and response analyses. Reports that were ultimately not deemed reliable enough for these purposes could be retained as a potential source of useful data in the future. This triage strategy would let us realize the advantages to be gained from anonymous reporting while reducing the risk of having anonymously provided information corrupt the data being relied upon for attribution and response analyses.

Sum: Modifying the Present System

My goal to this point has been to consider how we might modify our current approach to maintaining internal and external order to improve its ability to respond to cyberattacks. I have conducted the analysis on a broad, conceptual level for a very basic reason: We are speculating about more or less radical changes in systems we take for granted. Because we implicitly assume the "rightness"—the inevitability—of these systems, it is difficult to postulate how they can, and should, be altered. I have therefore limited my analysis to the critical issue: Are there structural modifications we ***could*** make while remaining within the constraints imposed by current law and policy? If we answer that question in the affirmative and elect to proceed, we can then analyze the micromechanics of implementing these modifications.

What we have seen to this point in our analysis is that as long as we continue to operate within those constraints, any modifications of the current approach necessarily are limited to (a) making information-sharing more permeable between the military and law enforcement; and (b) creating an initiative and a pliant organizational structure that will encourage, and allow, civilians to report information they acquire about cyberattacks. These modifications should enhance the effectiveness of the institutions we rely on to maintain order, both internally and externally; as we have seen, information about cyberattacks is essential, just as information about crimes, acts of terrorism and military assaults has been essential in the past. As we have also seen, the migration of these threats into cyberspace can make the collection of threat information far more difficult; online attackers can obscure not only their identities, but the nature and origins of their attacks, as well. Any effort that increases the quantum of attack data available to our formal institutional response mechanisms should improve nation-states' ability to control cyb3rchaos.

The civilian information-sharing effort outlined in the preceding section may seem a very modest contribution to this effort, but if it were to function as we have hypothesized it could be very helpful. In the section above, I noted the existence of two U.S. organizations that seek to encourage information-sharing among law enforcement personnel and between certain civilians and law enforcement personnel: The U.S. Secret Service's Electronic Crime Task Forces and the Federal Bureau of Investigation's InfraGard program. According to the InfraGard website, there are, as I write this in the summer of 2008, 23,682 members of InfraGard; I do not have figures for the Electronic Crimes Task Forces, but I suspect their membership is comparable to that of InfraGard. We will therefore assume for the purpose of analysis that these two valuable initiatives involve the participation of, say, 50,000 people in the United States.

That presumptive figure is in one sense very impressive: It has, I know, taken a great deal of effort to recruit this many people; participation in the Electronic Crime Task Forces and InfraGard chapters is selective. It is not as selective as, say, qualifying for a security clearance, but it is not open to the general public; both efforts informally recruit federal, state, and local law enforcement personnel plus civilians who are involved in the information security or the general corporate sector.[1007] Neither organization is designed to, nor aspires to, be ecumenical in its recruitment; they are intended to bring together people who can share information and expertise and, in so doing, improve the state's ability to respond to threats directly targeting aspects of the United States' critical infrastructure.

I admire the organizers of both groups, and I firmly believe both are playing an important role in dealing with cyberattacks. Both are innovative in that they are attempting to break down the associational and institutional barriers between individuals in various sectors of our society; both let law enforcement officers exchange information and share expertise with civilians from the information security and corporate sectors, and vice versa. As I explain in the next section, I believe that these barriers directly impede nation-state efforts to control cyb3rchaos.

1007 See, e.g., Become a Member of InfraGard, InfraGard, http://www.infragard.net/member.php: InfraGard is "dedicated to the protection of the United States. . . . [A]pplicants undergo a background check . . . by the FBI" and are "screened according to a defined criteria."

That is why I find our estimated membership figure for the Electronic Crimes Task Forces and the InfraGard chapters to be not so impressive. It is impressive given what the organizers of the efforts have set out to do; but it is also, I submit, not so impressive given what we need to do. In the next and final section, I explain what I believe we need to do and why.

New Processes, New Systems?

This section is the most difficult to write because it requires imagining what is not, and may never have been. As I noted earlier, this strategy encompasses two discrete alternatives: One is to retain the nation-state as our default governance structure but reconfigure our threat response processes so that they become effective against both offline and online threats. The other, more radical alternative is to replace the nation-state with one or more new governing configurations the legitimacy of which is not defined by controlling specific territory; the premise here is that the difficulties we increasingly experience in controlling cyb3rchaos are attributable to our continuing reliance on a system that evolved to deal with real-space and cannot adapt to an environment in which real- and virtual-space interact. We explore each option below.

Reconfigured Threat Response Processes

The first and most obvious change we could make in our threat response processes is to eliminate our insistence on partitioned response authority. The United States may be the most extreme example of this. As we saw in Chapter 6, U.S. law creates a rigid, essentially absolute partition between external threat response authority (the military) and internal threat response authority (law enforcement). Other countries rely on a similar approach to response authority, but the partition between the categories may not be as absolute as it is in the United States.

In Chapter 6, we also saw why this rigid partitioning of response authority exists in the United States. Partitioned response authority is the norm in other countries for similar reasons, the most important of which is a desire to retain civilian control of the process of maintaining order. Maintaining order has, at least to this point in our history, inevitably involved the use of physical force; as we saw in Chapter 6, the institutions we rely upon to maintain order have a monopoly on the legitimate use of force within a nation-state. Partitioning response authority according to the type of threat involved (internal versus external) divides those who are authorized to use physical force into two constituencies and, in so doing, presumptively reduces the likelihood that they will combine to depose civilian authority and assume control of a nation-state.

There were also other reasons to partition response authority: As we saw in Chapters 6 and 7, dividing authority among two distinct institutions made it possible for professionalism and specialization to develop within each institution, and that made each more effective in carrying out its designated tasks. A unified, essentially amateur response authority sufficed throughout the early millennia of our history, when human groupings were small and did not occupy fixed territories; an undifferentiated response authority is sufficient in small groups, because informal social controls are more effective in this context. As groupings expanded beyond the band and tribe into city-states, empires, and, finally, nation-states, it became necessary to supplement the eroding influence of informal social controls with the formal rules we examined in Chapter 2 and to create systems for enforcing those rules.

Our practice of partitioning response authority evolved as a response to two empirical developments: One was the constant increase in the population of discrete human groupings; the other was the linking of each grouping with a particular territorial area, a practice that was institutionalized by the Peace of Westphalia.[1008] By the seventeenth century, the focus was on securing the area controlled by a particular grouping from encroachments by other groupings and maintaining the baseline of internal order required for the populace to survive, reproduce and, perhaps, prosper.[1009] This divided focus quite pragmatically produced the initial partitioning of response authority. As we saw in Chapter 7, as human groupings coalesced around, and became defined by, territory, they found it necessary to fend off ever more dedicated, ever more sophisticated, efforts to compromise their respective dominion of particular territories. This, as we also saw in Chapter 7, resulted in the differentiation and professionalization of the cadre of individuals charged with fending off these threats—the military.

As nation-states became more distinct and more insular, the resulting concentration of discrete populations in specific territory led to a rise in urbanization. As increasing urbanization produced a rise in the number and

---

1008 *See* Chapter 7.

1009 *See* Chapters 2 & 7.

sophistication of internal threats, the military served as the model for a correlate cadre of individuals who became responsible for controlling these threats and maintaining internal order. As many have noted, law enforcement is in many respects a clone of the military: Both are authorized to use military force, both are hierarchically organized with a ranked division of authority, both tend to wear distinctive uniforms, and the members of both are, in varying degrees, distinct from the third cadre in any nation-state: the civilians.

This essentially dichotomous system functioned quite satisfactorily until toward the end of the last century. Its declining efficacy is, as we have seen, in large part attributable to the rise of computer technology and our immersion in cyberspace, but I suspect it began well before that. I believe the decline is in large part attributable to what Justice Cardozo described as "the tendency of a principle to expand itself to the limit of its logic."[1010]

The bifurcated system outlined above was quite adequate for the threat environment nation-states confronted throughout the nineteenth century and the early part of the twentieth century, but the threat landscape became more complex as the last century progressed. The increasing complexity of the threats nation-states confronted was primarily due to the evolving sophistication and proliferation of various technologies, including the automobile, the telephone, the airplane, and vast advances in military technology.[1011] As the threat environment became more complex, nation-states responded by replicating the strategy they had devised to deal with the core threats—crime and warfare.

So nation-states divided and sub-divided their militaries into distinct cadres, each of which was charged with implementing a particular aspect of the national response to external threats. By the middle of the last century, the U.S. military was divided into the Army, the Navy, the Marine Corps, the Air Force, and the Coast Guard.[1012] As I write this in 2008, the U.S. Army is subdivided into 23 distinct commands, each of which is subdivided in a series of organizational iterations.[1013] Like the Army, and like military organizations around the world, the other divisions of the U.S. military have the same fractionated, hierarchically tiered structure. They all have this structure for two reasons: One is purely historical: replicating what was effective in the past.

1010 Benjamin N. Cardozo, The Nature of the Judicial Process 51 (Yale University Press 1921).

1011 *See* Chapter 3.

1012 *See*, e.g., "Military of the United States," Wikipedia, http://en.wikipedia.org/wiki/Military_of_the_United_States.

1013 *See* U.S. Army, Organization: Units and Commands, http://www.army.mil/institution/organization/unitsandcommands/commandstructure/.

The other reason is that, as we saw in Chapter 7, a division of labor and the specialization it can allow have been effective principles for organizing activity, especially in the physical world.

For the same reasons, we see fractionated, hierarchically tiered structures in civilian law enforcement in the United States and other countries. In the United States, the basic division is between federal law enforcement and state law enforcement, each of which is subdivided with increasing levels of complexity. Federal law enforcement is divided into a number of agencies, including the Federal Bureau of Investigation; the U.S. Secret Service; the Bureau of Alcohol, Tobacco, Firearms and Explosives; the U.S. Marshals; the Drug Enforcement Agency; and the Immigration and Customs Enforcement agency. Each agency is segmented into divisions, subdivisions and other organizational increments.[1014] State law enforcement is initially apportioned into 50 units, one for each of the states; within each state, law enforcement is divided into state law enforcement[1015] and local law enforcement, which is subdivided into county and municipal law enforcement.[1016] Each law enforcement entity is complemented by a prosecutor's agency, which will display the same fractionated, hierarchically tiered structure and will interact with law enforcement personnel as they go about maintaining internal order.[1017] Military entities have their own complementary agencies, such as the

---

1014 *See*, e.g., "Federal Bureau of Investigation," Wikipedia, http://en.wikipedia.org/wiki/Federal_Bureau_of_Investigation#Organization (FBI is divided into five branches, each branch being subdivided into divisions or offices, and each division or office being further divided). There are also various other federal law enforcement agencies, such as the U.S. Mint Police.

1015 *See* "State police," Wikipedia http://en.wikipedia.org/wiki/State_police (e.g., State Police, Highway Patrol, Department of Public Safety, Department of Law Enforcement, State Bureau of Investigation, State Bureau of Narcotics, Motor Carrier Enforcement and Park Police).

1016 County law enforcement consists of a sheriff's office and/or a county police department. *See* "County police," Wikipedia, http://en.wikipedia.org/wiki/County_police#United_States. Municipal law enforcement consists of the police in each city, town or village. According to the Bureau of Statistics, there were nearly 13,000 local police departments in the United States in 2000. *See* U.S. Department of Justice Office of Justice Programs, Bureau of Justice Statistics, Local Police Departments 2000, http://www.ojp.usdoj.gov/bjs/abstract/lpd00.htm.

1017 *See*, e.g., Prosecuting Attorneys, District Attorneys, Attorneys General & US Attorneys, http://www.eatoncounty.org/prosecutor/proslist.htm. And there are over-arching agencies like the Department of Homeland Security, which also display the fractionated, hierarchically tiered structure described above and which interact with law enforcement entities of varying types. *See* About the Department of Homeland Security, http://www.dhs.gov/xabout/.

Department of Defense, the Department of State, the Central Intelligence Agency, and the National Security Agency, each of which also displays this structure.

This brings us back to the tendency of a principle to expand itself to the limit of its logic. The size and intricacy of the military and law enforcement institutions is no doubt exaggerated in the United States, a function of the relatively extensive territory and population it must protect. But the pattern I described above replicates itself in every modern nation-state for the reasons I noted above and because nation-states have a tendency to adopt strategies they believe are proving effective for other nation-states. The result is that we have consigned the processes of maintaining order to increasingly Byzantine, honeycombed organizational structures that often compete with each other. In the United States, for example, while law enforcement personnel are committed to working together at a macro level, rivalries manifest themselves at the micro level. State law enforcement often resent the intervention of federal law enforcement personnel, especially the FBI; state officers complain that federal agents tend to "take over" investigations and "claim the credit" for any successes that occur.[1018] There are also rivalries among federal law enforcement agencies[1019] and among the various branches of the military, in the United States and elsewhere.[1020] For the moment, though, I want to concentrate on law enforcement.

The rivalries are primarily a product of institutional loyalties: If we are going to create agencies and subdivisions within agencies, we should not be

---

1018 *See*, e.g., Lee Hammel, *Agent Closes the Case on a Long Career*, Worcester Telegram & Gazette (January 29, 2007), 2007 WLNR 1789485 ("The rivalry between state police and the FBI is legendary"). *See also* Timothy L. O'Brien & Lowell Bergman, *Law Enforcement Rivalry in U.S. Slowed Inquiry on Russian Funds*, New York Times (September 29, 1999), http://query.nytimes.com/gst/fullpage.html?res=9805E5DA173EF93AA1575AC0A96F958260; Martin Gottlieb, *The Twin Towers—Rivalry: Animus Surfaces Between Rival Agencies*, New York Times (March 7, 1993), http://query.nytimes.com/gst/fullpage.html?res=9F0CE5D7133BF934A35750C0A965958260.

1019 *See*, e.g., David Johnston, *Inquiry into CIA Tapes Seen as Payback for FBI*, International Herald Tribune (January 5, 2008), 2008 WLNR 310501; Edward M. Green & Michael Gill, *Focus on Explosives: Failures and Fees*, 22 Crowell & Moring Mining Law Monitor (Spring 2005), http://www.kph.com/NewsEvents/Article.aspx?id=241.

1020 *See*, e.g., Philip Gold, *The Essentials of Self-Preservation*, Policy Review 33 (2000); Bruce Carlson, *In It Together: Budget Decisions Should Be Immune to Interservice Rivalries*, Army Times, http://www.armytimes.com/community/opinion/airforce_opinion_carlson070319/. *See also* Brian Dollery, Zane Spindler, & Craig Parsons, *Nanshin: Budget-Maximizing Behavior, the Imperial Japanese Navy and the Origins of the Pacific War*, 4 Public Organization Review 135 (June 2004) (intense rivalry between World War II era Japanese army and navy).

surprised when they compete with each other, at some level. I am not suggesting that state and federal law enforcement personnel do not, and cannot work together on an investigation; they can and do. What I am pointing out is that there is always some sense of competitiveness, which is in large part due to commitment to one's particular agency or agency sub-subdivision. Like many endeavors, law enforcement is to a great extent a competitive effort: Law enforcement officers compete against the criminals they investigate, but they also compete with each other to see which entity can successfully conduct an investigation. That competitiveness is a function of pride in one's self and in one's colleagues, and it can enhance the effectiveness of an agency. It can also, though, become counterproductive, especially if the competition begins to be perceived as a matter of competing for the allocation of limited resources. Law enforcement agencies must report on their performance during a particular calendar or fiscal year, and their relative success can have a direct impact on the resources they are allocated for succeeding years.[1021]

That reality can have a negative effect on cooperation across agencies, as can law enforcement personnel's perception that officers from another agency are being less than fully cooperative in an effort to gain some advantage for their agency. About a year ago, officers from Federal Agency A told me that officers from Federal Agency B had a particularly effective software program they refused to share with Federal Agency A; indeed, according to the officers from Federal Agency A, officers from Federal Agency B would not even admit they had such a program. I was not sure whether to believe any of this, but that is not the point: The Federal Agency A officers believed there was a deliberate, strategic lack of cooperation from the other agency. That belief would inevitably have a corrosive effect on their willingness to cooperate with officers from Federal Agency B which, in turn, would no doubt erode the willingness of Federal Agency B officers to cooperate with officers from Federal Agency A. This is merely one, minor example of the tensions that exist among U.S. law enforcement agencies and among agencies in other countries, as well.

Tension among agency personnel is not the only factor that erodes the effectiveness of our threat response processes. The agency compartmentalization

---

1021 *See*, e.g., Edward M. Green & Michael Gill, *Focus on Explosives* ("ATF finds itself in a rivalry for funding, personnel, and attention with the much larger and much richer FBI). *See also* Bruce Carlson, *In It Together: Budget Decisions Should Be Immune to Interservice Rivalries*, Army Times (Springfield, VA: Army Times Publishing Company), http://www.armytimes.com/community/opinion/airforce_opinion_carlson070319/.

and segmentation I noted above also contributes to this. The Secret Service's Electronic Crime Task Forces (ECTFs) and the Federal Bureau of Investigation's InfraGard initiative are a good example of this: As we saw earlier, both are intended to facilitate networking and information-sharing among federal, state, and local law enforcement officers and representatives from the private sector. Their missions are functionally identical.[1022] Each initiative operates by establishing a presence (an ECTF or an InfraGard chapter) in a particular place, usually a city.[1023] Both have established a presence in a number of U.S. cities, including New York, Miami, and Chicago.[1024] Each has also established a unique presence in other cities, and neither has so far established a presence in many of the smaller cities in the U.S.[1025]

This means that: (a) in some cities, two independent federal agency-sponsored groups are operating simultaneously with the same agenda and for the same purpose; (b) branches of each group alternate among a number of other cities; and (c) neither has established a branch in many cities. Each outcome is to some extent inconsistent with the ultimate goal of facilitating information sharing among law enforcement and the private sector.

The first outcome seems to produce an unnecessary overlap in personnel, resources, and time: Are the same non-Secret Service, non-FBI people (state and local law enforcement officers and members of the private sector) joining both the ECTF and the InfraGard chapter in their city? If so, do they regularly attend the meetings of both? (Based on anecdotal evidence, I am assuming Secret Service agents only attend ECTF meetings, and FBI agents only attend InfraGard meetings.) If the non-Secret Service, non-FBI participants *do* attend the meetings of both, that would produce an unnecessary *duplication* of time and effort. As we saw earlier, both groups have the same mission; because both focus on promoting the sharing of information about

---

1022 *Compare* About the U.S. Secret Service Electronic Crimes Task Forces, http://www.ustreas.gov/usss/ectf_about.shtml *with* Goals and Objectives, InfraGard, http://www.infragard.net/about.php?mn=1&sm=1-0.

1023 The Secret Service's ECTFs are sometimes established in states, instead of cities. *See* Electronic Crimes Task Forces and Working Groups, http://www.ustreas.gov/usss/ectf.shtml (Oklahoma & South Caroline ECTFs). And some of the InfraGard chapters cover several states. *See, e.g.*, National Chapters, InfraGard, http://www.infragard.net/chapters/ (Minneapolis chapter covers South Dakota, Salt Lake City chapter covers Montana).

1024 *Compare* Electronic Crimes Task Forces and Working Groups, http://www.ustreas.gov/usss/ectf.shtml *with* National Chapters, InfraGard, http://www.infragard.net/chapters/.

1025 *Compare* Electronic Crimes Task Forces and Working Groups, http://www.ustreas.gov/usss/ectf.shtml *with* National Chapters, InfraGard, http://www.infragard.net/chapters/. InfraGard seems to have more chapters in smaller cities—like Dayton and Des Moines.

threats and response techniques, attending the meetings of both should result in one's receiving the same information twice.

The residual alternative, which I suspect is more likely to manifest itself, is even more problematic: If the non-federal law enforcement participants do not regularly attend the meetings of both groups, this should produce a *fragmentation* of effort because part of the local law enforcement-private sector constituency will be sharing information VIA the Secret Service-run ECTF and the remainder will be sharing information via the FBI-run InfraGard chapter. While a core of nationally and internationally derived threat information will be shared within each group, there is likely to be a disconnect when it comes to sharing locally derived threat information. Members of the ECTF may, for instance, immediately learn of an attack on a local company because a company representative attends ECTF meetings; since the representative does not attend InfraGard meetings, members of that group may not learn the details of the attack for some time. The obvious solution is to combine both groups into a single entity, but that is not likely to happen, at least not for the foreseeable future, because of the embedded partitioning that separates the FBI and Secret Service. Though their missions overlap when it comes to cyberthreats, each agency is institutionally and operationally distinct and, as I noted above, has various incentives to remain so.1026

The other outcomes can also have undesirable effects. The second outcome—ECTFs (only) in some cities, InfraGard chapters (only) in other cities—seems likely to produce gaps in information-sharing: Does the information collected by InfraGard chapters in InfraGard (only) cities make its

---

1026 Some interaction between the groups is developing. In April of 2007, for example, a Secret Service agent active in the ECTF initiative spoke at a New York InfraGard meeting. *See* New York Metro InfraGard Alliance, Security Summit—April Agenda, http://www.nym-infragard.us/nuke/downloads/agendas/Agenda%20NYM%20InfraGard%20Cybercrime%2004-13-07.pdf.

And in August of 2006, the "first-ever joint meeting" of the ECTF and InfraGard initiatives was held in Atlanta. ***See*** Government and Private Sector Experts Unite to Discuss CyberSecurity and Infrastructure, Business Wire (August 14, 2006), http://findarticles.com/p/articles/mi_m0EIN/is_2006_August_14/ai_n16621507. A press release said the meeting would "plot the ***periodic*** merger" of the organizations "to channel government and private sector expertise to address cyber security and infrastructure protection." ***Id.*** (emphasis added). From the agenda for the four-hour session, it seems to have focused on sharing information about two federal initiatives, the REALID program and the Department of Homeland Security's National Infrastructure Protection Plan. ***See id.*** The ECTF and InfraGard chapter in Philadelphia held a joint meeting a year later, so perhaps there is some general effort to have the initiatives combine forces on occasion. ***See*** Joint InfraGard and ECTF Meeting, Philadelphia InfraGard, http://infragardphl.org/calendar.php?mon=08&day=21&year=2007.

way to ECTFs in ECTF (only) cities, and vice versa? As I noted earlier, there will be a level of consistency in the information being shared within each initiative because much of it will pertain to nationally or internationally derived threats, such as new fraud tactics. Since this top-down information concerns threats the origins and significance of which transcend a city or geographical region, it should be a common, uniform component of the information sharing efforts in both initiatives. The potential problem lies with the bottom-up information generated by and shared within an initiative—information about threats at the local level. To the extent this information pertains to a local manifestation of a nationally or internationally derived threat, it should to a great extent duplicate top-down information being shared within both initiatives; this would mitigate the information sharing gap I noted above, but the local threat information could still be useful in identifying how the large-scale threat is evolving and adapting to different circumstances.

The real potential for a gap in information sharing lies in local threat information that does not pertain to an already identified, larger scale threat; as we saw in earlier chapters, what seems to be a localized, isolated attack on target in a medium-sized city could actually be the beginning of, or an increment in, a cyberwarfare campaign. If the victim of such an attack belonged to an ECTF and reported what had occurred to that ECTF, the Secret Service could share the information within its agency and with the other ECTFs. The question is whether, and how, the information would be shared outside the Secret Service-ECTF system. Because ECTFs and InfraGard chapters are organizationally distinct, that would require the recipient agency's (the Secret Service's, in this example) contacting the nonrecipient agency (the FBI, here) and advising them of what had occurred. It would then be up to the FBI to disseminate the attack information throughout its agency and through the InfraGard chapters.

In outlining this rather convoluted scenario, I do not mean to imply that the Secret Service would deliberately decline to share attack information with the FBI. I, for one, do not believe the institutional rivalries I noted above could have this effect, at least not in a post-9/11 world.[1027] My focus is not on intentional defaults but on the inadvertent but equally problematic failures that can result from having operationally distinct, hierarchically organized entities with overlapping missions operating in parallel. Information tends

---

1027 *Cf.* Abraham McLaughlin, ***Can FBI and CIA Cooperate?***, Christian Science Monitor (June 5, 2002), http://www.csmonitor.com/2002/0605/p01s01-uspo.html (lack of cooperation among these two agencies leading up to the 9/11 attacks).

to travel vertically in these entities; depending on its point of origin, it moves up or down. As we saw in the scenarios outlined above, higher-level information, such as national or international attack information, will move down through the organization, while lower-level, local threat information will move up. Information will also be disseminated laterally as it reaches a particular level of the organization, but that dissemination is only within the organization.[1028] The problem arises with regard to moving the information out of one organization and into another.

All organizations are insular to varying degrees. Every organization of whatever type—educational, charitable, professional, artistic, commercial, religious—competes at some level with other, similar organizations. Organizations have in effect become our tribes: I belong to my university's tribe and to the tribes of other organizations with which I am affiliated. We are loyal to our organizational tribes, and that makes us look inward. Our employer organization tends to be our primary tribe; when we are at work, we collaborate with other members of that tribe on advancing its—and our—interests. Our focus is internal, and our information-sharing patterns tend to be the same; we may share information with a competing organization, but only when, and to the extent that, we believe it will advance the interests of our tribe. The same is true to a heightened extent of law enforcement organizations. Law enforcement organizations tend to be particularly insular because they deal with information that often must be kept confidential and because of the interagency rivalries I noted earlier. This insularity and its attendant emphasis on sharing information only within an organization are not new phenomena in law enforcement, but they were less problematic in the past, when law enforcement entities focused primarily on discrete, localized threats.[1029] They are, as we have seen, becoming increasingly problematic in a world in which threats can transcend territory and categorization.[1030] We must therefore devise some way to overcome this institutional insularity without losing the residual benefits it continues to

---

1028 This phenomenon is known as the "stovepipe approach," and it is blamed, in part, for intelligence failures leading up to 9/11. *See id. See also* Remarks by the Vice President to Technology Industry Leaders—San Jose, California, The White House (February 21, 2002),http://www.whitehouse.gov/vicepresident/news-speeches/speeches/vp20020221.html (noting the problematic nature of "the stovepipe approach to information, in which data travels up and down within an organization, but is not shared with other agencies or jurisdictions").

1029 *See* Chapters 2 & 3.

1030 *See* Chapters 3–6.

provide with regard to real-world threats. We return to those issues in a moment, but first I need to note the consequences of the third outcome of the ECTF-IngraGard configurations we examined above.

The third, ultimately residual outcome—areas that have neither an ECTF nor an InfraGard chapter—means that in many parts of the country nonfederal law enforcement personnel and members of the private sector do not have the opportunity to participate in either initiative. This outcome is less a product of institutional partitioning than it is of scarce resources, but it still contributes to the gaps and uncertainties in information sharing that can impede the attribution and response processes for cyberthreats.

Before we consider how to overcome organizational insularity and otherwise improve these processes, I need to note one thing about the scope of the problem. The scenarios I outlined above are predicated on gaps in information sharing between two very similar federal law enforcement agencies. While their institutional autonomy exacerbates this problem, their operational similarity can mitigate it, at least to some extent; the personnel of each agency tend to speak the same language and understand the same threat dynamics. Neither is necessarily true of the many other law enforcement and law enforcement-related agencies that are involved in the various cyberthreat attribution and response processes, nor is either true of the civilian-based organizations that are also seeking to contribute to these processes.[1031]

The U.S. is often described as having a sectoral approach to privacy because we have no single, overarching privacy laws. I would argue that we, and every other nation-state, have a sectoral approach to security because

---

1031 In the United States, for example, we not only have the ECTFs and InfraGard Chapters, we also have Information Sharing and Analysis Centers that have emerged in various contexts and a number of other, similar initiatives. The mission of the ISACS is functionally analogous to that of the ECTFs and InfraGard chapters, though it focuses more on general security than on law enforcement. *See* About the Council, ISACCOUNCIL.ORG, http://www.isaccouncil.org/about/. As to the nature of the ISACS, there is a Communications ISAC, an Electricity Sector ISAC, a Financial Services ISAC, an Emergency Management and Response ISAC, a Highway ISAC, an Information Technology ISAC, a Multi-State ISAC, a Public Transit ISAC, a Surface Transportation ISAC, a Supply Chain ISAC and a Water ISAC. *See id.* There is also a Food and Agriculture ISAC, a Real Estate ISAC and a Research and Education Networking ISAC. *See* Food and Agriculture ISAC, http://www.fmi.org/isac/; Real Estate ISAC, http://www.reisac.org/; REN-ISAC, http://www.ren-isac.net/. And there are state ISACs. *See* Michigan ISAC, http://www.michigan.gov/cybersecurity/0,1607,7-217-43559-153667--,00.html; South Carolina ISAC, http://www.sc-isac.sc.gov/content.asp?contentid=579. *See also* New York Cyber Security and Critical Infrastructure Coordination, http://www.cscic.state.ny.us/partnership/. There are also more or less formal private sector information sharing initiatives.

we divide the task of securing ourselves from chaos in its real and virtual forms among an almost exponentially increasing number of entities, both public and private. The scenarios we examined above illustrate the excessive, redundant partitioning of responsibility for controlling internal threats and we see a similar phenomenon in the military. As noted earlier, the U.S. military is divided into a dizzying array of branches, divisions, subdivisions, sub-subdivisions, and so on. And the tribal insularity I noted above applies with equal validity in the military context, as well.

So how do we overcome all this? More precisely, how do we overcome all this while maintaining the benefits this system has provided and continues to provide? I, for one, believe that the fracturing of law enforcement and military entities has probably progressed beyond the point at which it makes sense, even for real world threats; if we continue along this path, we will expand the underlying principle far beyond the limits of its logic and, in so doing, undermine our ability to fend off chaos of varying types. At the same time, I think we would all concede that our militaries and law enforcement agencies do a more than adequate job of dealing with real-world threats. That discrepancy brings us to the critical issue: How can we retain the current system's efficacy against real-world threats—threats that will persist unless and until we transcend our essential nature and become placid cyborgs—while equipping our governing entities to deal effectively with cyb3rchaos? Logically, our options fall into two categories: The first, which we will examine here, is to reconfigure the threat response processes nation-states use to maintain internal and external order. The other, more drastic option is to move beyond nation-state governance into new systems that control real- and virtual-world threats with equal efficacy. We consider that option in the next section.

Our analysis of reconfiguring nation-state threat response processes will focus primarily on the United States, both because it is the system I know best and because it is probably the most intractable system.[1032] We assume that our current threat response processes remain in place as the default system for controlling real-world threats. Our task is to develop correlate processes for cyberthreats. Because they operate in the virtual world, these processes cannot be territorially based, that is, they cannot partition response authority according to whether a threat jeopardizes internal or external order. Since we are assuming nation-state governance, we also assume that the reconfiguration of these processes occurs at the nation-state level.

1032 *See* Chapter 6.

Because our goal is to implement nonterritorially based processes, we must design a system that integrates military and law enforcement functions into a single function: controlling threats emanating from cyberspace. Integrating these functions would involve both structural and doctrinal innovation. Doctrinally, we would have to develop a new threat vocabulary; because response authority is not segregated according to whether a threat is internal or external, the "crime," "terrorism," and "war" concepts would be inappropriate here. We might fuse them into a single, global concept—"cyberthreat"—but I suspect that level of generality would suffice only at the broadest operational levels. Because there are reasons to differentiate threats according to their source, their targets and the types of "harm" they inflict, I suspect this new threat response system would develop its own threat vocabulary, which might, or might not, employ the terms we currently rely upon.

Structurally, we would need to create a new institution—the Cyber Security Agency, say. There is, as we saw in Chapter 6, precedent for such a step: In 1947, Congress created a new institution—the Central Intelligence Agency—to respond to the unique realities of the Cold War. Many countries eventually followed suit.[1033] The Central Intelligence Agency and its equivalent in other countries are analogous to the institution we are postulating in at least one respect: Their mission deviates from the traditional division of responsibility for dealing with external threats by involving both military personnel and "pure" civilians in the response process.[1034]

The Central Intelligence Agency cannot, however, serve as legal precedent for creating our Cyber Security Agency because they differ in a critical respect: The creation of the Central Intelligence Agency did not directly contravene U.S. law, which would be the case here. As we saw in Chapter 6, the Posse Comitatus Act prohibits military personnel from playing an active role in civilian law enforcement. Creating a fused internal-external cyberthreat response agency would directly violate the Posse Comitatus Act.

Is that a fatal obstacle? Probably not, as far as the law is concerned. The Posse Comitatus Act is merely a statute, and statutes can be revised or even repealed. As we saw in Chapter 6, it was adopted in response to a specific evil—the egregious misuse of federal troops in the Reconstruction South. As we also saw in Chapter 6, prior to the adoption of the Act, law enforcement officers routinely used military personnel in maintaining

---

1033 *See*, e.g., "List of intelligence agencies," Wikipedia, http://en.wikipedia.org/wiki/List_of_intelligence_agencies.

1034 *See* Chapter 6.

internal order. And as we again saw in that chapter, in the 1980s, Congress adopted legislation excluding certain activities—primarily activities associated with the war on drugs—from the scope of the Posse Comitatus Act. The Act is clearly malleable.

Would we need to, and be willing to, repeal the Posse Comitatus Act to create our new Cyber Security Agency? I think not. I suspect that most Americans, myself included, would not be willing to repeal the Posse Comitatus Act in its entirety, even if doing so would increase our security from cyberthreats. The prospect of having armed military personnel patrolling streets to maintain order is one I, and I suspect many others, find distasteful and even frightening. But that reaction is, I also suspect, limited to the physical world and to the peculiar circumstances

I, for one, do not find the prospect of an internal-external response institution operating in cyberspace particularly discomfiting. I think there are two reasons why I, and I suspect many others, would have much less difficulty with this scenario: One is that the activity is to a great extent invisible because it occurs in cyberspace. While I may be aware, at some level, that agents of the Cyber Security Agency are online and are conducting activities that may or may not impact on my own online activity, I do not "see" them in the way I would see armed military personnel patrolling my community. Their presence and their activities are subtle, essentially invisible and therefore far less intimidating. The other reason I think I, and others, would have less difficulty with this scenario is that it would not literally involve sending military personnel into cyberspace to act as law enforcement officers. The analogy between armed military personnel patrolling the streets of my community and the online activities of Cyber Security Agents is, in other words, imperfect. We are, after all, exploring the possibility of creating a new institution, one that would combine the internal and external threat response processes but that would do so by creating a unique institution, one that is neither military nor law enforcement, as such. The Cyber Security Agents who would patrol the online world would be neither fish nor fowl, neither military nor police; they would be something entirely new. Because they are not actually members of the military, we should not find their participation in the process of maintaining order to be as disconcerting, as objectionable, as the prospect of military personnel patrolling our streets.

It seems, then, that we *could* create our Cyber Security Agency. We could modify the Posse Comitatus Act so that it neither explicitly nor implicitly bars the creation and operation of a fused internal-external threat institution *in cyberspace*. And, as we saw above, that limitation should overcome

Americans' concerns about eradicating the partition between the military and law enforcement. There is still, though, a residual problem: Although the activities of the Cyber Security Agency and its agents would be limited to cyberspace, the effects of those activities would not. For the foreseeable future, anyway, cyberthreats, like real-world threats, will be the product of human activity. Our Cyber Security Agency, like our real-world militaries and law enforcement agencies, is both an attribution and response agency; all three control threats by identifying the nature of the threat and the architects of the threat and then directing a response at the architects of the threat. The response, as we saw earlier, is intended to incapacitate the architects of this threat from implementing further threats and to provide a disincentive that will deter others from implementing similar threats.

Our Cyber Security Agency could use online response techniques to achieve these ends, at least in part. But as long as humans are responsible for the consummation of threats, we will, I believe, also need to rely on response techniques—sanctions—that are directed at the specific individuals whose activities undermine order. This is why the effects of the activities of our Cyber Security Agency will bleed out of cyberspace and into the real-world. To the extent that our cyber security effort depends on individual apprehension and sanctioning, it will require either the participation of traditional military and law enforcement personnel or the creation of a real-world adjunct to the Cyber Security Agency that would act as a surrogate for military and/or law enforcement personnel. Because there seem to be neither any legal necessity nor practical reasons for creating a new, surrogate institution,[1035] I assume that the Cyber Security Agency would interact at some level with real-world law enforcement and military agencies. That could create problems, of law and perception.

Legally, we might have a Posse Comitatus Act issue if a non-law enforcement agency (or, more accurately, a non-exclusively law enforcement agency) collected evidence implicating someone in the commission of a crime and turned that evidence over to law enforcement. Law enforcement, of course, would use the evidence to apprehend the suspect, who could then be prosecuted, convicted, and sanctioned. We might be able to anticipate,

1035 Creating such an institution should not serve to overcome any Posse Comitatus Act issues because it would need to combine law enforcement and military functions. Depending on the particular threat and threat architects involved, the real-world response to a cyberthreat might necessitate either a law enforcement response or a military response or, perhaps, both. *See* Chapters 3–5.

and eliminate, this issue when we revised the Posse Comitatus Act to allow us to create the Cyber Security Agency. One option would be to revise the Act so that it allowed the Cyber Security Agency to share evidence and other information with law enforcement as long as the data being shared was relevant to a legitimate law enforcement function. This alternative would simply ignore the question of the Cyber Security Agency's institutional status, that is, whether it was in part a military agency. The other, I think less satisfactory option would be definitional: In revising the Posse Comitatus Act, we could include a section that specifically defined the Cyber Security Agency as not being a military agency within the compass of the Act. I find that alternative less satisfactory because I suspect such a definition would be challenged as being both legally and factually problematic.

In other respects, our Cyber Security Agency would come within the information-sharing scheme I outlined earlier in the chapter. That is, it would be able to share information freely with military agencies, both providing information to such agencies and receiving information from them. It should also be able to receive information freely from law enforcement agencies because that should not create any Posse Comitatus Act or other issues.[1036]

I am not going to attempt to outline the structural and operational contours of our Cyber Security Agency because these are issues we would have to resolve once we embarked upon the process of creating what is to some extent a twenty-first century analogue of the CIA. I do, though, want to note what I think should be an essential aspect of its structure: Hierarchical organizations function very well in the real-world (at least, until they begin to expand beyond the limits of their inherent logic), but, as we have seen, are not well adapted for the virtual world. [1037] I think the greatest challenge those charged with creating a virtual Cyber Security Agency would have is conceptualizing a threat response agency that was not overtly, exclusively hierarchical in nature. We think of hierarchies because that is what we know; indeed, it is all we know.

What might a nonhierarchical Cyber Security Agency look like? It obviously needs some kind of command structure, but I would propose that this command structure be as flat—as lateral—as possible. Cyberspace is a network of networks. It follows, I submit, that an online threat response organization should adopt a lateral, networked organizational model.

1036 *See* Chapter 6.

1037 For more on this, *see* Susan W. Brenner, ***Toward A Criminal Law for Cyberspace: Distributed Security***, 10 Boston University Journal of Science & Technology Law 1, 76–83 (2004).

Hierarchical organizations are effective in concentrating resources—human and otherwise—to overcome real-world challenges.[1038] The resources entities like the Cyber Security Agency will employ in cyberspace are digital resources and, as such, have no fixed, tangible form. They are the product of computer technology manipulated by an individual. Concentrating resources and directing them at a challenge in cyberspace is, therefore, primarily an automated process. Tiered, partitioned organizational structures not only become irrelevant in this context, they become counterproductive. My hope is that we will devise and implement alternate organizational models—lateral, network models—for order-sustaining processes in cyberspace.

New Governing Configurations

Postulating the new governing configurations that may replace nation-states is a very difficult task because it requires imagining what is not, and probably has never been. It is, of course, possible that we might regress—abandoning the nation-state and returning to the pre-Westphalian system of personal sovereignty perhaps buttressed by an overlying conceptual principle analogous to the ***Respublica Christiana***.[1039] I, for one, find that possibility to be quite unlikely, given the influence of history and the accelerating impact of technology. History has made us skeptical of personal sovereigns; and technology breaks down cultural and territorial barriers and, in so doing, makes us skeptical of provincial solutions.

1038 *See*, e.g., Susan W. Brenner, ***Toward a Criminal Law for Cyberspace: Product Liability and Other Issues***, 5 University of Pittsburgh Journal of Technology Law and Policy 2, 4 (2005):

> The . . . model of law enforcement is . . . analogous to the . . . model of military order: Both are concerned with organizing and concentrating human and other forces to respond to physical activity . . . by human beings . . . situated in an identifiable physical environment. Both models therefore feature a hierarchical structure that uses a chain of command to concentrate . . . resources on those to whom a response is necessary; this is an appropriate approach to concentrating . . . resources to achieve an objective in the real-world. . . .
>
> Technology eliminates the need . . . to focus on specific, localized activity. Communication technologies free us from the constraints of the empirical world; we can communicate instantaneously with anyone from anywhere. This produces a new type of social organization. . . . Networks are displacing hierarchies in every sector of society, including the military because hierarchical organization is not an effective means of organizing technologically mediated activities.

1039 *See* Chapter 7.

I will, therefore, assume that the governing configurations to come will neither constitute a return to personal sovereignty nor continue to take the form of territorially based nation-states. It seems to me that the possibilities logically fall into either of two categories: territorially based models that transcend the nation-state, and corporate governance.

Territorially based configurations that transcend the limitations of the nation-state are the most obvious possibility for two reasons. One is that they do not require us to devise a new model of governance; we continue to rely on territorially based governance, but increase the amount of territory encompassed by a governing unit. The other, related reason is that they are probably the simplest, and certainly the most literal, solution to the problems that currently plague nation-states.

As we have seen, the various cyberthreats challenge nation-states not because they constitute new modalities of inflicting "harm,"[1040] but because cyberspace gives those who inflict "harm" the ability to do so remotely. This challenges nation-states because their bipartite threat response processes cannot deal effectively with externally generated threats that do not conform to the traditional model of warfare.[1041] The premise here is that moving to a higher order of territorially based governance systems would resolve these problems by making it more difficult, or even impossible, for perpetrators to inflict "harm" remotely.

The remote infliction of "harm" should become a factual impossibility if we were to move to the highest order of territorially based governance system: a global governance structure. If there were only one governance structure, and if that governance structure exercised dominion over all of the Earth's territory, it would presumably be impossible for anyone to launch threats remotely, that is, from outside the governance structure's jurisdiction.[1042] The problem here lies not with the solution as such, since it does seem to eliminate the problem of extra-territorial threats. It lies, instead,

---

1040 As we saw in Chapter 3, most cyberthreats involve inflicting traditional "harms" in new ways. There is, as I noted there, at least one new "harm"—that "harm" resulting from a distributed denial of service attack—which has required us to define a new criminal offense. We may find it necessary to address other new "harms" as our experience with cyberspace progresses, but the nature of the "harm" is not, as we have seen, the critical issue.

1041 *See* Chapters 3–6.

1042 We are assuming, for the purposes of this analysis, that humanity has not colonized planets or satellites other than Earth, so extraterrestrial threats are not a consideration.

with the viability of the solution: It is exceedingly unlikely that nation-states will, in the foreseeable future, surrender their respective identities to fuse into a single, global entity. Like the personal sovereigns who were their predecessors, nation-states are jealous of their prerogatives and vigilant in securing their survival. It is equally unlikely that the people of the world would be willing to accept the demise of the nation-state and its replacement by a global governance structure. Although evolved transportation and communication technologies are making us more cosmopolitan, the vast majority of people strongly identifies with their nation-state of origin and are therefore disinclined to become part of a vast, undifferentiated mass.[1043]

The same obstacles would impede any effort to move to a higher order, but less than global, governance system. This would presumably involve creating supranational entities analogous to, but more autonomous than, the European Union. Though the European Union has been notably successful in creating a collaborative political and economic community, it has not sought to absorb its constituent nation-states.[1044] The higher order governance structure we are hypothesizing would do precisely that; here, instead of combining all of the nation-states of the world into a single, global governance structure, we would fuse geographically related states into regional governance structures. So, we could have the North American Union, the South American Union, the European Union, the African Union, the Middle East Union, and the Arab Union with, perhaps, a few outliers. This approach would be as problematic to implement as the global governance system because it also involves eliminating the nation-states to which we are so accustomed. And it would not be as effective in eliminating the problems we seek to resolve because it would still divide the world up into discrete territorial enclaves. In so doing, it would preserve one's ability to inflict "harm" remotely, that is, from outside the territory controlled by a particular regional governance structure.

Another, less drastic possibility would be to implement a variation of the strategy we examined in the previous section: creating a single entity—a Cyber Security Agency—that would respond to cyberthreats of whatever type. The premise there was that we can improve nation-state level responses to cyberthreats by fusing what we now consider to be military and law

---

1043 This is evident in the idiosyncratic paranoia that exists as to the United Nations' plan to become a global governing system. *See*, e.g., Carl S. Parnell, ***The United Nations' Secret Agenda: Globalization and the Elimination of America's Sovereignty***, Free Republic (January 24, 2008), http://www.freerepublic.com/focus/f-news/1958703/posts.

1044 *See*, e.g., "European Union," Wikipedia, http://en.wikipedia.org/wiki/European_Union.

enforcement functions into a single agency. We could implement a version of this strategy at the supranational or even at the global level. It would involve parsing the order-maintaining functions into two categories: those directed at controlling threats emanating from the real, physical world; and those directed at controlling threats emanating from the cyberworld. We would then leave the first function to be performed at the nation-state level, and consign the second either to regional Cyber Security Agencies or to a single, Global Cyber Security Agency.

Though this approach has a certain, superficially appealing logic, I do not believe it is a viable alternative. The primary problem with it is that threats cannot really be divided between those that originate in real-space and those that originate in cyberspace. As we have seen, the threats that originate in cyberspace directly or indirectly inflict "harm" in real-space. If we were to implement this approach, we would have higher-order governance structures (supranational or global) directly participating in the order-maintaining process within a nation-state. We would in effect create a threat response overlay: When citizens of the United States were "harmed" in real-space, they would rely on U.S. law enforcement for assistance. When they were "harmed" in cyberspace, they would rely on the North American Union Cyber Security Agency or on the Global Cyber Security Agency. Because the "harms" that emanate from cyberspace reverberate in the real-world, agents of the higher-order response authority would certainly interact with, and would also probably compete with, the local, U.S. law enforcement agents. We would replicate the problem of duplicate, overlapping response authorities we in the United States are already grappling with, but we would exacerbate the problem by involving "outsiders," that is, non-United States nationals. I see no reason to pursue this strategy.

This brings us to the residual alternative: corporate governance. Some believe we are in the process evolving into a system of corporate governance.[1045] As I understand this theory, it begins with the premises (a) that corporate entities are increasingly transnational in nature and (b) that this trend will accelerate. It then postulates that corporate entities are more efficient at using modern technologies, are becoming wealthier than many nation-states, and are taking over certain of the functions that have been carried out exclusively by nation-states.[1046] The extrapolation seems to be

1045 *See*, e.g., van Creveld, The Rise and Decline of the State, *supra* at 415–418.

1046 *See id.*

that as corporate entities become increasingly wealthy and powerful, they will eclipse the nation-state, which will suffer a correlate decline in its power and prestige.

That extrapolation may be extraordinarily prescient. It is, after all, very difficult to realize when things—including governance systems—begin to change. As we saw in Chapter 7, the architects of the Peace of Westphalia did not realize that they had set into motion the rise of the nation-state and the decline, and eventual disappearance, of the personal sovereignties which were all they knew. We may well be on the cusp of a shift to private, corporate governance. I could see a system—a twenty-first century feudal system—in which we all become clients of one or more corporate governance systems. I would be, say, a primary client of the university corporate structure that employs me; I could be a secondary client of the corporate system that is responsible for maintenance and security in the suburb where I live and, perhaps, of other corporate systems, as well.

My only reservations about corporate systems as our new governance model may reveal how wedded I am to the nation-state model. I wonder how they would handle the processes of maintaining order in the real-world and in the cyberworld. How would they carve up the world's territory? Would there be internecine corporate wars? Or would war become a quaint artifact of times past, an antedated relic that had no place in a world driven by profit and loss statements?

If we do evolve into a system of exclusive corporate governance, I assume the problem of maintaining order—online and offline—would be assigned to private contractors. Discrete corporate governing entities could, I suppose, create their own threat response organizations—their own militaries and law enforcement agencies—but that seems inconsistent with the notion of evolving into a corporate, functionality-rather-than-territorially based governance structure.

Final Thoughts

My primary goal in this book has been to explain how and why our use of cyberspace and related computer technologies challenges the order-maintaining systems we have relied on for hundreds of years. I believe—I hope—I have achieved that goal.

As to how we address those challenges, I have explained what I think are the critical issues we must address and I have offered some short-term solutions for addressing those issues. I may be wrong, but I suspect the challenges emerging in this area are analogous to pre-shocks that signal an impending earthquake. Our law enforcement and military institutions evolved in a world controlled by territorially defined nation-states. As the principle of territoriality becomes increasingly less dispositive in human governance, and in human affairs generally, these systems—and the nation-states they serve—become increasingly problematic.

The solution for the challenges we confront and for those yet to come may ultimately lie in abandoning our adherence to territoriality and developing new governance systems. I have outlined the contours of what those successor systems *might* look like. In a sense, though, each of the alternatives I outlined in the previous section is merely an extrapolation from the nation-state governance model that is all we know.

Earlier, I criticized that model for exemplifying a principle's tendency to expand itself to the limit of its logic. If we really want to speculate creatively about successor systems, we should abandon any adherence to territoriality and attempt to extrapolate the defining principles of cyberspace—lateral, occasional organization predicated on communication—to the problems of organizing human endeavors and maintaining order in human social systems. I, for one, find that a daunting task, far too difficult to undertake here. It would presumably involve flattening our organizational models, eradicating hierarchies and replacing them with lateral congeries, more or less adventitious associations of like-minded individuals. I can conceptualize such a model; what I cannot comprehend is how the members of those associations would prevent certain of their colleagues from preying on the others while discouraging hostile advances from other associations. Cyberspace lets us do many things, but it has not, so far, anyway, allowed us to transcend greed, anger, bigotry, or a number of other traits that inevitably import elements of instability into human organizations.

Index

Note: page numbers following by an *n* or *nn* and a second number refer to a numbered note or notes on the indicated pages.